C0-ASO-538

HANDBOOK OF PLANT-BASED BIOFUELS

Handbook of Plant-Based Biofuels

Edited by
Ashok Pandey

CRC Press is an imprint of the
Taylor & Francis Group, an **informa** business

CRC Press
Taylor & Francis Group
6000 Broken Sound Parkway NW, Suite 300
Boca Raton, FL 33487-2742

Printed in the United States of America on acid-free paper
10 9 8 7 6 5 4 3 2 1

International Standard Book Number-13: 978-1-56022-175-3 (Hardcover)

Library of Congress Cataloging-in-Publication Data

Handbook of plant-based biofuels / editor, Ashok Pandey.
p. cm.
Includes bibliographical references and index.
ISBN 978-1-56022-175-3 (alk. paper)
1. Biomass energy. 2. Biodiesel fuels. 3. Alcohol as fuel. I. Pandey, Ashok. II. Title.

TP339.H37 2008
662'.88--dc22 2008022722

Visit the Taylor & Francis Web site at
http://www.taylorandfrancis.com

and the CRC Press Web site at
http://www.crcpress.com

Contents

SECTION III Production of Biodiesel

Preface

With the depletion of oil resources as well as negative environmental impact associated with the use of fossil fuels, there is a renewed interest in alternate energy sources. As the world reserves of fossil fuels and raw materials are limited, active research interest has been stimulated in nonpetroleum, renewable, and nonpolluting fuels. Biofuels are the only alternate energy source for the foreseeable future and can still form the basis of sustainable development in terms of socioeconomic and environmental concerns. Biodiesel and bioethanol, derived from plant sources, appear to be promising future energy sources. It is against this background that this book was conceived and prepared.

The book has three sections. Section 1 has four chapters. Chapter 1 is introductory and gives a profile of plant-based biofuels. Chapter 2 deals with the world biofuel scenario and provides an overview of the production of biofuels from biomass materials by thermochemical and biochemical methods, as well as trends of utilization of the products in the world. Chapter 3 deals with the thermochemical conversion of biomass to liquids and gaseous fuels and focuses on pyrolysis and other conventional thermochemical processes. It describes various types of pyrolysis processes, namely, slow, fast, flash, and catalytic processes in detail to give the reader better insight into these thermochemical processes. Chapter 4 describes the production of biofuels, with special emphasis on biodiesel.

Section 2 of the book deals with the production of bioethanol and has seven chapters. Chapter 5 is titled "Fuel Ethanol: Current Perspectives and Future," and analyses the current status of biomass-to-ethanol programs. It summarizes the ways in which rapid increase in world demand for fuel ethanol and the state of the oil market may notably influence the international market price of ethanol and provide opportunities for large-scale production in other regions such as Europe and Asia. In the long term, lignocellulose to ethanol conversion is the most viable pathway, from a sustainability point of view. However, its production cost must be reduced significantly in order for this process to have a chance to drive forward the strategy of biomass-to-ethanol conversion worldwide. Chapter 6 on molasses ethanol provides an overview of the status of ethanol fermentation from molasses and process practices applied for the improvement of ethanol production by ethanologenic microorganisms such as yeasts *Saccharomyces*, *Kluyveromyces*, and the bacterium *Zymomonas mobilis*. Chapters 7 and 8 deal with the topic of ethanol from starchy biomass. Chapter 7 focuses on the production of starch saccharifying enzymes. The enzymes involved in the hydrolysis of starch include α-amylase, α-amylase, glucoamylase, and pullulanase. These enzymes can be obtained from plant and microbial sources but industrial demand is met through the latter. This chapter presents a brief description of the sources, applications, and production of these enzymes. Chapter 8 is on hydrolysis using these enzymes and fermentation. The remaining chapters in this section cover the production of bioethanol from lignocellulosic biomass (Chapter 9), the pretreatment of the substrates and production of cellulases and hemicellulases (Chapter 10), and hydrolysis and fermentation (Chapter 11).

Section 3 of the book is on the production of biodiesel from plant sources and has eight chapters. Chapter 12 in this section discusses current perspectives and the future of biodiesel production. It argues that opportunities for the future for biodiesel include improvements in the conversion technology, which appears promising, and expanding the amount of available feedstock through various plans to increase oil yields or oilseed production. Chapter 13 describes biodiesel production technologies and substrates. Other chapters describe the lipase catalyzed preparation of biodiesel (Chapter 14), biodiesel production with supercritical fluid technologies (Chapter 15), and the production of biodiesel from various plant sources, such as palm oil (Chapter 16), rice bran oil (Chapter 17), karanja and jatropha seed oils (Chapter 18), mahua oil (Chapter 19), and rubber seed oil (Chapter 20).

Each of the chapters incorporates state-of-art information. It is our hope that readers will find the book useful.

The Editor

Professor Ashok Pandey was born in 1956. He graduated from the University of Kanpur in 1974 (Biology) and received his master's degree with first class honors in organic chemistry (1976) and PhD in microbiology (1979) from the University of Allahabad. During 1979 to 1985, he worked as a postdoctoral fellow and scientist in India and Germany. In 1987, he joined the CSIR's National Institute for Interdisciplinary Science and Technology (formerly Regional Research Laboratory) at Trivandrum as scientist, where currently he holds the position of deputy director. He heads the Biotechnology R&D department. Professor Pandey has strong research interest in the area of industrial biotechnology. He has published more than three hundred papers and book chapters. He has edited the *Encyclopedia of Bioresource Technology* (Haworth Press, USA), has written two popular science books and edited sixteen books published by Springer, Kluwer, Asiatech Inc., IBH & Oxford, Wiley Eastern, Doehring Druck, among others. He has acted as guest editor for eleven special issues of journals, which include *Food Technology and Biotechnology, Journal of Scientific & Industrial Research, Applied Biochemistry and Biotechnology, Indian Journal of Biotechnology,* and *Biochemical Engineering Journal.* He is currently editorial board member of seven international journals, four Indian journals and editor of *Bioresource Technology.* He has won several national and international awards, which include the Honorary Doctorate Degree from Blaise Pascal University, Clermont-Ferrand, France in 2007 and the Thomson Scientific India Laureate Award in 2006.

Contributors

Ramakrishnan Anish, MSc
Research Scholar
Biochemical Sciences Division
National Chemical Laboratory
Pune, India

Ajay Kumar Dalai, PhD
Professional Engineer and Canada Research Chair
Department of Chemical Engineering
Catalysis and Chemical Reaction Engineering Laboratories
University of Saskatchewan
Saskatoon, Saskatchewan, Canada

Ayhan Demirbas, PhD
Department of Chemical Engineering
Selcuk University
Konya, Turkey

Muhammed F. Demirbas, PhD
Renewable Energy Researcher
Sila Science
University Mahallesi
Trabzon, Turkey

Hideki Fukuda, PhD
Organization of Advanced Science and Technology
Kobe University
Kobe, Japan

Edgard Gnansounou, PhD
Head, Energy Planning Group
Laboratory of Energy Systems
LASEN-ICARE-ENAC
Ecole Polytechnique Fédérale de Lausanne (EPFL) 1015
Lausanne, Switzerland

Hari Bhagwan Goyal, PhD
Scientist
Indian Institute of Petroleum
Dehradun, India

Paramasamy Gunasekaran, PhD
Head, Department of Genetics
Center for Excellence in Genomic Sciences
School of Biological Sciences
Madurai Kamaraj University
Madurai, India

Ramang bin Hajar, BEng
Research Assistant
Department of Mechanical Engineering
University of Malaya
Kuala Lumpur, Malaysia

Milford A. Hanna, PhD
Director of Industrial Agricultural Products Center
Kenneth E. Morrison Professor of Biological Systems Engineering
University of Nebraska
Lincoln, Nebraska

Masjuki Hj Hassan, PhD
Department of Mechanical Engineering
University of Malaya
Kuala Lumpur, Malaysia

Loren Isom
Technical Assistance Coordinator
Industrial Agricultural Products Center
University of Nebraska
Lincoln, Nebraska

Simon Jayaraj, PhD
Mechanical Engineering Department
National Institute of Technology
Calicut, India

Yi-Hsu Ju, PhD
Department of Chemical Engineering
National Taiwan University of Science and Technology
Taipei, Taiwan

Abul Kalam, MEngSc
Project Manager
Engine Tribology Laboratory
Department of Mechanical Engineering
University of Malaya
Kuala Lumpur, Malaysia

Akihiko Kondo, PhD
Department of Chemical Science and Engineering
Faculty of Engineering
Kobe University
Kobe, Japan

Anil H. Lachke, PhD
Senior Scientist
Division of Biochemical Sciences
National Chemical Laboratory
Pune, India

Ryali Seeta Laxman, PhD
Senior Scientist
Division of Biochemical Sciences
National Chemical Laboratory
Pune, India

Indra Mahlia, PhD
Senior Lecturer
Department of Mechanical Engineering
University of Malaya
Kuala Lumpur, Malaysia

Lekha Charan Meher, PhD
Research Scholar
Centre for Rural Development and Technology
Indian Institute of Technology Delhi,
Hauz Khas,
New Delhi, India

Eiji Minami, PhD
Postdoctoral Fellow
Department of Socio-Environmental Energy Science
Graduate School of Energy Science
Kyoto University
Kyoto, Japan

Chandrashekaran Muraleedharan, PhD
Assistant Professor
Mechanical Engineering Department
National Institute of Technology
Calicut, India

Malaya Kumar Naik, MSc
M Tech Scholar
Centre for Rural Development and Technology
Indian Institute of Technology Delhi,
Hauz Khas,
New Delhi, India

Satya Narayan Naik, PhD
Associate Professor
Centre for Rural Development and Technology
Indian Institute of Technology Delhi,
Hauz Khas,
New Delhi, India

Subhash U. Nair, PhD
Department of Microbiology
Institute of Chemical Technology
University of Mumbai
Mumbai, India

Ashok Pandey, PhD
Deputy Director
Head, Biotechnology Division
National Institute for Interdisciplinary Science and Technology
(formerly Regional Research Laboratory), CSIR
Trivandrum, India

Binod Parameswaran, MSc
Senior Research Fellow
Biotechnology Division
National Institute for Interdisciplinary Science and Technology
(formerly Regional Research Laboratory), CSIR
Trivandrum, India

Rachapudi Badari Narayana Prasad, PhD
Head, Lipid Science & Technology Division
Indian Institute of Chemical Technology
Hyderabad, India

Sukumar Puhan, BSc, ME
Senior Research Fellow
Chemical Engineering Division
Central Leather Research Institute
Chennai, India

Boppana Venkata Ramabrahmam, B Tech, MBA
Chemical Engineering Division
Central Leather Research Institute
Chennai, India

Sumitra Ramachandran, MSc
Laboratoire de Génie Chimique et Biochimique
Polytech'Clermont-Ferrand
University Blasie Pascal
Aubiere, France

Arumugam Sakunthalai Ramadhas, PhD
Research Officer
Engine Testing - Fuels and Hydrogen Department
Indian Oil Corporation Ltd.
Faridabad, India

Bhamidipati Venkata Surya Koppeswara Rao, PhD
Scientist
Lipid Science & Technology Division
Indian Institute of Chemical Technology
Hyderabad, India

Mala Rao, PhD
Scientist
Biochemical Sciences Division
National Chemical Laboratory
Pune, India

Andrea C. M. E. Rayat, MS
Lecturer
Department of Chemical Engineering
University of San Carlos – Technological Center
Cebu City, The Philippines

Shiro Saka, PhD
Director
Department of Socio-Environmental Energy Science
Graduate School of Energy Science
Kyoto University
Kyoto, Japan

Ramesh Chandra Saxena, MSc
Technical Office
Indian Institute of Petroleum
Dehradun, India

Diptendu Seal, MSc
Chemist
Reliance Industries Limited
Jalore, India

Velusamy Senthilkumar, PhD
Postdoctoral Fellow
Department of Genetics
Center for Excellence in Genomic Sciences
School of Biological Sciences
Madurai Kamaraj University
Madurai, India

Reeta Rani Singhania, MSc
Senior Research Fellow
Biotechnology Division
National Institute for Interdisciplinary Science and Technology
(formerly Regional Research Laboratory), CSIR
Trivandrum, India

Tamalampudi Sriappareddy, PhD
Postdoctoral Fellow
Department of Molecular Science and Material Engineering
Faculty of Engineering
Kobe University
Kobe, Japan

Rajeev K. Sukumaran, PhD
Scientist
Biotechnology Division
National Institute for Interdisciplinary Science and Technology
(formerly Regional Research Laboratory), CSIR
Trivandrum, India

Muhammad Redzuan bin Umar, BEng
Research Assistant
Department of Mechanical Engineering
University of Malaya
Kuala Lumpur, Malaysia

Nagarajan Vedaraman, PhD
Technical Officer
Chemical Engineering Division
Central Leather Research Institute
Chennai, India

Muhd Syazly bin Yusuf, BEng
Research Assistant
Department of Mechanical Engineering
University of Malaya
Kuala Lumpur, Malaysia

Section I

General

1 Plant-Based Biofuels
An Introduction

Reeta Rani Singhania, Binod Parameswaran, and Ashok Pandey

CONTENTS

ABSTRACT

With the depletion of oil resources as well as the negative environmental impact associated with the use of fossil fuels, there is a renewed interest in alternate energy sources. As world reserves of fossil fuels and raw materials are limited, it has stimulated active research interest in nonpetroleum, renewable, and nonpolluting fuels. Biofuels are the only viable source of energy for the foreseeable future and can still form the base for sustainable development in terms of socioeconomic and environmental concerns. Biodiesel and bioethanol appear to be promising future energy sources.

1.1 INTRODUCTION

Self-sufficiency in energy requirement is critical to the success of any developing economy. With the depletion of oil resources and the negative environmental impact associated with the use of fossil fuels, there is a renewed interest in alternate energy sources. Apart from the search for alternatives, there is a need to achieve energy independence, directing much focus on biofuels. Biofuels are renewable fuels that are produced predominantly from domestic biomass feedstock, or as a by-product from the industrial processing of agricultural or food products, or from the recovery and reprocessing of products such as cooking and vegetable oil. Bioethanol and biodiesel are the most widely recognized biofuel sources for the transport sector. Biofuel does not contain petroleum, but it can be blended in any proportion with petroleum fuel to create a biofuel blend. It can be used in conventional heating equipment or diesel engines with no major modification.

Biofuel is in the process of acquiring a cult status. It does not provide an opportunity to address issues of energy security and climate change. It is simple to use, biodegradable, nontoxic and essentially free of sulfur and aromatics. By themselves or as blends, biofuels such as bioethanol can help cut substantially both oil imports and carbon emissions. There is a need to evolve the right biofuel model in terms of the feedstock and technology, and a plan of action to ensure availability over the long term. Efforts to identify the right varieties of feedstock and to put in place the research and development, production, and processing facilities on a national scale have been unreliable. Among the crop options, *Jatropha curcas*, a shrub that can sustain itself under difficult climatic and soil conditions, is being considered as a good source for biodiesel.

1.2 THE WORLD ENERGY SCENARIO

The present energy scenario has stimulated active research interest in nonpetroleum, renewable, and nonpolluting fuels. The world reserves of primary energy and raw materials are, obviously, limited. According to an estimate, the reserves will last another 218 years for coal, 41 years for oil, and 63 years for natural gas, under a business-as-usual scenario (Agarwal, 2007). Oil has no equal as an energy source for its intrinsic qualities of extractability, transportability, versatility, and cost. Being the product of the burial and transformation of biomass over the last 200 million years, the amount of underground oil is finite. Hence, there is an urgent need to understand the world energy crisis and the underlying science behind it, and of course, to transition to sustainable energy sources. Concerns have arisen in recent years about the relationship between the growing consumption of oil and the availability of oil reserves, as well as the impact of the potentially dwindling supplies and rising prices on the world's economy and social welfare. Oils can be derived from conventional and nonconventional sources of energy. A conventional source is one that uses the present mainstream technologies, whereas nonconventional sources are those that require more complex or more expensive technologies. The additional cost and technological challenges surrounding the production of the nonconventional sources make these resources more uncertain.

Oil accounts for approximately one-third of all the energy used in the world. Following the record oil prices associated with the Iranian revolution in 1979 to 1980 and with the start of the Iran-Iraq war in 1980, there was a drop in the total world oil consumption, from about 63 million barrels per day in 1980 to 59 million barrels per day in 1983. Since then, however, world consumption of petroleum products has increased, totaling about 84 million barrels per day in 2005 (GAO-07-283). Future world demand for oil is uncertain because it depends on economic growth and government policies throughout the world. Rapid economic growth in China and India could significantly increase world demand for oil, while environmental concerns, including oil's contribution to global warming, may spur conservation or the adoption of alternative fuels that would reduce future demand for oil. Being the fifth largest energy consumer, India imported nearly 70% of its crude oil requirement (90 million tonnes) during 2003–04. Estimates indicate that this figure will rise to 95% by 2030 (*World Energy Outlook* 2005)

Reserves of petrol or gasoline, which is a complex mixture of hundreds of different hydrocarbons, are finite. The NO_x, SO_2, CO_2, and particulate matter that cause pollution are emissions from engines using gasoline. Tetraethyl lead (TEL) improves the antiknocking rating of gasoline when used as an additive. The addition of oxygenated compounds helps in antiknocking. TEL and benzene have been banned because of the harmful effects of the lead in the former and the carcinogenic property of the latter. Although natural gas remains available, it is more efficient to use methane directly and less carbon for the useful energy obtained is released. The production of hydrogen from natural gas does not appear to be worthwhile. A better option, if sufficient quantities of liquefied natural gas (LNG) will be imported, would be to transfer it directly from a tanker to containers on road vehicles. This would avoid liquefying the gas a second time.

The growth in demand for oil and gas is rising exponentially. The combined world production of oil and gas in equivalent units rose by around 1.50% per annum from 1990 to 2000, but from 1999 to 2000, it rose by 4%. As the reserves approach exhaustion, demand is accelerating, bringing the emptying of reserves ever nearer. Over the last five years, oil consumption has increased by 11% and gas consumption by 14%. The world has large coal resources, with a R/P ratio in 2005 of 155 (compare oil 41 and gas 65). As gas and oil production rates approach their Hubbert peaks in 2010 and 2020, respectively, coal resources will be utilized at a higher rate as substitutes for liquid fuels and chemical products. Also, the thermal efficiency of the coal liquefaction process is around half that of gas-to-liquids equivalents, increasing its demand enormously. If, hypothetically, coal were provided for the combined world 2005 production rates for oil, gas, and coal, its production would peak around 2040 to 2050, tailing off thereafter, indicating that fossil fuels will not outlast the century. Petroleum prices, kept low for political reasons, are responsible for the lack of enthusiasm for private investment in alternative energy sources. Rather than subsidizing alternative energy sources, it would be better to increase the fuel tax which in turn will attract private funding of alternatives. The tax on alternatives should be lower than that for petroleum-based fuels or removed altogether as a further incentive. More fundamental assessment should be done while considering alternative sources, that is, the ratio of energy inputs required to build, supply, and maintain the system to the energy output over the plant life cycle.

Thus, biofuels from feedstock are apparently the only foreseeable alternative sources of energy that can efficiently replace petroleum-based fuels in the long term.

1.3 RENEWABLE ENERGY

Before proceeding further, let us look at the different kinds of renewable energy sources available. Renewable energy is the energy derived from resources that are regenerative or, for all practical purposes, cannot be depleted. For this reason, renewable energy sources are fundamentally different from fossil fuels, and do not produce as many greenhouse gases and other pollutants as fossil fuel combustion. Renewable energy sources like wind, solar, geothermal, hydrogen, and biomass play an important role in the future of our energy demand.

1.3.1 Hydroelectricity

Hydroelectricity is electricity produced by hydropower. It is the world's leading form of renewable energy, accounting for over 63% of the total in 2005. The long-term technical potential is believed to be nine to twelve times the current hydropower production, but environmental concerns increasingly block new dam construction. There are also growing interests in mini-hydro projects, which avoid many of the problems of the larger dams.

1.3.2 Solar Power

Solar power (also known as solar energy) is the technology of obtaining usable energy from sunlight. It has been used in many traditional technologies for centuries, and has come into widespread use where other sources of power are absent, such as in remote locations and in space. Commercial solar cells can presently convert about 20% of the energy of incident sunlight to electrical energy. Solar energy is currently used in a number of applications, such as for the generation of heat for boiling water, heating rooms, cooking, electricity generation in photovoltaic cells and heat engines, and for desalination of seawater.

1.3.3 Wind Power

Wind power is the conversion of wind energy into electricity using wind turbines. It is one of the most cost-competitive renewable energy sources today. Its long-term technical potential is believed to be five times the current global energy consumption, or 40 times current electricity demand. It currently produces less than 1% of worldwide electricity use and accounts for approximately 20% of electricity use in Denmark, 9% in Spain, and 7% in Germany. The wind energy is ample, renewable, widely distributed, clean, and reduces toxic atmospheric and greenhouse gas emissions if used to replace fossil-fuel-derived electricity. The intermittency of wind seldom creates problems when using wind power at low to moderate penetration levels.

1.3.4 Geothermal Power

Geothermal power is the use of geothermal heat to generate electricity. Geothermal power and tidal power are the only renewable energy sources not dependent on the sun but are today limited to special locations. Geothermal power has a very large potential if all the heat existing inside the earth is considered, although the heat flow from the interior to the surface is only 1/20,000 as great as the energy received from the sun or about two to three times that from tidal power. At the moment, Iceland and New Zealand are two of the largest users of geothermal energy, although many others also have potential.

1.3.5 Tidal Power

Tidal power is the power achieved by capturing the energy contained in the moving water in tides and ocean currents. It is classified as a renewable energy source, because tides are caused by the orbital mechanics of the solar system and are considered inexhaustible. The root source of the energy is the orbital kinetic energy of the earth-moon system, and also the earth-sun system. The tidal power has great potential for future power and electricity generation because of the essentially inexhaustible amount of energy contained in these rotational systems. All the available tidal energy is equivalent to one-fourth of the total human energy consumption today. Tidal power is reliably predictable (unlike wind energy and solar power). In Europe, tide mills have been used for nearly a thousand years, mainly for grinding grains. Several smaller tidal power plants have recently started generating electricity in Canada and Norway. They all exploit the strong periodic tidal currents in narrow fiords using subsurface water turbines.

1.3.6 Biofuels

Biomass (produced by burning biological materials to generate heat), biofuels (produced by processing biological materials to generate fuels such as biodiesel and bioethanol), and biogas (using anaerobic digestion to generate methane from biodegradable materials and wastes) are other renewable sources of energy to produce an array of energy-related products, including electricity, liquid, solid, and gaseous fuels, heat, chemicals, and other materials. The term *biomass* refers to any plant-derived organic matter available on a renewable basis, including dedicated energy crops and trees, agricultural food and feed crops, agricultural crop wastes and residues, wood wastes and residues, aquatic plants, animal wastes, municipal wastes, and other waste materials. There are several reasons for biofuels to be considered as relevant technologies by both developing and industrialized countries. These include energy security, environmental concerns, foreign exchange savings, and socioeconomic issues related to the rural sector (Demirbas 2007).

1.4 BIOFUELS FOR THE TRANSPORTATION SECTOR

The transportation sector is one of the major consumers of fossil fuels and the biggest contributor to environmental pollution, which can be reduced by replacing the mineral-based fuels with bio-origin renewable fuels. There are a variety of biofuels potentially available, but the main biofuels being considered globally are biodiesel and bioethanol for the transport sector. Bioethanol can be produced from a number of crops, including sugarcane, corn (maize), wheat, and sugar beet. Biodiesel is a fuel that can be produced from straight vegetable oils, edible and nonedible, recycled waste vegetable oils, and animal fat.

In the past few years, biofuel programs have gained new momentum, as a result of rising prices of petroleum fuels as well as the advent of flex-fuel vehicles which can utilize different percentages of ethanol blended with gasoline. The governments of many countries grant subsidies and tax reductions to promote the assimilation of bioethanol. The cultivation, processing, and use of liquid fuels emit less climate-relevant CO_2 than the fossil fuels. Biofuel also has the advantage of being biodegradable. The seed oils are combustibles that have great potential to be used as biofuels. They essentially comprise the triglycerides of the long chain saturated and unsaturated fatty acids.

Pure ethanol is rarely used for transportation; instead, it is usually mixed with gasoline. The most popular blend for light-duty vehicles is E85, which is 85% ethanol and 15% gasoline. Ethanol contains more oxygen than gasoline; its use favors more complete combustion, as a consequence of which emissions of hydrocarbons and particulate matter, which result from incomplete combustion of gasoline, is reduced. In fact, bioethanol demand will grow very fast until 2015 owing to the ban on methyl tert-butyl ether (MTBE) in gasoline, new legislations favoring biofuels and the evolution of flexible-fuel vehicles in Brazil and other parts of the world. However, it is apparent that on a volume basis bioethanol is not presently competitive, apart from Brazil, assuming that the production cost of gasoline is around US$0.25/l. Due to its ecological merits, the share of bioethanol in the automobile fuel market will peak.

Ethanol can be blended with gasoline to save the use of fossil fuels that cause greenhouse gas emission resulting in climate changes. According to the Federation of Indian Chambers of Commerce and Industry, India can save nearly 80 million liters of petrol annually if it is blended with ethanol by 10%. It is one of the possible fuels for diesel replacement in compression ignition (CI) engines. The alcohols have higher octane number than the gasoline. A fuel with a higher octane number can endure higher compression ratios before the engine starts knocking, thus giving the engine the ability to deliver more power efficiently and economically. The alcohol burns cleaner than regular gasoline and produces less carbon monoxide, hydrocarbons (HCs), and oxides of nitrogen. It has higher heat of vaporization; therefore, it reduces the peak temperature inside the combustion chamber leading to lower NO_x emissions and increased engine power. However, the aldehyde emissions go up significantly, which play an important role in the formation of photochemical smog.

Blends of ethanol in gasoline are commonly used in vehicles designed to operate on gasoline; however, vehicle modification is required for alcohol fueling because its properties are different from those of gasoline. It has low stoichiometric air-fuel ratio and high heat of vaporization that require carburetor recalibration and increased heating of the air-fuel mixture to provide satisfactory driveability. Ethanol-diesel blends up to 20% can be used in present-day constant-speed CI engines without modification. Up to a 62% reduction in CO emission is possible with the use of ethanol-diesel blends as compared to diesel alone. NO_x emissions are also reduced (up to 24%) when using ethanol-diesel blends.

Biodiesel and bioethanol are eco-friendly, even when considered on a life cycle basis. They have the lowest life cycle greenhouse gas (GHG) emissions (in grams GHG per kilometer traveled). In fact, both emit larger quantities of CO_2 than the conventional fuels, but as most of this is from renewable carbon stocks that fraction is not counted towards the GHG emissions from the fuel. CO, formed by the incomplete combustion of the fuels, is produced most readily from the petroleum fuels, which contain no oxygen in their molecular structure. Because ethanol and other "oxygenated" compounds contain oxygen, their combustion in automobile engines is more complete. The result is a substantial reduction in CO emissions. Because of its high octane rating, adding ethanol to gasoline leads to reduction or removal of aromatic HCs (such as benzene), and other hazardous high-octane additives commonly used to replace TEL in gasoline. Because of its effect in reducing the HC and CO in exhaust, adding ethanol to gasoline results in an overall reduction in the ozone-forming potential of exhaust.

Using ethanol as a fuel additive to unleaded gasoline causes an improvement in engine performance and exhaust emissions. The ethanol addition results in an improvement in brake power, brake thermal efficiency, volumetric efficiency, and fuel consumption; however, the brake specific fuel consumption and equivalence air-fuel ratio decrease because of the lower calorific value of the gasohol.

Biofuels can still form the basis for sustainable development in terms of socio-economic and environmental concerns. Currently, bioethanol and biodiesel have already reached commercial market, to be used as blends with petro-fuels.

1.5 STATUS OF BIOFUEL

Several initiatives have been taken in recent years to link poor and marginal farmers in arid lands in different parts of the world, especially the developing countries, with the global biofuels revolution without compromising food security. The main motive is to benefit these farmers. The rising price of the fuel could provide opportunity to serve the purpose. Innovative research on bioethanol from sweet sorghum and biodiesel from *Jatropha* and *Pongamia* ensures energy, livelihood and food security to farmers in arid regions, as well as reduces the use of fossil fuel, which can help in extenuating climate change. These crops do not require much water, can withstand stress, are not expensive to cultivate, and are thus suitable crops to be cultivated by farmers in dry lands. There is a need to develop partnerships between the public and private sectors in such initiatives so that they can be economically beneficial to poor farmers.

India has a rich biomass resource which can be converted into renewable energy. The Planning Commission of the government of India has launched an ambitious National Mission on biodiesel to be implemented by a number of government agencies and coordinated by the Ministry of Rural Development. The mission focuses on the cultivation of the physic nut, *Jatropha curcas*, a shrubby plant of the castor family. The seed contains 30 to 40% oil and can be mixed with diesel after transesterification.

1.6 BIOETHANOL

Ethanol is an alcohol-based fuel produced by fermenting plant sugars. It can be made from many agricultural products and food wastes if they contain sugar, starch, or cellulose, which can then be fermented and distilled into ethanol. The technology for producing ethanol, at least from certain feedstocks, is generally well established, and ethanol is currently produced in many countries around the world, but current efforts are underway mainly to develop methods for producing ethanol from lignocellulosic biomass, including forest trimmings and agricultural residues (cellulosic ethanol).

Ethanol is produced mainly from sugar as it is the cheapest means. In Brazil, which is the largest ethanol producer, ethanol is produced from sugarcane. Brazil has the lowest cost of production worldwide and is in a position to capture a large share of the international market in the future. India is one of the largest producers as well as consumers of sugar, hence, it is not envisaged to use sugarcane for the production of ethanol, at least in the near future. In India, molasses is the major raw material for ethanol production, but it cannot fulfill the demand when used for the automotive sector.

Sweet sorghum competes with sugarcane for ethanol production. Sweet sorghum possesses some advantages over sugarcane as it can be grown in dry conditions, requiring one-seventh the amount of water required by sugarcane. Though the ethanol yield per unit weight of feedstock is less, the lower production cost from sweet sorghum compensates for the loss. Sweet sorghum has a competitive cost advantage. The production cost of ethanol from sweet sorghum and sugarcane is about US$0.29 and 0.33 per liter, respectively (ICRISAT 2007).

Currently, efforts are being made to produce ethanol from a variety of agricultural products, including trees, grasses, and forestry residues, which are of considerable interest, as the lignocellulosic materials are being seen as the only foreseeable source of energy (Sukumaran, Reeta, and Pandey 2005). The U.S. Departments of Energy (DOE) and Agriculture (USDA) project that more cellulosic ethanol will ultimately be produced than corn ethanol because cellulosic ethanol can be produced from a variety of feedstocks, but more fundamental reductions in the production costs will be needed to make cellulosic ethanol commercially viable (Gnansounou and Dauriat 2005). The production of ethanol from cellulosic feedstocks is currently more costly than the production of corn ethanol because the cellulosic material must first be broken down into fermentable sugars that can be converted into ethanol. The production costs associated with this additional processing would have to be reduced in order to make cellulosic ethanol cost-competitive with gasoline at today's prices.

However, corn and cellulosic ethanol are more corrosive than gasoline, and the widespread commercialization of these fuels would require substantial retrofitting of the refueling infrastructure—pipelines, storage tanks, and filling stations. To store the ethanol, gasoline stations may have to retrofit or replace their storage tanks, at an estimated cost of $100,000 per tank. The DOE also reported that some private firms consider significant capital investment in ethanol refineries to be risky, unless the future of alternative fuels becomes more certain. Finally, the widespread use of ethanol would require a turnover in the vehicle fleet because the most current vehicle engines cannot effectively burn ethanol in high concentrations.

The current cost of producing ethanol from corn is between US$0.90 and $1.25 per gallon, depending on the size of the production plant, transportation cost for the corn, and the type of fuel used to provide the steam and other energy needs for the facility. The current cost of producing ethanol from biomass is not cost competitive, but the cost is expected to drop significantly, to about $1.07 per gallon by 2012. The The key infrastructure costs associated with ethanol include retrofitting refueling stations to accommodate E85 (estimated at between $30,000 and $100,000) and constructing or modifying the pipelines to transport the ethanol. The 2005 production of ethanol in the United States was approximately 4 billion gallons. By 2014–15, corn ethanol production is expected to peak at approximately 9 billion to 18 billion gallons annually. Assuming success with cellulosic ethanol technologies, the experts project cellulosic ethanol production levels of over 60 billion gallons by 2025–30. Corn ethanol is produced commercially today and production continues to expand rapidly.

For cellulosic ethanol, the economic challenges are the initial capital investment and high feedstock and production costs. Several technical challenges still remain, including improving the enzymatic pretreatment, fermentation, and process integration.

1.7 BIODIESEL

Biodiesel is the methyl or ethyl ester of the fatty acid made from virgin or used vegetable oils (both edible and nonedible) and animal fat. The main resources for biodiesel production can be nonedible oils obtained from plant species such as *Jatropha curcas* (ratanjyot), *Pongamia pinnata* (karanj), *Calophyllum inophyllum* (nagchampa), *Hevea brasiliensis* (rubber), etc. (Swarup 2007). Biodiesel can be blended in any proportion with mineral diesel to create a biodiesel blend or can be used in its pure form. Just like petroleum diesel, biodiesel operates in the compression ignition (diesel) engine, and essentially requires very little or no engine modifications because the biodiesel has properties similar to mineral diesel. It can be stored just like mineral diesel and hence does not require separate infrastructure. The use of biodiesel in conventional diesel engines results in substantial reduction in the emission of unburned hydrocarbons, carbon monoxide, and particulates.

Biodiesel is currently used in small quantities in the United States, but is not cost-competitive with gasoline or diesel. The cost of the biodiesel feedstocks, which in the United States largely consist of soybean oil, is the largest component of the production costs. The price of soybean oil is not expected to decrease significantly in the near future owing to competing demands from the food industry and from soap and detergent manufacturers. These competing demands, as well as the limited land available for the production of feedstocks, also are projected to limit biodiesel's capacity for large-volume production, according to the DOE and USDA. As a result, the experts believe that the total production capacity of biodiesel is ultimately limited, compared with other alternative fuels.

Like petroleum diesel, biodiesel operates in compression ignition engines. Blends of up to 20% biodiesel (B20) can be used in nearly all diesel equipment and are compatible with most storage and distribution equipment. These low-level blends generally do not require any engine modifications. Higher blends and 100% biodiesel (B100) may be used in some engines with little or no modification, although the transportation and storage of B100 requires special management.

The current wholesale cost of pure biodiesel (B100) ranges from about US$2.90 to $3.20 per gallon, although recent sales have also been reported at $2.75 per gallon (GAO-07-283). To date, there has been limited evaluation of the projected infrastructure costs required for biodiesel. However, it is acknowledged that there are infrastructure costs associated with the installation of manufacturing capacity, distribution, and blending of the biodiesel.

In 2005, the U.S. production of biodiesel was 75 million gallons, and the DOE projected about 3.6 billion gallons per year by 2015 (GAO-07-283). Under a more speculative scenario requiring major changes in land use and price supports, the experts project that it would be possible to produce 10 billion gallons of biodiesel per year. Although biodiesel is commercially available, in many ways it is still in the development and demonstration stages. Key areas of focus for development and demonstration include quality, warranty coverage, and impact of air pollutant emissions and compatibility with advanced control systems. The experts project (GAO-07-283) that with adequate resources, the key remaining difficulties could be resolved in the next five years. Initial capital costs are significant and the technical learning curve is steep, which deters many potential investors. The economic challenges are significant for biodiesel. The DOE is currently collaborating with the biodiesel and automobile industries in funding research and development efforts on biodiesel use, and the USDA is conducting research on feedstocks.

REFERENCES

Agarwal, A. K. 2007. Biofuels (alcohols and biodiesel) applications as fuels for internal combustion engines. *Progress in Energy and Combustion Science* 33: 233–271.

Demirbas, A. 2007. Progress and recent trends in biofuels. *Progress in Energy and Combustion Science* 33: 1–18.

GAO-07-283. n.d. Crude Oil-Uncertainty about Future Oil Supply Makes It Important to Develop a Strategy for Addressing a Peak and Decline in Oil Production. http://www.gao.gov/cgi-bin/getrpt?GAO-07-283

Gnansounou, E. and A. Dauriat. 2005. Ethanol fuel from biomass: A review. *Journal of Scientific and Industrial Research* 64: 809–821.

ICRISAT. 2007. ICRISAT promotes pro-poor biofuels initiatives. *Advanced Biotech* 15(10): 8–10.

Sukumaran, R. K., R. S. Reeta, and A. Pandey. 2005. Microbial cellulases: Production, applications and challenges. *Journal of Scientific and Industrial Research* 64: 832–844.

Swarup, R. 2007 (April). Biofuels: Breathing new fire. *Biotech News* II(2): 14.

World Energy Outlook. 2005. Paris: International Energy Agency.

2 World Biofuel Scenario

Muhammed F. Demirbas

CONTENTS

ABSTRACT

The term biofuel refers to liquid or gaseous fuels mainly for the transport sector that are predominantly produced from plant biomass. There are several reasons for biofuels to be considered as relevant technologies by both developing and industrialized countries. These include energy security, environmental concerns, foreign exchange savings, and socioeconomic issues, mainly related to the rural sector. A large number of research projects in the field of thermochemical and biochemical conversion of biomass, mainly on liquefaction, pyrolysis, and gasification, have been carried out. Liquefaction is a thermochemical conversion process of biomass or other organic matters into primarily liquid oil products in the presence of a reducing reagent, for example, carbon monoxide or hydrogen. Pyrolysis products are divided into a volatile fraction, consisting of gases, vapors, and tar components, and a carbon-rich solid residue. The gasification of biomass is a thermal treatment, which results in a high production of gaseous products and small quantities of char and ash. Bioethanol is a petrol additive/substitute. It is possible that wood, straw, and even household wastes may be economically converted to bioethanol. Bioethanol is derived from alcoholic fermentation of sucrose or simple sugars, which are produced from biomass by hydrolysis process. There has been renewed interest in the use of vegetable oils for making biodiesel due to its less polluting and renewable nature as against the conventional petroleum diesel fuel. Methanol is mainly manufactured from natural gas, but biomass can also be gasified to methanol. Methanol can be produced from hydrogen-carbon oxide mixtures by means of the catalytic reaction of carbon monoxide and some carbon dioxide with hydrogen. Bio-synthesis gas (bio-syngas) is a gas rich in CO and H_2 obtained by gasification of biomass. Biomass sources are preferable for biomethanol, than for bioethanol because bioethanol is a high-cost and low-yield product. The aim of this chapter is to present an overview of the production of biofuels from biomass materials by thermochemical and biochemical methods and utilization trends for the products in the world.

2.1 INTRODUCTION

The term biofuel refers to liquid or gaseous fuels for the transport sector that are predominantly produced from biomass. Biofuels are important because they replace petroleum fuels. Biofuels are generally considered as offering many priorities, including sustainability, reduction of greenhouse gas emissions, regional development, social structure and agriculture, and security of supply (Reijnders 2006). Worldwide energy consumption has increased seventeen-fold in the last century and emissions of CO_2, SO_2, and NO_x from fossil-fuel combustion are primary causes of atmospheric pollution. Known petroleum reserves are estimated to be depleted in less than 50 years at the present rate of consumption (Sheehan et al. 1998). In developed countries there is a growing trend toward employing modern technologies and efficient bioenergy conversion using a range of biofuels, which are becoming cost competitive with fossil fuels (Puhan et al. 2005). The demand for energy is increasing at an exponential rate due to the exponential growth of the world's population. Advanced energy-efficiency technologies reduce the energy needed to provide energy services, thereby reducing environmental and national security costs of using energy and potentially increasing its reliability.

Biomass is composed of organic carbonaceous materials such as woody or lignocellulosic materials, various types of herbage, especially grasses and legumes, and crop residues. Biomass can be converted to various forms of energy by numerous technical processes, depending upon the raw material characteristics and the type of energy desired. Biomass energy is one of humanity's earliest sources of energy. Biomass is used to meet a variety of energy needs, including generating electricity, heating homes, fueling vehicles, and providing process heat for industrial facilities. Biomass is the most important renewable energy source in the world and its importance will increase as national energy policies and strategies focus more heavily on renewable sources and conservation. Biomass power plants have advantages over fossil-fuel plants, because their pollution emissions are less. Energy from biomass fuels is used in the electric utility, lumber and wood products, and pulp and paper industries. Biomass can be used directly or indirectly by converting it into a liquid or gaseous fuel.

The aim of this chapter is to present an overview of the production of biofuels from biomass materials by thermochemical and biochemical methods and utilization trends for the products in the world.

2.2 BIOMASS LIQUEFACTION

Liquefaction is a thermochemical conversion process of biomass or other organic matters into primarily liquid oil products in the presence of a reducing reagent, for example, carbon monoxide or hydrogen. Liquefaction is usually conducted in an environment of moderate temperatures (from 550 to 675 K) and high pressures. Aqueous liquefaction of lignocellulosic materials involves disaggregation of the wood ultrastructure followed by partial depolymerization of the constitutive families (hemicelluloses, cellulose, and lignin). Solubilization of the depolymerized material is then possible (Chornet and Overend 1985).

During liquefaction, hydrolysis and repolymerization reactions occur. At the initial stage of liquefaction, biomass is thermochemically degraded and depolymerized to small compounds, and then these compounds may rearrange through condensation, cyclization, and polymerization to form new compounds in the presence of a suitable catalyst. With pyrolysis, on the other hand, a catalyst is usually unnecessary, and the light decomposed fragments are converted to oily compounds through homogeneous reactions in the gas phase (Demirbas, 2000). The differences in operating conditions for liquefaction and pyrolysis are shown in Table 2.1.

The alkali (NaOH, Na_2CO_3, or KOH) catalytic aqueous liquefaction of wood to oils may be a promising process to make good use of them. Liquid products obtained from the wood samples could eventually be employed as fuels or other useful chemicals after suitable refining processes.

Liquefaction was linked to hydrogenation and other high-pressure thermal decomposition processes that employed reactive hydrogen or carbon monoxide carrier gases to produce a liquid fuel from organic matter at moderate temperatures, typically between 550 and 675 K. Direct liquefaction involves rapid pyrolysis to produce liquid tars and oils and/or condensable organic vapors. Indirect liquefaction involves the use of catalysts to convert noncondensable, gaseous products of pyrolysis or gasification into liquid products. In the liquefaction process, the carbonaceous materials are converted to liquefied products through a complex sequence of physical structure and chemical changes. The changes involve all kinds of processes such as solvolysis, depolymerization, decarboxylation, hydrogenolysis, and hydrogenation. Solvolysis results in micellar-like substructures of the biomass. The depolymerization of biomass leads to smaller molecules. It also leads to new molecular rearrangements through dehydration and decarboxylation. When hydrogen is present, hydrogenolysis and hydrogenation of functional groups, such as hydroxyl groups, carboxyl groups, and keto groups also occur (Chornet and Overend 1985). The micellar-like broken down fragments produced by hydrolysis are then degraded

TABLE 2.1
Comparison of Liquefaction and Pyrolysis

Thermochemical Process	Temperature (K)	Pressure (MPa)	Drying
Liquefaction	525–600	5–20	Unnecessary
Pyrolysis	650–800	0.1–0.5	Necessary

to smaller compounds by dehydration, dehydrogenation, deoxygenation, and decarboxylation (Demirbas 2000).

The heavy oil obtained from the liquefaction process is a viscous tarry lump, which sometimes caused troubles in handling. For this reason, organic solvents are added to the reaction system. Among the organic solvents tested, propanol, butanol, acetone, methyl ethyl ketone, and ethyl acetate were found to be effective for the formation of heavy oil having low viscosity.

Alkaline degradation of whole biomass or of its separate constituent components (cellulose and lignin) leads to a very complex mixture of chemical products. In turn, these compounds, due to their greater variance in structure, must involve extensive and complex mechanistic pathways for their production. Clarification of these mechanisms should lead to a better understanding of the conversion process. Several distinctly different classes of compounds, including mono- and dinuclear phenols, cycloalkanones and cycloalkanols, and polycyclic and long chain alkanes and alkenes, were identified by Eager, Pepper, and Roy (1983).

2.3 BIOMASS PYROLYSIS

Pyrolysis seems to be a simple and efficient method to produce gasoline and diesel-like fuels. Hydrocarbons from biomass materials were used as raw materials for gasoline and diesel-like fuel production in a cracking system similar to the petroleum process now used. Pyrolysis is the thermal decomposition of biomass by heat in the absence of oxygen, which results in the production of char, bio-oil, and gaseous products. Thermal decomposition in an oxygen-deficient environment can also be considered to be true pyrolysis as long as the primary products of the reaction are solids or liquid. Three-step mechanism reactions for describing the kinetics of the pyrolysis of biomass can be proposed:

$$\text{Virgin biomass} \rightarrow \text{Char}_1 + \text{Volatile}_1 + \text{Gases}_1 \tag{2.1}$$

$$\text{Char}_1 \rightarrow \text{Char}_2 + \text{Volatile}_2 + \text{Gases}_2 \tag{2.2}$$

$$\text{Char}_2 \rightarrow \text{Carbon-rich solid} + \text{Gases}_3 \tag{2.3}$$

The most interesting temperature range for the production of the pyrolysis products from biomass is between 625 and 775 K. The charcoal yield decreases as the temperature increases. The production of the liquid products has a maximum at temperatures between 625 and 725 K. The main pyrolysis applications and their variants are listed in Table 2.2. Conventional pyrolysis is defined as pyrolysis that occurs at a slow rate of heating. The first stage of biomass decomposition, which occurs between 395 and 475 K, can be called pre-pyrolysis. During this stage some internal rearrangement, such as water elimination, bond breakage, appearance of free radicals, and the formation of carbonyl, carboxyl, and hydroperoxide groups, takes place. The second stage of the solid decomposition corresponds to the main pyrolysis process. It proceeds at a high rate and leads to the formation of the pyrolysis products. During the third stage, the char decomposes at a very slow rate and carbon-rich residual solid forms.

TABLE 2.2
Main Pyrolysis Applications and Their Variants

Method	Residence Time	Temperature (K)	Heating Rate	Products
Carbonization	Days	675	Very low	Charcoal
Conventional	5–30 min	875	Low	Oil, gas, char
Fast	0.5–5 s	925	High	Bio-oil
Flash-liquid	<1 s	<925	Very high	Bio-oil
Flash-gas	<1 s	<925	Very high	Chemicals, gas
Hydro-pyrolysis	<10 s	<775	High	Bio-oil

Biomass is a mixture of structural constituents (hemicelluloses, cellulose, and lignin) and minor amounts of extractives which each pyrolyse at different rates and by different mechanisms and pathways. It is believed that as the reaction progresses the carbon becomes less reactive and forms stable chemical structures, and consequently the activation energy increases as the conversion level of biomass increases. Lignin decomposes over a wider temperature range compared to cellulose and hemicelluloses, which degrade rapidly over narrower temperature ranges, hence the apparent thermal stability of lignin during pyrolysis.

In the thermal depolymerization and degradation of biomass, cellulose, hemicelluloses, and products are formed, as well as a solid residue of charcoal. The mechanism of the pyrolytic degradation of structural components of the biomass samples were separately studied (Demirbas 2000). If wood is completely pyrolysed, the resulting products are about what would be expected by pyrolysing the three major components separately. The hemicelluloses break down first, at temperatures of 470 to 530 K and cellulose follows in the temperature range 510 to 620 K, with lignin being the last component to pyrolyse, at temperatures of 550 to 770 K (Demirbas 2000).

The pyrolysis of lignin has been studied widely (Demirbas 2000). Its pyrolysis products, of which guaiacol is that chiefly obtained from coniferous wood, and guaiacol and pyrogallol dimethyl ether show the aromatic nature of lignin from deciduous woods. Lignin gives higher yields of charcoal and tar from wood although lignin has a threefold higher methoxyl content than wood. Cleavage of the aromatic C-O bond in lignin leads to the formation of one-oxygen atom products and the cleavage of the methyl C-O bond to form two-oxygen atom products is the first reaction to occur in the thermolysis of 4-alkylguaiiacol at 600 to 650 K. Cleavage of the side chain C-C bond occurs between the aromatic ring and α-carbon atom.

The liquid fraction of the pyrolysis products consists of two phases: an aqueous phase containing a wide variety of organo-oxygen compounds of low molecular weight and a nonaqueous phase containing insoluble organics (mainly aromatics) of high molecular weight. This phase is called bio-oil or tar and is the product of greatest interest. The ratios of acetic acid, methanol, and acetone of the aqueous phase are higher than those of the nonaqueous phase. If the purpose were to maximize the yield of liquid products resulting from biomass pyrolysis, a process involving low temperature, high heating rate, and short gas residence time would be required. For a high char production, a low temperature, low heating rate process would be chosen.

If the purpose was to maximize the yield of fuel gas resulting from pyrolysis, a high temperature, low heating rate, long gas residence time process would be preferred.

2.4 BIOMASS GASIFICATION

Gasification describes the process in which oxygen-deficient thermal decomposition of organic matter primarily produces noncondensable fuel or synthesis gases. The gasification of biomass is a thermal treatment, which results in a high production of gaseous products and small quantities of char and ash. Gasification generally involves pyrolysis as well as combustion to provide heat for the endothermic pyrolysis reactions. Gasification of biomass is well-known technology that can be classified depending on the gasifying agent: air, steam, steam-oxygen, air-steam, O_2-enriched air, etc. The main gasification reactors are designed as fixed-bed, fluidized-bed, or moving-bed reactors. Fixed-bed gasifiers are the most suitable for biomass gasification. Fixed-bed gasifiers are usually fed from the top of the reactor and can be designed in either updraft or downdraft configurations. The gasification of biomass in fixed-bed reactors provides the possibility of combined heat and power production in the power range of 100 kWe up to 5 MWe. With fixed-bed updraft gasifiers, the air or oxygen passes upward through a hot reactive zone near the bottom of the gasifier in a direction counter-current to the flow of solid material. Fixed-bed downdraft gasifiers were widely used in World War II for operating vehicles and trucks. During operation, air is drawn downward through a fuel bed; the gas in this case contains relatively less tar compared with the other gasifier types.

Fluidized-bed gasifiers are a more recent development that takes advantage of the excellent mixing characteristics and high reaction rates of this method of gas-solid contacting. The fluidized bed gasifiers are typically operated at 1075 to 1275 K. Heat to drive the gasification reaction can be provided in a variety of ways in fluidized-bed gasifiers. Direct heating occurs when air or oxygen in the fluidizing gas partially oxidizes the biomass and heat is released by the exothermic reactions that occur. At temperatures of approximately 875 to 1275 K, solid biomass undergoes thermal decomposition to form gas-phase products that typically include hydrogen, CO, CO_2, methane, and water. In most cases, solid char plus tars that would be liquids under ambient conditions are also formed. The product distribution and gas composition depends on many factors, including the gasification temperature and the reactor type.

Assuming a gasification process using biomass as a feedstock, the first step of the process is a thermochemical decomposition of the lignocellulosic compounds with production of char and volatiles. Further, the gasification of char and some other equilibrium reactions occur as shown in Equations 2.4 to 2.7.

$$C + H_2O \text{ D } CO + H_2 \tag{2.4}$$

$$C + CO_2 \text{ D } 2CO \tag{2.5}$$

$$CO + H_2O \text{ D } H_2 + CO_2 \tag{2.6}$$

$$CH_4 + H_2O \text{ D } CO + 3H_2 \tag{2.7}$$

2.5 GREEN DIESEL FUEL FROM BIO-SYNGAS VIA FISHER-TROPSCH SYNTHESIS

Gasification processes provide the opportunity to convert renewable biomass feedstocks into clean fuel gases or synthesis gases. The synthesis gas includes mainly hydrogen and carbon monoxide (H_2 + CO) which is also called bio-syngas. To produce bio-syngas from a biomass fuel, the following procedures are necessary: (1) gasification of the fuel, (2) cleaning of the product gas, (3) use of the synthesis gas to produce chemicals, and (4) use of the synthesis gas as energy carrier in fuel cells. Bio-syngas is a gas rich in CO and H_2 obtained by gasification of biomass. In the steam-reforming reaction of a biomass material, steam reacts with hydrocarbons in the feed to predominantly produce bio-syngas. Figure 2.1 shows the production of diesel fuel from bio-syngas by Fisher-Tropsch synthesis (FTS).

The Fischer–Tropsch synthesis was established in 1923 by German scientists Franz Fischer and Hans Tropsch. The main aim of FTS is the synthesis of long-chain hydrocarbons from CO and H_2 gas mixture. The use of iron-based catalysts is attractive due to their high FTS activity as well as their water-gas shift reactivity, which helps make up the deficit of H_2 in the syngas from modern energy-efficient coal gasifiers (Rao et al. 1992). The interest in using iron-based catalysts stems from its relatively low cost and excellent water-gas shift reaction activity, which helps to make up the deficit of H_2 in the syngas from coal gasification (Jothimurugesan et al. 2000).

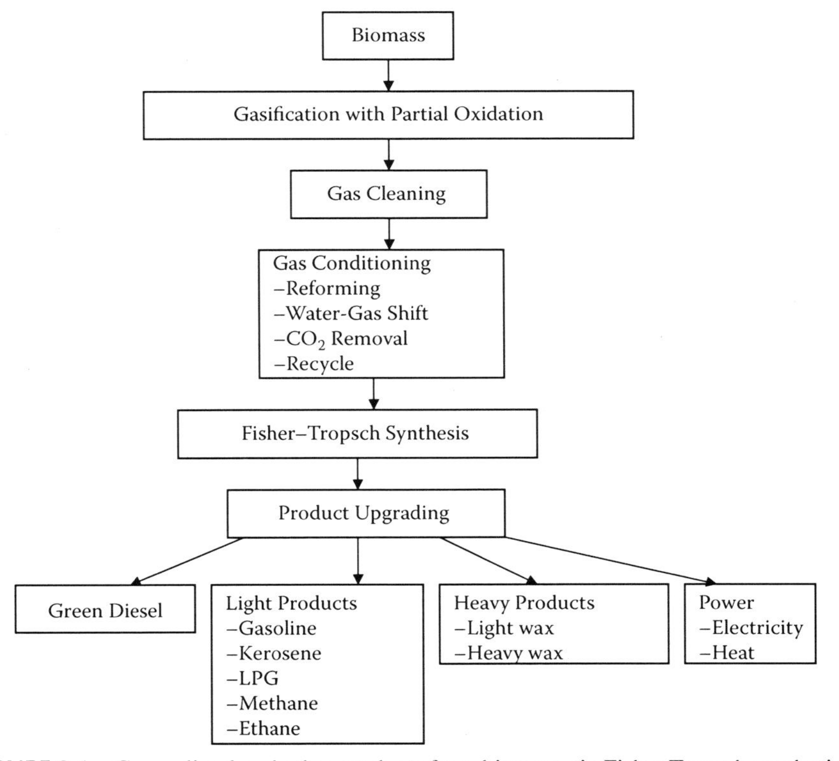

FIGURE 2.1 Green diesel and other products from biomass via Fisher-Tropsch synthesis.

The FTS-based gas to liquids technology includes three processing steps, namely syngas generation, syngas conversion, and hydroprocessing. It has been estimated that the FTS should be viable at crude oil prices of about $20 per barrel (Dry 2004). The current commercial applications of the FT process are geared to the production of the valuable linear alpha olefins and of fuels such as liquefied petroleum gas (LPG), gasoline, kerosene, and diesel. Since the FT process produces predominantly linear hydrocarbons the production of high quality diesel fuel is currently of considerable interest (Dry 2004). The most expensive section of an FT complex is the production of purified syngas and so its composition should match the overall usage ratio of the FT reactions, which in turn depends on the product selectivity.

The Al_2O_3/SiO_2 ratio has significant influences on iron-based catalyst activity and selectivity in the process of FTS. Product selectivities also change significantly with different Al_2O_3/SiO_2 ratios. The selectivity of low-molecular-weight hydrocarbons increases and the olefin to paraffin ratio in the products shows a monotonic decrease with increasing Al_2O_3/SiO_2 ratio. Table 2.3 shows the effects of Al_2O_3/SiO_2 ratio on hydrocarbon selectivity (Jothimurugesan et al. 2000). Jun et al. (2004) studied FTS over Al_2O_3 and SiO_2 supported iron-based catalysts from biomass-derived syngas. They found that Al_2O_3 as a structural promoter facilitated the better dispersion of copper and potassium and gave much higher FTS activity. The reaction results from FTS with balanced syngas are given in Table 2.4.

There has been some interest in the use of FTS for biomass conversion to synthetic hydrocarbons. Biomass can be converted to bio-syngas by noncatalytic, catalytic, and steam gasification processes. The bio-syngas consists mainly of H_2, CO, CO_2, and CH_4. The FTS has been carried out using $CO/CO_2/H_2/Ar$ (11/32/52/5 vol.%) mixture as a model for bio-syngas on co-precipitated Fe/Cu/K, Fe/Cu/Si/K, and Fe/Cu/Al/K catalysts in a fixed-bed reactor. Some performances of the catalysts that depended on the syngas composition are also presented (Jun et al. 2004).

TABLE 2.3
Effects of Al_2O_3/SiO_2 Ratio on Hydrocarbon Selectivity

Hydrocarbon Selectivities (wt%)	100Fe/ 6Cu/5K/ $25SiO_2$	100Fe/6Cu/ $5K/3Al_2O_3/$ $22SiO_2$	100Fe/6Cu/ $5K/5Al_2O_3/$ $20SiO_2$	100Fe/6Cu/ $5K/7Al_2O_3/$ $18SiO_2$	100Fe/6Cu/ $5K/10Al_2O_3/$ $15SiO_2$	100Fe/ 6Cu/5K/ $25Al_2O_3$
CH_4	6.3	8.7	10.4	10.7	14.3	17.3
C_{2-4}	24.5	27.8	30.8	29.9	33.4	46.5
C_{5-11}	26.8	27.6	32.2	33.9	40.0	31.0
C_{12-18}	21.9	21.2	15.8	15.0	6.0	4.9
C_{19+}	20.5	14.4	11.0	10.6	6.1	0.4

Reaction condition: 523 K, 2.0 MPa, H2/CO = 2.0, and gas stream velocity: 2000 h^{-1}.
*From Jothimurugesan, K. et al. 2000. Catal. Today 58:*335–344. With permission.

TABLE 2.4
Reaction Results from FTS With Balanced Syngas (H_2-Supplied Bio-Syngas)

Conversion (%)				Hydrocarbon Distribution (C mol%)				Olefin Selectivity (%) in C2–C4
CO	CO2	CO +	CO2	CH4	C2–C4	C5–C7	C8+	
82.9	0.3	21.2		12.6	39.2	21.9		84.9
88.2	28.9	43.6		26.3				84.0
				13.8	37.7	22.2		
				26.4				

Reaction conditions: Fe/Cu/Al/K (100/6/16/4), $CO/CO_2/Ar/H_2$ (6.3/19.5/5.5/69.3), 1 MPa, 573 K, 1800 mL/(g_{cat} h).

From Jun, K. W. et al. 2004. Appl. Catal. A 259: 221–226. With permission.

2.6 BIO-ALCOHOLS FROM BIOMASS

The alcohols are oxygenates, fuels in which the molecules have one or more oxygen, which decreases the combustion heat. Practically, any of the organic molecules of the alcohol family can be used as a fuel. The alcohols that can be used for motor fuels are methanol (CH_3OH), ethanol (C_2H_5OH), propanol (C_3H_7OH), and butanol (C_4H_9OH). However, only methanol and ethanol fuels are technically and economically suitable for internal combustion engines (ICEs). Ethanol (ethyl alcohol, grain alcohol, CH_3-CH_2-OH or ETOH) is a clear, colorless liquid with a characteristic, agreeable odor. Ethanol can be blended with gasoline to create E85, a blend of 85% ethanol and 15% gasoline. E85 and blends with even higher concentrations of ethanol, such as E95, are being explored as alternative fuels in demonstration programs. Ethanol has a higher octane number (108), broader flammability limits, higher flame speeds, and higher heats of vaporization than gasoline. These properties allow for a higher compression ratio, shorter burn time, and leaner burn engine, which lead to theoretical efficiency advantages over gasoline in an ICE. Disadvantages of ethanol include its lower energy density than gasoline, its corrosiveness, low flame luminosity, lower vapor pressure, miscibility with water, and toxicity to ecosystems.

The components of lignocellulosic biomass include cellulose, hemicelluloses, lignin, extractives, ash, and other compounds. Cellulose, hemicelluloses, and lignin are three major components of a plant biomass material. Cellulose is a remarkable pure organic polymer, consisting solely of units of anhydro glocose held together in a giant straight chain molecule. Cellulose must be hydrolyzed to glucose before fermentation to ethanol. Conversion efficiencies of cellulose to glucose may be dependent on the extent of chemical and mechanical pretreatments to structurally and chemically alter the pulp and paper mill wastes. The method of pulping, the type of wood, and the use of recycled pulp and paper products also could influence the accessibility of cellulose to cellulase enzymes. Hemicelluloses (arabinoglycuronoxylan and galactoglucomammans) are related to plant gums in composition, and occur in much shorter molecule chains than cellulose. The hemicelluloses, which are present in deciduous woods chiefly as pentosans and in coniferous woods almost entirely as hexosanes, undergo thermal decomposition very readily. Hemicelluloses are derived mainly from chains of pentose sugars, and act as the cement material holding together the cellulose micells and fiber. Lignins are polymers of aromatic compounds. Their functions are to provide structural strength, provide sealing of the water conducting system that links roots with leaves, and protect plants against degradation. Lignin is a macromolecule that consists of alkylphenols and has a complex three-dimensional structure. Lignin is covalently linked with xylans in the case of hardwoods and with galactoglucomannans in softwoods. Even though mechanically cleavable to a relatively low molecular weight, lignin is not soluble in water. It is generally accepted that free phenoxyl radicals are formed by thermal decomposition of lignin above 525 K and that the radicals have a random tendency to form a solid residue through condensation or repolymerization. Cellulose is insoluble in most solvents and has a low accessibility to acid and enzymatic hydrolysis. Hemicelluloses are largely soluble in alkali and, as such, are more easily hydrolysed. Table 2.1 shows the relative abundance of individual sugars in the carbohydrate fraction of wood.

Bioethanol is derived from alcoholic fermentation of sucrose or simple sugars, which are produced from biomass. Bioethanol is a fuel derived from renewable sources of feedstock, typically plants such as wheat, sugar beet, corn, straw, and wood. By contrast, petrol, diesel, and the road fuel gases LPG and compressed natural gas (CNG) are fossil fuels in finite supply. Bioethanol is a petrol additive/substitute. It is possible that wood, straw, and even household wastes may be economically converted to bioethanol. Bioethanol can be used as a 5% blend with petrol under the EU quality standard EN 228. This blend requires no engine modification and is covered by vehicle warranties. With engine modification, bioethanol can be used at higher levels, for example, E85 (85% bioethanol).

A large amount of ethanol can be produced from ethylene (a petroleum product). Catalytic hydration of ethylene produces synthetic ethanol.

$$\underset{\text{Ethylene}}{C_2H_4} + \underset{\text{Steam}}{H_2O} \rightarrow \underset{\text{Ethanol}}{C_2H_5OH} \tag{2.8}$$

Bioethanol can be produced from a large variety of carbohydrates with a general formula of $(CH_2O)_n$. Fermentation of sucrose is performed using commercial yeast such as *Saccharomyces cerevisiae*. Chemical reaction is composed of enzymatic hydrolysis of sucrose followed by fermentation of simple sugars (Gnansounou, Dauriat , and Wyman 2005). First, invertase enzyme in the yeast catalyzes the hydrolysis of sucrose to convert it into glucose and fructose.

$$\underset{\text{Sucrose}}{C_{12}H_{22}O_{11}} \rightarrow \underset{\text{Glucose}}{C_6H_{12}O_6} + \underset{\text{Fructose}}{C_6H_{12}O_6} \tag{2.9}$$

Second, zymase, another enzyme also present in the yeast, converts the glucose and the fructose into ethanol.

$$C_6H_{12}O_6 \rightarrow 2C_2H_5OH + 2CO_2 \tag{2.10}$$

Glucoamylase enzyme converts the starch into D-glucose. The enzymatic hydrolysis is then followed by fermentation, distillation, and dehydration to yield anhydrous bioethanol. Corn (60 to 70% starch) is the dominant feedstock in the starch-to-bioethanol industry worldwide.

Carbohydrates (hemicelluloses and cellulose) in lignocellulosic materials can be converted to bioethanol. The lignocellulose is subjected to delignification, steam explosion, and dilute acid prehydrolysis, which is followed by enzymatic hydrolysis and fermentation into bioethanol. A major processing step in an ethanol plant is enzymatic saccharification of cellulose to sugars through treatment by enzymes; this step requires lengthy processing and normally follows a short pretreatment step.

Hydrolysis breaks down the hydrogen bonds in the hemicellulose and cellulose fractions into their sugar components, pentoses and hexoses. These sugars can then be fermented into bioethanol. The most commonly applied methods can be classified in two groups: chemical hydrolysis (dilute and concentrated acid hydrolysis) and enzymatic hydrolysis. In chemical hydrolysis, pretreatment and hydrolysis may be carried out in a single step.

There are two basic types of acid hydrolysis processes commonly used: dilute acid and concentrated acid. The biggest advantage of dilute acid processes is their fast rate of reaction, which facilitates continuous processing. Since five-carbon sugars degrade more rapidly than six-carbon sugars, one way to decrease sugar degradation is to have a two-stage process. The first stage is conducted under mild process conditions to recover the five-carbon sugars while the second stage is conducted under harsher conditions to recover the six-carbon sugars. Concentrated sulfuric or hydrochloric acids are used for hydrolysis of lignocellulosic materials. The concentrated acid process uses relatively mild temperatures, and the only pressures involved are those created by pumping materials from vessel to vessel. Reaction times are typically much longer than for dilute acid. This process provides a complete and rapid conversion of cellulose to glucose and hemicelluloses to five-carbon sugars with little degradation. The critical factors needed to make this process economically viable are to optimize sugar recovery and cost effectively recover the acid for recycling. The solid residue from the first stage is dewatered and soaked in a 30 to 40% concentration of sulfuric acid for 1 to 4 hours as a pre-cellulose hydrolysis step. The solution is again dewatered and dried, increasing the acid concentration to about 70%. After reacting in another vessel for 1 to 4 hours at low temperatures, the contents are separated to recover the sugar and acid. The sugar/acid solution from the second stage is recycled to the first stage to provide the acid for the first stage hydrolysis.

The primary advantage of the concentrated acid process is the potential for high sugar recovery efficiency. The acid and sugar are separated via ion exchange and then acid is reconcentrated via multiple effect evaporators.

Before modern production technologies were developed in the 1920s, methanol was obtained from wood as a co-product of charcoal production and, for this reason, was commonly known as wood alcohol. Methanol is currently manufactured worldwide by conversion or derived from syngas, natural gas, refinery off-gas, coal, or petroleum. Methanol can be produced from essentially any primary energy source. Thus, the choice of fuel in the transportation sector is to some extent determined by the availability of biomass. Methanol is currently made from natural gas but can also be made using biomass via partial oxidation reactions. Biomass can be considered as a potential fuel for gasification and further bio-syngas production and methanol synthesis. Adding sufficient hydrogen to the synthesis gas to convert all of the biomass into methanol can double the methanol produced from the same biomass base. Bio-syngas is altered by catalyst under high pressure and temperature to form methanol. The gases produced can be steam reformed to produce hydrogen and followed by water-gas shift reaction to further enhance hydrogen production.

The use of methanol as a motor fuel received attention during the oil crises of the 1970s due to its availability and low cost. Problems occurred early in the development of gasoline-methanol blends. As a result of its low price some gasoline marketers over blended. Many tests have shown promising results using 85 to 100% percent by volume methanol as a transportation fuel in automobiles, trucks, and buses. Methanol can be used as one possible replacement for conventional motor fuels. Methanol has been seen as a possible large volume motor fuel substitute at various times during gasoline shortages. It was often used in the early part of the twentieth century to power automobiles before inexpensive gasoline was widely introduced.

2.7 BIODIESEL FROM VEGETABLE OILS

Biodiesel is a fuel consisting of long chain fatty acid alkyl esters made from renewable vegetable oils, recycled cooking greases, or animal fats (ASTM D6751). Vegetable oil (m)ethyl esters, commonly referred to as "biodiesel," are prominent candidates as alternative diesel fuels. Biodiesel is technically competitive with or offer technical advantages compared to conventional petroleum diesel fuel.

Methyl esters of vegetable oils have several outstanding advantages among other new-renewable and clean engine fuel alternatives, as the physical characteristics of fatty acid (m)ethyl esters are very close to those of diesel fuel and the production process is relatively simple. Furthermore, the (m)ethyl esters of fatty acids can be burned directly in unmodified diesel engines, with very low deposit formation. There are more than 350 oil bearing crops identified, among which only sunflower, safflower, soybean, cottonseed, rapeseed, and peanut oils are considered as potential alternative fuels for diesel engines. Dilution, micro-emulsification, pyrolysis, and transesterification are the four techniques applied to solve the problems encountered with the high fuel viscosity. The purpose of the transesterification process is to lower the viscosity of the oil. Ethanol is a preferred alcohol in the transesterification process compared to methanol because it is derived from agricultural products and is renewable and biologically less objectionable in the environment (Demirbas 2003).

The properties of biodiesel are close to those of diesel fuels. The biodiesel was characterized by determining its viscosity, density, cetane number, cloud and pour points, characteristics of distillation, flash and combustion points and higher heating value (HHV) according to ISO norms.

2.8 THE FUTURE OF BIOMASS

Biomass provides a number of local environmental gains. Energy forestry crops have a much greater diversity of wildlife and flora than the alternative land use, which is arable or pasture land. Energy crops may also offer a corridor for wildlife between woodland habitats. Energy crops that are carefully sited and designed will enhance local landscapes and provide a new habitat for recreation. Provision of recreation habitat is important near urban centers.

It is important to underline here that the collection of fuel from European forestry and agriculture and the use of energy crops is a sustainable activity that does not deplete future resources. By the year 2050, it is estimated that 90% of the world population will live in developing countries (Ramage and Scurlock 1996).

In industrialized countries, the main biomass processes utilized in the future are expected to be direct combustion of residues and wastes for electricity generation, bioethanol and biodiesel as liquid fuels and combined heat and power production from energy crops. The future of biomass electricity generation lies in biomass integrated gasification/gas turbine technology, which offers high energy conversion efficiencies. Biomass will compete favorably with fossil mass for niches in the chemical feedstock industry. Biomass is a renewable, adaptable resource. Crops can be grown to satisfy changing end use needs.

In the future, biomass has the potential to provide a cost-effective and sustainable supply of energy, while at the same time aiding countries in meeting their greenhouse gas reduction targets.

2.9 GLOBAL BIOFUEL SCENARIO

Renewable resources are more evenly distributed than fossil and nuclear resources, and energy flows from renewable resources are more than three orders of magnitude higher than current global energy use. Today's energy system is unsustainable because of equity issues as well as environmental, economic, and geopolitical concerns that have implications far into the future (UNDP 2000). According to the International Energy Agency (IEA), scenarios developed for the United States and the European Union indicate that near-term targets of up to 6% displacement of petroleum fuels with biofuels appear feasible using conventional biofuels, given available cropland. A 5% displacement of gasoline in the EU requires about 5% of available cropland to produce ethanol whereas in the United States 8% is required. A 5% displacement of diesel requires 13% of U.S. cropland, 15% in the EU (IEA 2004).

The recent commitment by the U.S. government to increase bio-energy threefold in 10 years has added impetus to the search for viable biofuels. The advantages of biofuels are the following: (1) biofuels are easily available from common biomass sources, (2) they are carbon dioxide neutral fuels, (3) they have a considerable environmentally friendly potential, (4) there are many benefits to the environment, economy, and consumers in using biofuels, and (5) they are biodegradable and contribute to sustainability (Puppan 2002).

The dwindling fossil fuel sources and the increasing dependence of the United States on imported crude oil have led to a major interest in expanding the use of bio-energy. The recent commitment by the U.S. government to increase bio-energy threefold in 10 years has added impetus to the search for viable biofuels. The EU has also adopted a proposal for a directive on the promotion of the use of biofuels with measures ensuring that biofuels account for at least 5.75% of the market for gasoline

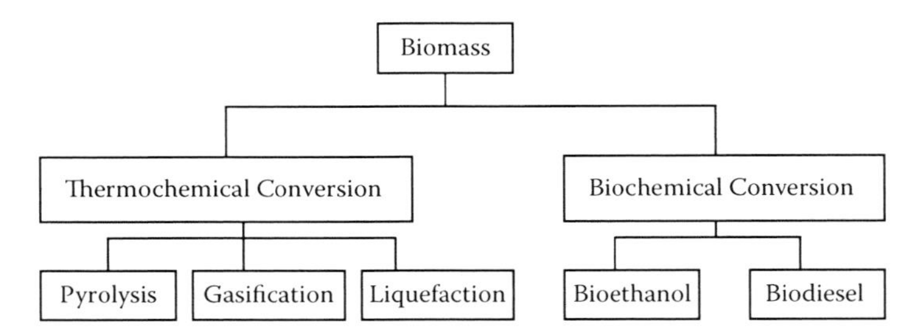

FIGURE 2.2 Main biomass conversion processes.

and diesel sold as transport fuel by the end of 2010 (Hansen, Zhang, and Lyne 2005). Figure 2.2 shows the main biomass conversion processes. Biomass can be converted to biofuels such as bioethanol and biodiesel and thermochemical conversion products such as syn-oil, bio-syngas, and biochemicals. Bioethanol is a fuel derived from renewable sources of feedstock, typically plants such as wheat, sugar beet, corn, straw, and wood. Bioethanol is a petrol additive/substitute. Biodiesel is better than diesel fuel in terms of sulfur content, flash point, aromatic content, and biodegradability (Bala 2005).

If the biodiesel is used for transportation fuel, it would benefit the environment and the local population: job creation, provision of modern energy carriers to rural communities, avoiding urban migration, and reduction of CO_2 and sulfur levels in the atmosphere. Biofuels are useful for energy security reasons, environmental concerns, foreign exchange savings, and socioeconomic issues related to the rural sector.

Figure 2.3 shows the share of alternative fuels compared to the total automotive fuel consumption in the world as a futuristic view. Hydrogen is currently more expensive than conventional energy sources. There are different technologies presently being practiced to produce hydrogen economically from biomass. Biohydrogen

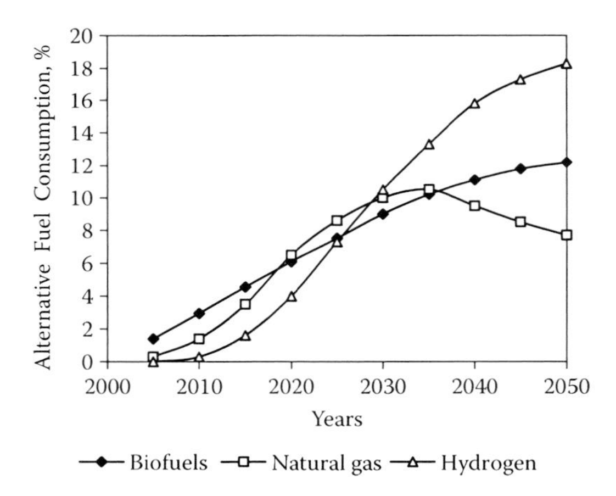

FIGURE 2.3 Shares of alternative fuels compared to the total automotive fuel consumption in the world. (From Demirbas, A. 2006. *Energy Edu. Sci. Technol.* 17: 32–63. With permission.)

technology will play a major role in the future because it can utilize the renewable sources of energy (Nath and Das 2003).

Hydrogen for fleet vehicles will probably dominate in the transportation sector. To produce hydrogen via electrolysis and the transportation of liquefied hydrogen to rural areas with pipelines would be expensive. The production technology will be site specific and include steam reforming of methane and electrolysis in hydropower rich countries. In the long run, when hydrogen is a very common energy carrier, distribution by pipeline is probably the preferred option. The cost of hydrogen distribution and refueling is very site specific.

2.10 CONCLUSIONS

Due to its environmental merits, the share of biofuel such as bioethanol and biodiesel in the automotive fuel market will grow rapidly in the next decade. There are several reasons for biofuels to be considered as relevant technologies by both developing and industrialized countries. The strategy is based on producing biofuel from biomass liquefaction and pyrolysis using a co-product strategy to reduce the cost of biofuel. It can be concluded that only this strategy could compete with the cost of the commercial hydrocarbon-based technologies. This strategy will demonstrate how biofuel is economically feasible and can foster the development of rural areas when practiced on a larger scale.

Bioethanol is a petrol additive/substitute. It is possible that wood, straw, and even household wastes may be economically converted to bioethanol. Biodiesel is an environmentally friendly alternative liquid fuel that can be used in any diesel engine without modification. Biodiesel is better than diesel fuel in terms of sulfur content, flash point, aromatic content, and biodegradability.

REFERENCES

Bala, B. K. 2005. Studies on biodiesels from transformation of vegetable oils for diesel engines. *Energy Edu. Sci. Technol.* 5: 1–45.

Chornet, E. and R. P. Overend. 1985. Biomass liquefaction: An overview. In *Fundamentals of Thermochemical Biomass Conversion*, ed. R. P. Overend, T. A. Milne and L. K. Mudge, 967–1002. New York: Elsevier Applied Science.

Demirbas, A. 2000. Mechanisms of liquefaction and pyrolysis reactions of biomass. *Energy Convers. Mgmt.* 41: 633–646.

Demirbas, A. 2003. Biodiesel fuels from vegetable oils via catalytic and non-catalytic supercritical alcohol transesterifications and other methods: a survey. *Energy Convers. Mgmt.* 44: 2093–2109.

Demirbas, A. 2006. Global biofuel strategies. *Energy Edu. Sci. Technol.* 17: 32–63.

Dry, M. E. 2004. Present and future applications of the Fischer–Tropsch process. *Appl. Catal. A*: General 276: 1–3.

Eager, R. L., J. M. Pepper, and J. C. Roy. 1983. Chemical studies on oils derived from aspen poplar wood, cellulose, and an isolated aspen poplar lignin. *Can. J. Chem.* 61: 2010–2015.

Gnansounou, E., A. Dauriat, C. E. Wyman. 2005. Refining sweet sorghum to ethanol and sugar: economic trade-offs in the contex of North China. *Biores. Technol.* 96: 985–1002.

Hansen, A. C., Q. Zhang, and P. W. L. Lyne. 2005. Ethanol–diesel fuel blends: A review. *Biores. Technol.* 96: 277–285.

IEA (International Energy Agency). 2004. Biofuels for Transport: An International Perspective. Paris: IEA (available from www.iea.org).

Jothimurugesan, K., J. G. Goodwin, S. K. Santosh, and J. J. Spivey. 2000. Development of Fe Fischer–Tropsch catalysts for slurry bubble column reactors. *Catal. Today* 58: 335–344.

Jun, K. W., H. S. Roh, K. S. Kim, J. S. Ryu, and K. W. Lee. 2004. Catalytic investigation for Fischer–Tropsch synthesis from bio-mass derived syngas. *Appl. Catal.* A 259: 221–226.

Nath, K. and D. Das. 2003. Hydrogen from biomass. Current Sci. 85: 265–271.

Puhan, S., N. Vedaraman, B. V. Rambrahaman, and G. Nagarajan. 2005. Mahua (Madhuca indica) seed oil: A source of renewable energy in India. J. Sci. Ind. Res. 64: 890-896.

Puppan, D. 2002. Environmental evaluation of biofuels. *Periodica Polytechnica Ser. Social Management Sci.* 10: 95–116.

Ramage, J. and J. Scurlock. 1996. Biomass. In *Renewable Energy-Power for a Sustainable Future*, ed. G. Boyle. Oxford: Oxford University Press, pp. 137–182.

Rao, V. U. S., G. J. Stiegel, G. J. Cinquergrane, and R. D. Srivastava. 1992. Iron-based catalysts for slurry-phase Fischer-Tropsch process: Technology review. *Fuel Proc. Technol.* 30: 83-107.

Reijnders, L. 2006. Conditions for the sustainability of biomass based fuel use. *Energy Policy* 34: 863–876.

Sheehan, J., V. Cambreco, J. Duffield, M. Garboski, and H. Shapouri. 1998. An Overview of Biodiesel and Petroleum Diesel Life Cycles. Washington, D.C.: U.S. Department of Agriculture and U.S. Department of Energy.

UNDP (United Nations Development Programme). 2000. World Energy Assessment. *Energy and the Challenge of Sustainability.* New York: UNDP.

3 Thermochemical Conversion of Biomass to Liquids and Gaseous Fuels

Hari Bhagwan Goyal, Rakesh Chandra Saxena, and Diptendu Seal

CONTENTS

ABSTRACT

Energy management will be difficult for the coming generations due to increasing demand for energy caused by rapid industrialization and social growth. The emerging alternative and renewable energy resources are expected to play an important role in energy consumption scenarios of the future, particularly to handle environmental concerns. Biomass is renewable, clean, and abundantly available and its thermochemical conversion to liquid and gaseous fuels is one of the prospective approaches. Various thermochemical conversion processes, namely, combustion, gasification, liquefaction, hydrogenation, and pyrolysis have been considered for converting biomass into liquid and gaseous fuels. The pyrolysis process has received considerable attention as it converts biomass directly into solid, liquid, and gaseous products by thermal decomposition of biomass in the absence of oxygen.

This chapter focuses on pyrolysis; other conventional thermochemical processes are discussed briefly. Various types of pyrolysis processes, namely, slow, fast, flash, and catalytic processes are discussed in detail to provide better insight into these thermochemical processes. Besides properties of biomass, the composition and use of pyrolysis products are also discussed in detail. In addition, various thermochemical processes, such as pyrolysis and supercritical water extraction gasification for hydrogen-rich gas production are also highlighted.

3.1 INTRODUCTION

The demand for energy is growing faster due to rapid industrialization and social growth. Conventional energy sources, such as coal, oil, and natural gas, have limited reserves that are expected not to last for an extended period. Consequently, energy management will be difficult for the coming generations (Adhikari et al. 2006). In addition, environment-related problems associated with conventional energy sources are continuously increasing. Over the last half century, a trend toward continuous increases in the average atmospheric temperature has been observed, totalling a half degree centigrade (Goyal et al. 2005) This trend may lead to natural calamities such as excessive rainfall and consequent floods, droughts, and local imbalances. With increasing energy demand, the emerging alternative and renewable energy resources are expected to play an increasing role in future energy consumption, at least in order to reduce the environmental concerns.

In contrast to conventional energy sources, nonconventional energy sources such as wind, sunlight, water, and biomass have been used since ancient times. Biomass is now being considered as an important energy resource all over the world and is being used to meet a variety of energy needs, including generating electricity, fueling vehicles, and providing process heat for industrial facilities. Among all the renewable sources of energy, biomass is unique as it effectively stores solar energy. It is the only renewable source of carbon that can be converted into convenient solid, liquid, and gaseous fuels. Biomass is the fourth largest source of energy in the world, accounting for about 15% of the world's primary energy consumption and about 38% of the primary energy consumption in developing countries (Chen, Andries, and Spliethoff 2003).

The term *biomass* generally refers to any living matter available on the earth. However, in the present context, only plant materials would be considered as biomass. Plants produce carbohydrates through the process of photosynthesis using CO_2, water, minerals, sunlight, and chlorophyll. Carbohydrates, which make up the bulk of tissues, trap the solar energy in their chemical bonds. It is these energy-rich bonds, when broken via different processes, that produce energy (Goyal et al. 2006). There is a need for an environmentally benign, cheap, and efficient process that can extract stored solar energy from waste biomass.

Biomass resources that can be used for energy production cover a wide range of materials, which include wood and wood wastes, agricultural crops and their waste by-products, waste from food processing, municipal solid waste, aquatic plants and algae, etc. Plant-based biomass can be divided into three major categories, residues from agriculture production, forest products, and energy crops. Energy crops include short rotation woody crops, herbaceous woody crops, grasses, starch crops, sugar crops, oilseed crops, etc. The choice of biomass for the production of energy and the type of process to be used for its conversion depends on the chemical and physical properties of the large molecules from which it is made. The major components present in any biomass are:

1. Cellulose: Cellulose is a linear chain polymer of (1, 4)-D-glucopyranose units. The units are linked 1-4 in the β-configuration. It has an average molecular weight of around 100,000 Da and general formula of $C_6H_{10}O_5$.
2. Hemicellulose: These are the complex polysaccharides that exist in association with the cellulose in the cell wall and consist of branched structures, which vary considerably with different biomass. It is a mixture of monosaccharides such as glucose, mannose, xylose, and arabinose, as well as methlyglucoronic and galaturonic acids, having an average molecular weight of <30,000 Da.
3. Lignin: Lignins are highly branched, substituted, mononuclear aromatic polymers in the cell walls of certain biomass, especially woody species, and are often adjacent to cellulose fibers to form a lignocellulose complex. Lignin is regarded as a group of amorphous, high-molecular-weight, chemically related compounds.

In most kinds of biomass, cellulose is generally the largest fraction, about 40 to 50% by weight, followed by hemicellulose about 20 to 40%. Besides these, biomass also has other components (Bridgwater, 1999). Table 3.1 highlights different types of biomass with their compositions.

Due to its renewable and environmentally friendly nature, the use of biomass for production of energy would result in a net reduction in greenhouse gas emission. Furthermore, biomass-derived fuels have negligible sulfur content and, therefore, do not contribute to the emission of sulfur dioxide, which causes acid rain. The combustion of biomass produces less ash than coal combustion. Moreover, the ash produced can be used as soil additive on farms.

TABLE 3.1
Biomass Components

	Percent, weight		
Biomass	**Cellulose**	**Hemicellulose**	**Lignin**
Sugarcane bagasse	38	27	20
Sugarcane leaves	36	21	16
Napier grass	32	20	09
Sweet sorghum	36	16	10
Eucalyptus saligna	45	12	25
Municipal solid waste	33	09	17
Newspaper	62	16	21

3.2 BIOMASS CONVERSION PROCESSES

Biomass can be converted into useful forms of energy using various processes. The choice of conversion process depends on the type and quantity of the biomass feedstock, the desired form of the energy, that is, end use requirements, environmental standards, economic conditions, and specific factors for the project (Manuel, Abdelkader, and Roy 2002).

The two main processes for the conversion of biomass are thermochemical processes and biochemical/biological processes (Saxena, Adhikari, and Goyal 2007).

3.2.1 Thermochemical Conversion Processes

The thermochemical conversion processes involve heating of biomass at high temperatures. There are two basic approaches. The first is gasification of biomass and its conversion to hydrocarbons. The second approach is to liquefy biomass directly by high-temperature pyrolysis, high-pressure liquefaction, ultra-pyrolysis, or supercritical extraction. Various thermochemical conversion processes are described below.

3.2.1.1 Combustion

Combustion is the burning of biomass in air. It converts the chemical energy stored in the biomass into heat, mechanical power, or electricity using different process equipment, for example, stoves, furnaces, boilers, steam turbines, turbo generators, etc. Combustion produces hot gases at temperatures around 800 to 1000°C. This is an older method of utilizing biomass for obtaining energy. Combustion has been used on a small scale for domestic purposes and on a large scale for industries. On a small scale, it can be used to provide energy for cooking, space heating, etc. Large-scale uses include combustion in boilers or furnaces to get heat, generation of steam for turbines, etc. Co-combustion of biomass with coal is a good option for use in the production of power on a larger scale. Complete combustion is actually a chemical reaction of the biomass and oxygen, giving CO_2, water, and heat. Combustion equipment is available with various designs of the combustion chambers, operating temperatures, etc. Examples are refractory lined furnaces, water wall incinerators, etc.

The choice of the specific reactor depends on the type of biomass, quantity, and the required form of final energy.

The process of combustion has many drawbacks. The biomass rarely exists naturally in an acceptable form for burning. Straw, wood, and some other types of biomass require primary treatment such as compressing, chopping, and grinding for better combustion, which can be expensive (McKendry 2002). Small-scale applications, such as domestic cooking and space heating can be very inefficient, with heat transfer loss of 30 to 90%. Large biomass power generation systems involve higher cost due to the presence of moisture in the biomass.

3.2.1.2 Gasification

Gasification is the conversion of the biomass into a combustible gas mixture by the partial oxidation of the biomass at high temperature, in the range of 800 to 900°C. The following reaction takes place in the reactor during the gasification reaction.

$$C + O_2 \rightarrow CO_2$$

$$C + \frac{1}{2} O_2 \rightarrow CO$$

$$CO + \frac{1}{2} O_2 \rightarrow CO_2$$

$$CO_2 + C \leftrightarrow 2\,CO$$

Methane and hydrogen formed simultaneously by the thermal splitting of the organic material may also be combusted and carbon may also be reduced by the hydrogen present in the gaseous mixture:

$$CO_2 + 4\,H_2 \rightarrow CH_4 + 2\,H_2O$$

The resulting gas, known as *producer gas,* is a mixture of carbon monoxide, hydrogen, and methane along with carbon dioxide and nitrogen. Various gasifiers are known to run on different types of biomass, such as rice husk, coconut shells, charcoal, wood, etc. The low calorific value gas produced (about 4–6 MJ/Nm^3) can be burnt directly, or can be used as fuel for gas engines and gas turbines. The production of synthesis gas from biomass allows the production of methanol and hydrogen, each of which may have a future as fuels for transportation (Asadullah et al. 2002; Demirbas 2004).

3.2.1.3 Pyrolysis

Pyrolysis is the heating of biomass in an inert atmosphere. Pyrolysis generally starts at 300°C and continues up to 600–700°C. The biomass is converted into useful liquid, gaseous, and solid products. Details of the pyrolysis process are covered in Section 3.3.

3.2.1.4 Liquefaction

Liquefaction is a low-temperature, high-pressure, thermochemical process using a catalyst with the addition of hydrogen and producing a marketable liquid product. High pressure is employed to assure good heat transfer, or to maintain a liquid-phase system at high temperatures. Interest in liquefaction is low because the reactors and fuel feeding systems are more complex and more expensive than for the pyrolysis and gasification processes (Demirbas 2001). The heavy oil obtained from the liquefaction process is a viscous tarry lump, which sometimes causes trouble in handling.

3.2.1.5 Hydrogenation

Hydrogenation is mainly employed for the production of methane by hydro-gasification. In one of the routes, synthesis gas is produced in the first step, followed by reaction with hydrogen to yield methane. In the other route, the feed reacts directly with the hydrogen; the shredded biomass is converted with the hydrogen-containing gas to a gas containing relatively high methane concentrations in the first-stage reactor. The product char from the first stage is used in a second-stage reactor to generate the hydrogen-rich synthesis gas. Presently, more attention is focused on two kinds of processes, pyrolysis to produce liquid fuel, and gasification to produce hydrogen, as these are environmentally benign and produce a better quality product. These processes are discussed here in detail.

3.3 PYROLYSIS OF BIOMASS TO LIQUID FUELS

Pyrolysis is the thermal decomposition of an organic material in the absence of oxygen, leading to the formation of liquid, gases, and a highly reactive carbonaceous char. The quantity and quality of the products depend on various parameters, such as reaction temperature, pressure, heating rate, reaction time, etc. Chars, organic liquids, gases, and water are formed in varying amounts, depending particularly on the biomass composition, heating rate, pyrolysis temperature, and residence time in the pyrolysis reactor. Lower process temperature and longer vapor residence times favor the production of charcoal, whereas high temperature and long vapor residence time increase the gas yields. Moderate temperature and short vapor residence time are optimum for higher liquid yields (Bridgwater 1994).

The basic phenomena that take place during pyrolysis are as follows:

1. Heat transfer from a heat source, leading to an increase in the temperature inside the fuel.
2. Initiation of the pyrolysis reactions due to increased temperature, leading to release of the volatiles and the formation of char.
3. Outflow of the volatiles, resulting in heat transfer between the hot volatiles and cooler unpyrolysed fuel.
4. Condensation of some of the volatiles in the cooler parts of the fuel to produce tar.
5. Autocatalytic secondary pyrolysis reactions.

The pyrolysis process may be endothermic, or exothermic, depending on the temperature of the reacting system. The process steps include drying the feed, grinding the feed to sufficiently small particles for rapid reaction, pyrolysis reaction, and the separation of the products (bio-oil). The pyrolysis processes are of the following types.

3.3.1 Slow Pyrolysis

This is a conventional process whereby the heating rate is kept slow (approximately 5–7°C/min) (Ozbay et al. 2001). This slow heating rate leads to higher char yields than the liquid and gaseous products. Different kinds of biomass, such as wood samples, safflower seeds, sugarcane bagasse, sunflower seeds, municipal wastes, etc., are generally subjected to slow pyrolysis.

3.3.2 Fast Pyrolysis

Fast pyrolysis is considered a better process than conventional, slow pyrolysis. In this, the heating rates are kept high, about 300 to 500°C/min and the liquid product yield is higher. Fluidized-bed reactors are best suited for this process as they offer high heating rates, rapid devolatilization and also are easy to operate. Reactors such as entrained flow reactors, circulating fluidized-bed reactors, rotating reactors, etc. are used for this purpose (Table 3.2).

3.3.3 Flash Pyrolysis

This is an improved version of fast pyrolysis, whereby high reaction temperature is obtained within a few seconds. The heating rates are very high, about 1000°C/min with reaction times of few to several seconds. This is carried out at atmospheric pressure. Entrained flow and fluidized-bed reactors are the best reactors for this purpose. Because there is rapid heating of the biomass, for better yields this process requires smaller particle size (-60+140 mesh) compared to other processes. Flash pyrolysis can be categorized as:

1. Flash hydro-pyrolysis: The pyrolysis is carried out in the presence of hydrogen. It involves a pressure of 20 MPa.
2. Solar flash pyrolysis: Solar energy is used for the pyrolysis process. It is stored in conventional devices and is used to increase the temperature of the reaction system.
3. Rapid thermal process: The rapid thermal process involves very short residence time of 30 ms to 1.5 s and is carried out at temperatures between 900 and 950°C. Rapid heating eliminates the side reactions in the system, with high yields of the desired product.
4. Vacuum flash pyrolysis: Vacuum pyrolysis incorporates a vacuum in the pyrolysis system. This stops the secondary decomposition reactions, giving higher liquid yields, and reduces gas production. The vacuum facilitates quick removal of the liquid from the system.
5. Catalytic biomass pyrolysis: Catalytic pyrolysis is done to improve the quality of the oil (the oil from pyrolysis processes is generally unsuitable

TABLE 3.2
Pyrolysis Reactors and Processes

Biomass Type	Reactor Type	Process Type
Safflower seed	Fixed bed	Slow
Rapeseed	Heinz retort	Slow
Rapeseed	Fixed bed	Fast
Rapeseed	Tubular transport reactor	Flash
Sugarcane bagasse	Cylindrical SS reactor	Vacuum
Sugarcane bagasse	Pilot plant	Vacuum
Sugarcane bagasse	Wire mesh reactor	Flash
Euphorbia rigida	Fixed bed	Hydropyrolysis
Sunflower bagasse	Fixed bed	–
Hazelnut shells	Tubular reactor, fixed bed	Slow
Wood	Fixed bed	Slow
Wood	Electric screen heater reactor	Fast
Wood	Fluid bed	Fast
Rice husks	Fluidized bed	Catalytic
Rice husks	Fluidized bed	Slow
Cotton seed	Tubular reactor, Heinz retort	–
Cotton seed cake	Fixed bed, tubular reactor	Slow
Cotton cocoon	Tubular reactor	Fast, flash
Pinewood	Static batch reactor	Slow
Softwood	Vacuum pyrolysis reactor	Vacuum
Hardwood	Electric screen heater reactor	Fast
Softwood and hardwood	Fluidized bed	Fast
Lignocell HBS	Circulating fluid bed reactor	Flash, catalytic
Sewage sludge	Fluidized bed	Slow
Coconut shell	Packed bed	Slow
Wheat straw	Packed bed	Slow
Groundnut shell	Packed bed	Slow
Pine wood	Static batch reactor	Slow
Forest and agricultural residue	Fixed bed reactor	Slow, steam
Eucalyptus and pine bark	—	Slow and flash
Canadian oil shales	Cylindrical SS retort reactor	Vacuum

for use in transportation). Catalysts such as zeolites and basic materials are used for carrying out these reactions (Williams and Nugranad 2000). The product from catalytic pyrolysis does not require costly techniques to upgrade the quality of the product.

3.3.4 Mechanism of Pyrolysis of Biomass

The mechanism of pyrolysis consists indirectly of the mechanisms of pyrolysis of its components, that is, cellulose, hemicellulose, and lignin.

3.3.4.1 Pyrolysis of the Cellulose

Cellulose degradation starts at temperatures lower than 325 K and is characterized by decreasing degrees of polymerization. The two basic reactions take place in the thermal degradation of the cellulose: (a) degradation, decomposition, and charring on heating at lower temperature; and (b) a rapid volatilization accompanied by the formation of levoglucosan on pyrolysis at higher temperatures.

The glucose chains in the cellulose are first cleaved to glucose, followed by the splitting of one molecule of water to give glucosan ($C_6H_{10}O_5$). The initial degradation reactions include depolymerization, hydrolysis, oxidation, dehydration, and decarboxylation. The mechanism of the pyrolysis of the cellulose is as follows:

$$\underset{\text{Cellulose}}{(C_6H_{10}O_5)_x} \overset{535\ K}{\leftrightarrow} \underset{\text{Levoglucosan}}{x\ C_6H_{10}O_5}$$

$$\underset{\text{Levoglucosan}}{C_6H_{10}O_5} \leftrightarrow \underset{\text{(Methyl glyoxal)}}{H_2O + 2\ CH_3 - CO - CHO}$$

$$2\ CH_3\text{-}CO\text{-}CHO + 2\ H_2 \leftrightarrow \underset{\text{(Acetal)}}{2\ CH_3\text{-}CO\text{-}CH_2OH}$$

$$2\ CH_3\text{-}CO\text{-}CH_2OH + 2\ H_2 \leftrightarrow 2\ CH_3\text{-}CHOH\text{-}CH_2OH$$

(Propylene glycol)

$$CH_3\text{-}CHOH\text{-}CH_2OH + H_2 \leftrightarrow CH_3\text{-}CHOH - CH_3 + H_2O$$

(Isopropyl alcohol)

3.3.4.2 Pyrolysis of the Hemicellulose

The hemicelluloses, which are present in deciduous woods chiefly as pentosans and in coniferous woods almost entirely as hexosans, undergo thermal decomposition very readily. Hemicellulose reacts more readily than cellulose during heating. The thermal degradation of hemicelluloses begins above 373 K.

3.3.4.3 Pyrolysis of Lignin

Lignin is a complex, naturally occurring polymer characterized by the general empirical formula: $C_9\ H_{8\text{-}x}\ O_2\ [H_2]\ [< 1.0\ [OCH_3]\ _x]$. It is susceptible to high-yield depolymerization/upgrading, leading to reformulated gasoline compositions as the final products. Two different processes have been developed for the degradation of lignin.

The first is a two-stage process comprised of base catalyzed depolymerization (BCD) and deoxygenative hydroprocessing (DHP). BCD is carried out in supercritical methanol reaction medium. This is followed by DHP, resulting in the reformulated hydrocarbons, gasoline, multibranched paraffins, C_6-C_{11} mono-, di-, tri-, and polyalkylated naphthenes, and C_7 –C_{11} alkyl benzenes.

Another two-stage process comprises mild BCD, followed by nondeoxygenative hydrotreatment/mild hydrocracking hydrotreatment (HT). This yields a reformulated, partially oxygenated gasoline.

The different procedures for converting lignin into fuel are (1) base-catalyzed depolymerization, (2) hydroprocessing, (3) selective hydrocracking, and (4) etherification.

3.3.5 Pyrolysis Reactors

Pyrolysis reactor designs include fixed beds, moving beds, suspended beds, fluidized beds, entrained-feed solids reactors, stationary vertical shaft reactors, inclined rotating kilns, high-temperature electrically heated reactors with gas blanketed walls, single and multi-hearth reactors, circulating fluidized-bed reactors, ablative pyrolysis type reactors, entrained flow reactors, vacuum pyrolysis reactors, rotating cone reactors, ultra-pyrolysis entrained flow reactors, wire mesh reactors, etc. Different pyrolysis processes with different types of reactor and biomass used are listed in Table 3.2.

3.3.6 Properties of Bio-Oils

The liquid, or bio-oil, or bio-crude is a micro-emulsion containing many reactive species, which contribute to its unusual properties. It is composed of a complex mixture of oxygenated compounds that provide both potential and challenges for its utilization. The pyrolysis oil is a dark brown, free-flowing liquid. Depending on the type of biomass and the mode of pyrolysis, the color can be almost black through dark red-brown to dark green. The liquid has an acrid, smoky, irritating smell due to the presence of low-molecular-weight aldehydes and acids. The bio-oils are comprised of different sized molecules along with significant amounts of water. In contrast to petroleum fuels, bio-oil contains a large amount of oxygen (45–50 wt%) in the form of different compounds. The other major groups of components identified are hydroxy-aldehydes, hydroxy-ketones, sugars, carboxylic acids, and phenolic compounds.

Bio-oil is immiscible with water but soluble in polar solvents such as methanol, acetone, etc. It is totally immiscible with petroleum-derived fuels. The density of the liquid is very high at around 1.2 kg/l compared to light fuel oil at around 0.85 kg/l. This means that the liquid has about 42% of the energy content of fuel oil on a weight basis, but 61% on a volumetric basis. The viscosity of bio-oil varies from as low as 100 cSt (measured at 40°C). This depends on the biomass type, the water content of the oil, the amounts of lighter ends that have been collected, and the storage conditions.

Bio-oil contains substantial amounts of organic acids (acetic acid and formic acid). It results in a pH of 2 to 3 and an acid number of 50 to 100 mg KOH/g. Bio-oils can be corrosive to common construction materials, such as carbon, steel, and aluminum, due to the presence of these acidic components. They contain molecules of different sizes, ranging from water to oligomeric phenolic compounds and the molecular weight depends mostly on the process conditions.

The complexity and nature of bio-oil causes some unusual behavior; specifically, properties that change with time are increase in viscosity, decrease in volatility, phase separation, and the deposition of gums. The degradation products from the cellulose include organics acids such as formic acid and acetic acid, which give the bio-oil its low pH. The high heating value of bio-oil is about 17 MJ/kg at 25% water, which is about 40% of that of fuel oil/diesel in weight terms. This means that 2.5 kg bio-oil is required for the same energy input as 1 kg fossil fuel oil, which amounts to 1.5 liters per liter of fossil fuel oil due to its high density. The low heating value (LHV) of bio-oils is in the range of 14 to 18 MJ/kg, which is similar to that for biomass, and is only 40 to 45% of that for hydrocarbon fuels. Proton nuclear magnetic resonance (HNMR) spectra of bio-oils indicate that the aromaticity of bio-oil obtained by steam pyrolysis techniques is higher than that of bio oil obtained by other techniques. Most of the biomass oils seem to be homogeneous, though some have a frothy top layer, which usually represents less than 10% of the oil. These oils are immiscible with hydrocarbon fuels but can be emulsified with diesel oil using surfactants.

3.3.7 Composition of Bio-Oil

Various types of compounds are found to be present in bio-oil. The composition depends on factors such as feedstock type and composition, process conditions (temperature, residence time, etc.), reactor type, storage conditions, etc. According to the National Renewable Energy Laboratory, approximately 400 types of organic compounds can be present in bio-oil. The components can be divided into organic and inorganic.

The organic components include:

- Acids: formic, acetic, propanoic, hydroxyacetic, 2-butenic, pentanoic, 2-me-butanoic, 4-oxypentanoic, 4-hydroxypentanoic, hexanoic, benzoic, heptanoic, etc.
- Esters: methyl formate, methyl acetate, methyl propionate, butyrolactone, methyl crotonate, methyl *n*-butyrate, velerolactone, angelicalactone, methyl valerate, etc.
- Alcohols: methanol, ethanol, 2-propene-1-ol, isobutanol, 3-methyl-1-butanol, ethylene glycol, etc.
- Ketones: acetone, 2-butanone, 2-butanone (mek), 2,3-butandione, cyclopentanone, 2-pentanone, 3-pentanone, 2-cyclopentanone, 2,3-pentenedione, 3-me-2-cyclopenten-2-ollone, me-cyclopentanone, 2-hexanone, cyclohexanone, methylcyclohexanone, 2-Et-cyclopentanone, dimethylcyclopentanone, trimethylcyclopentanone, etc.
- Aldehydes: formaldehyde, acetaldehyde, 2-propenal, 2-butenal, 2- methyl-2-butenal, pentanal, ethanedial.
- Phenols: phenol, 2-methyl phenol, 3-methyl phenol, 4-methyl phenol, 2,3-dimethyl phenol, 2,4-dimethyl phenol, 2,5-dimethyl phenol, 2,6-dimethyl phenol, etc.
- Alkenes: 2-methyl propene, dimethylcyclopentene, alpha-pinene, dipentene.
- Aromatics: benzene, toluene, xylenes, naphthalenes, phenanthrene, fluoranthrene, chrysene, etc.
- Nitrogen compounds: ammonia, methylamine, pyridine, methylpyridine, etc.

- Furans: furan, 2-methyl furan, 2-furanone, furfural, 3-methyl-2(3h)furanone, furfural alcohol, furoic acid, methyl furoate, 5-methylfurfural, dimethylfuran, etc.
- Guaiacols: 2-methoxy phenol, 4-methyl guaiacol, ethyl guaiacol, eugenol, isoeugenol, 4-propylguaiacol, acetoguiacone, propioguiacone, etc.
- Syringols: 2,6-DiOMe phenol, methyl syringol, 4-ethyl syringol, propyl syringol, syringaldehyde, 4-propenylsyringol, etc.
- Sugars: levoglucosan, glucose, fructose, D-xylose, D-arabinose, cellobiosan, 1,6-anhydroglucofuranose, etc.
- Miscellaneous oxygenates: hydroxyacetaldehyde, hydroxyacetone, methylal, dimethyl acetal, acetal, acetoxy-2-propanone, methyl cyclopentenolone, 1-acetyloxy-2-propanone, 2-methyl-3-hydroxy-2-pyrone, 2-methoxy-4-methylanisole, 4-OH-3-methoxybenzaldehyde, maltol, etc. (Diebold 2000).

Inorganic components can be present in bio-oil in the following forms: associated with counter ions, connected to organic acids, and related to various enzymatic compounds. These include Ca, Si, K, Fe, Al, Na, S, P, Mg, Ni, Cr, Zn, Li, Ti, Mn, Ln, Ba, V, Cl, etc.

3.3.8 Composition of Pyrolysis Gas and Char

The combustible gas from the pyrolysis process contains mostly CO, CO_2, and CH_4. Other components present are H_2, propane, propylene, butane, butenes, C_5, ethane, etc. The gas can be used for obtaining energy by combustion. The char obtained contains around 85% elemental carbon and 3% hydrogen. The typical analysis of charcoal is given in Table 3.3.

3.3.9 Uses of Bio-Oil and Char

The oil obtained from the pyrolysis of biomass has a variety of uses, which include as fuel for combustion, power generation, the production of chemicals, as a fuel for transportation and synthetic fossil fuels, for the preparation of phenol formaldehyde resole resins, as liquid smoke, for the production of anhydro-sugars such as levoglucosan, as binders for palletizing and briquetting of combustible organic waste materials, as preservatives, for example, wood preservative, in the preparation of adhesives, as diesel engine fuels (after blending suitably with diesel oil). The oil obtained from sewage sludge pyrolysis can be used directly in diesel-fuelled engines. The oil may be stored and transported, hence does not have to be used at or near the process plant only.

The char obtained from the pyrolysis of biomass has several uses. As the yield of char is between 20 and 26 wt%, this solid fuel can be used in boilers where bagasse or other biomass is presently burnt. This can also be converted into brickets alone, or mixed with the biomass and can be used as high efficiency fuel in boilers. It can be used as feedstock for the production of activated carbon and for the gasification process to obtain hydrogen-rich gas. Also, the possibility of using this carbon feedstock for making carbon-nano-tubes may be explored.

TABLE 3.3
Charcoal Proximate and Elemental Analysis (wt%)

	Laboratory Scale	Pilot Plant Scale
	Proximate Analysis	
Volatile matter	18.9	15.4
Fixed carbon	74.4	79.1
Ash	6.7	5.5
	Elemental Analysis	
C	85.6	81.5
H	2.9	3.1
N	1.3	0.8
S	<0.1	<0.1
O + ash (by difference)	10.2	14.6

3.4 HYDROGEN FROM BIOMASS

The production of hydrogen from biomass is also being studied as it would be environmentally benign. Hydrogen is either manufactured from fossil fuels such as natural gas, naphtha and coal, or from nonfossil energy resources like water electrolysis, photolysis, and thermolysis (Momirlan and Vziroglu 1999). Different thermochemical routes of hydrogen production from biomass are described below.

3.4.1 Pyrolysis

Conventional pyrolysis methods (Section 3.3) produce hydrogen in amounts that are not significant. Catalytic pyrolysis could be a useful method in this regard, which could be achieved by (1) the catalytic steam reforming of pyrolysis liquids to produce hydrogen, (2) pyrolysis at 700°C, with removal of the tar content of the gas and improving the quality of the product gas, and (3) pyrolysis at a lower temperature (<750°C) and incorporation of catalyst in the same reactor. Some types of reactor used in this process are Waterloo fast pyrolysis unit, free fall reactor, reactor for $RhKeO_2/SiO_2$ catalyst, and dual bed gasifier reactor.

3.4.2 Supercritical Water Extraction

Biomass can be converted into fuel gases rich in hydrogen by supercritical water extraction. The water is attractive as a potential medium for industrial chemical reactions because it is environmentally benign. The supercritical water acts as a homogeneous, nonpolar solvent of high diffusivity and high transport properties, able to dissolve organic compounds and gases (Feng et al. 2004). In such a process, hydrogen can be produced at thermodynamic equilibrium.

3.4.3 Gasification

Gasification converts biomass into a combustible gas mixture (CO, CO_2, CH_4, H_2, and H_2O) by the partial oxidation of the biomass at high temperatures, typically in the range of 800 to 900°C. The main reaction steps in biomass gasification are:

- Heating and pyrolysis of the biomass, converting biomass into gas, char, and primary tar.
- Cracking of the primary tar to gases and secondary and ternary tars.
- Cracking of secondary and tertiary tars.
- Heterogeneous gasification reactions of the char formed during the pyrolysis and the homogeneous gas phase reactions.
- The combustion of char formed during pyrolysis and oxidation of combustible gases.

In a gasification process, the solid fuels are completely converted (except the ashes in the feed) to gaseous products having different compositions. The gasification process is attractive because of the production of cleaner gaseous fuel as well as almost complete conversion of biomass. The gasification processes can be categorized as below:

1. Air gasification. Air gasification is the most widely used technology, as a single product is formed at high efficiency and without requiring oxygen. Temperatures of 900 to 1100°C are achieved.
2. Oxygen gasification. This gives a better quality gas of 10 to 15 MJ/ Nm^3 and temperatures of 1000 to 1400°C are achieved. However, this process requires an O_2 supply, with concomitant problems of cost and safety.
3. Steam gasification. Steam gasification of biomass results in the conversion of carbonaceous material to permanent gases (H_2, CO, CO_2, CH_4, and light hydrocarbons), char, and tar (Kim 2003).

For thermochemical processes, different reactor designs are needed. One way to characterize reactor types is based on the method of transport of fluids or solids through the reactor. The four main types are quasi-non-moving or self-moving feedstock, mechanically moved feedstock, fluid-moved feedstock, and special reactors. Some of the reactors used include multi-stage circulating fluidized-bed reactors, pressurized fluidized-bed air-blown gasifiers, two-stage gasification reactors, atmospheric bubbling fluidized-bed reactors, countercurrent fixed-bed gasifiers, open top reburn downdraft gasifiers, down draft fixed-bed, etc. (Dasappa et al. 2004; Narvaez et al. 1996; Padban et al. 2000; Saxena et al. 2007). There is no significant advantage to using either the fixed-bed or the fluidized-bed reactor (Warnecke 2000).

3.5 CONCLUSION

Visualizing the present energy requirements worldwide and concerns for environmental problems, it is necessary to develop a process that can convert biomass into useful energy products, especially to liquids that can be good substitutes for the depleting fossil fuels. Thermochemical conversion methods like pyrolysis have received a significant amount of interest. The fast and the flash pyrolysis processes give higher yields of liquid product as compared to slow pyrolysis. Catalytic pyrolysis, which is another way to upgrade the quality of the products, is also a process of interest. It may also be applied to improve the quantity and the quality of the gaseous products to make them more useful for obtaining energy.

Apart from liquid fuels, hydrogen from biomass also is a potential fuel that may meet the energy requirement without creating problems for the environment. Various thermochemical conversion technologies can be applied for conversion of renewable biomass into hydrogen-rich gas.

ACKNOWLEDGMENTS

The authors gratefully acknowledge the continuous encouragement given by the Director, Indian Institute of Petroleum, Dehradun during the preparation of this chapter.

REFERENCES

Adhikari, D. K., Seal Diptendu, R. C. Saxena, and H. B. Goyal. 2006. Biomass based fuel/energy. *Journal of Petrotech Society* 3(1): 28–42.

Asadullah, M., T. Miyazawa, S. Ito, K. Kunimori, and K. Tomishige. 2002. Role of catalysis and its fluidisation in the catalytic gasification of biomass to syngas at low temperature. *Industrial Engineering & Chemistry Research* 41: 4567–4575.

Bridgwater, A. V. 1994. Catalysis in thermal biomass conversion. *Applied Catalysis A: General* 116: 5–47.

Bridgwater, A. V. 1999. Principles and practice of biomass fast pyrolysis processes for liquids. *Journal of Analytical and Applied Pyrolysis* 51: 3–22.

Chen, G., J. Andries, and H. Spliethoff. 2003. Catalytic pyrolysis of biomass for hydrogen rich fuel gas production. *Energy Conversion and Management* 44: 2289–2296.

Dasappa, S. P. J. Paul, H. S. Mukunda, N. K. S. Rajan, G. Giridhar, and H. V. Sridhar. 2004. Biomass gasification technology: A route to meet energy needs. *Current Sciences* 87: 908–916.

Demirbas, A. 2001. Biomass resources facilities and biomass conversion processing for fuels and chemicals. *Energy Conversion and Management* 42: 1357–1378.

Demirbas, A. 2004. Hydrogen-rich gas from fruit shale via supercritical water extraction. *International Journal of Hydrogen Energy* 29: 1237–1243.

Diebold, J. P. 2000. *A review of the chemical and physical mechanism of the storage stability of fast pyrolysis bio-oils.* Golden, CO: National Renewable Energy Laboratory.

Feng, W., J. Hedzer, K. Vander, and A. Jakob de Swan. 2004. Phase equilibria for biomass conversion process in subcritical and supercritical water. *Chemical Engineering Journal* 98: 105–113.

Goyal, H. B., A. K. Gupta, A. K. Bhatnagar, P. S. Verma, and K. V. Padmaja. 2005. Liquid fuels from biomass. Paper presented at 6th International Petroleum Conference and Exhibition (Petrotech-2005), January 16–19, India.

Goyal, H. B., Seal Diptendu, and R. C. Saxena. 2006. Bio-fuels from thermochemical conversion of renewable resources: A review. *Renewable & Sustainable Energy Reviews*, Available online at www.sciencedirect.com.

Kim, H. Y. 2003. A low cost production from cabonaceous waste. *International Journal of Hydrogen Energy* 28: 1179–1186.

Manuel, G. P., Abdelkader Chaala, and C. Roy. 2002. Vacuum pyrolysis of sugarcane bagasse. *Journal of Analytical and Applied Pyrolysis* 65: 111–136.

McKendry, P. 2002. Energy production from biomass. II. Conversion technologies. *Bioresource Technology* 83: 47–54.

Momirlan, M. and T. Vziroglu. 1999. Recent directions of world hydrogen production. *Renewable and Sustainable Energy Reviews* 3: 219–231.

Narvaez, I., A. Orio, M. P. Aznar, and J. Corella. 1996. Biomass gasification with air in an atmospheric bubbling fluidized bed. Effects of six operational variables on the quality of the produced raw gas. *Industrial Engineering & Chemistry Research* 35: 2110–2120.

Ozbay, N., A. E. Putun, B. B. Uzun, and E. Putun. 2001. Biocrude from biomass: pyrolysis of cottonseed cake. *Renewable Energy* 24: 615–625.

Padban, N.; W. Wang, Z. Ye, I. Bjerle, and I. Odenbrand. 2000. Tar formation in pressurised fluid bed air gasification of woody biomass. *Energy and Fuels* 14: 603–611.

Saxena, R. C., D. K. Adhikari, and H. B. Goyal. In press. Biomass based energy fuel through biochemical routes: A review. *Renewable & Sustainable Energy Reviews.* Available online at www.sciencedirect.com.

Saxena, R. C., Seal Diptendu, Kumar Satinder, and H. B. Goyal. In press. Thermochemical routes for hydrogen rich gas from biomass: A review. *Renewable & Sustainable Energy Reviews.* Available online at www.sciencedirect.com.

Warnecke, R. 2000. Gasification of biomass; comparison of fixed bed and fluidized bed gasifier. *Biomass and Bioenergy* 18: 489–497.

Williams, P. T. and N. Nugranad. 2000. Comparison of products from pyrolysis and catalytic pyrolysis of rice husks. *Energy* 25: 493–513.

4 Production of Biofuels with Special Emphasis on Biodiesel

Ayhan Demirbas

CONTENTS

ABSTRACT

Biodiesel, an alternative biodegradable diesel fuel, is a renewable fuel that can be produced from vegetable oils, animal fats, and used cooking oil, including triglycerides. It is derived from triglycerides by transesterification with methanol and ethanol. Concerns about the depletion of diesel fuel reserves and the pollution caused by the continuously increasing energy demands make biodiesel an attractive alternative motor fuel for compression ignition engines. There are four different ways of modifying vegetable oils and fats to use them as diesel fuel, such as pyrolysis/cracking, dilution with hydrocarbons blending, emulsification, and transesterification. The most commonly used process is the transesterification of vegetable oils and animal fats.

4.1 INTRODUCTION

The term biofuel generally refers to liquid or gaseous fuels for the transport sector that are predominantly produced from biomass. Biomass (all plants and living organisms) can be converted into liquid and gaseous fuels through thermochemical and biological routes. Biofuel is a nonpolluting, locally available, accessible, sustainable, and reliable fuel obtained from renewable sources. The liquid biofuels being considered the world-over fall into the following categories: (1) vegetable oils and biodiesels, (2) alcohols, and (3) biocrude and synthetic oils (Demirbas 2006). Figure 4.1 shows the resources of the main liquid biofuels for automotives. Figure 4.2 shows the main biomass conversion processes.

4.2 VEGETABLE OILS AS DIESEL FUELS

The use of vegetable oils as alternative renewable fuel competing with petroleum was proposed at the beginning of the 1980s. The advantages of vegetable oils as diesel fuel are their liquid nature, which implies portability, ready availability, renewability, higher heat content (about 88% of No. 2 petroleum diesel fuel), lower sulfur content, lower aromatic content, and biodegradability. A sustainable biofuel has two favorable properties, which are availability from renewable raw material and lower negative environmental impact than that of the fossil fuels. As an alternative fuel, vegetable oil is one of the renewable fuels. Vegetable and animal oils/fats, which

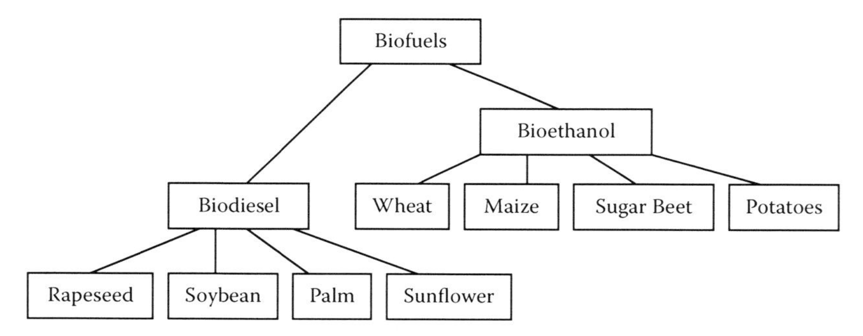

FIGURE 4.1 Resources of main liquid biofuels for automotives.

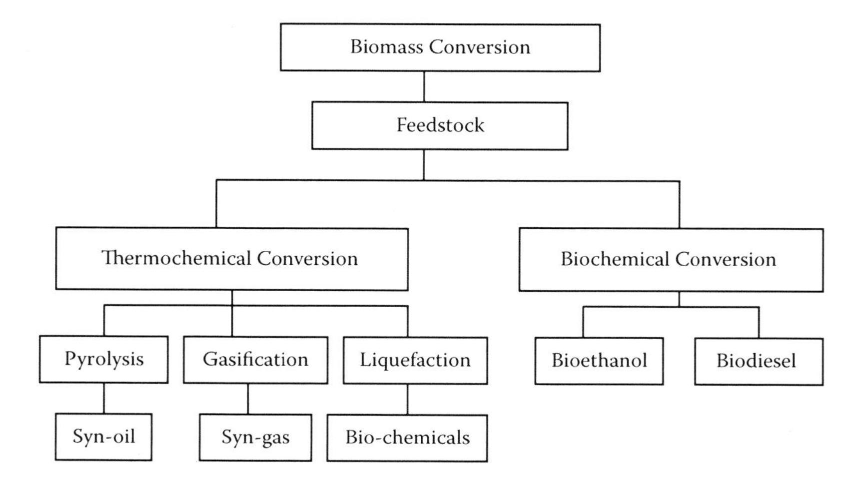

FIGURE 4.2 Main biomass conversion processes.

mainly consist of the triglyceride of the straight chain fatty acid, are organic chemicals made from carbon dioxide and water using solar energy. Vegetable oil is a potentially inexhaustible source of energy, with an energy content close to that of diesel fuel. Vegetable oils, such as palm, soybean, sunflower, peanut, and olive oils, can be used as alternative fuels for diesel engines. Vegetable oils used as alternative engine fuels are all extremely viscous, with viscosities ranging from 10 to 20 times higher than petroleum diesel fuel. The major problem associated with the use of pure vegetable oils as fuels for diesel engines is caused by the high fuel viscosity in compression ignition engines. The vegetable oils pose many problems when used directly in diesel engines. These include (1) coking and trumpet formation on the injectors, to such an extent that fuel atomization does not occur properly or even is prevented as a result of plugged orifices; (2) carbon deposits; (3) oil ring sticking; (4) thickening or gelling of the lubricating oil as a result of contamination by vegetable oils; and (5) lubricating problems.

There are different ways of modifying vegetable oils and fats to use them as diesel fuel, and such methods as pyrolysis, dilution with hydrocarbons, and emulsification have been considered. Figure 4.3 shows the use of vegetable oils as petroleum alternative fuels (Demirbas 2003). Emulsification with alcohols has been proposed to overcome the problem of high viscosity of vegetable oils (Madras, Kolluru, and Kumar 2004).

4.3 BIODIESEL

The diesel made from oils/fats has been noted as an ecological fuel because the oils/fats are a sustainable energy resource. Thus, alkyl esters of fatty acids that meet transportation fuel standards are called "biodiesel." It has been also defined as the monoalkyl esters of the long chain fatty acids derived from renewable feedstocks, such as vegetable oils or animal fats, for use in compression ignition (diesel) engines. Biodiesel has become more attractive because of its environmental benefits and the fact that it is made from renewable resources (Ma and Hanna 1999).

One popular process for producing biodiesel from the fats/oils is *trans*-esterification of triglyceride by methanol (methanolysis) to make methyl esters of the straight chain fatty acid. Alcohols are primary or secondary monohydric aliphatic alcohols having one to eight carbon atoms. Among the alcohols that can be used in the transesterification reaction are methanol, ethanol, propanol, butanol, and amyl alcohol. Methanol and ethanol are used most frequently; ethanol is a preferred alcohol in the transesterification process compared to methanol because it is derived

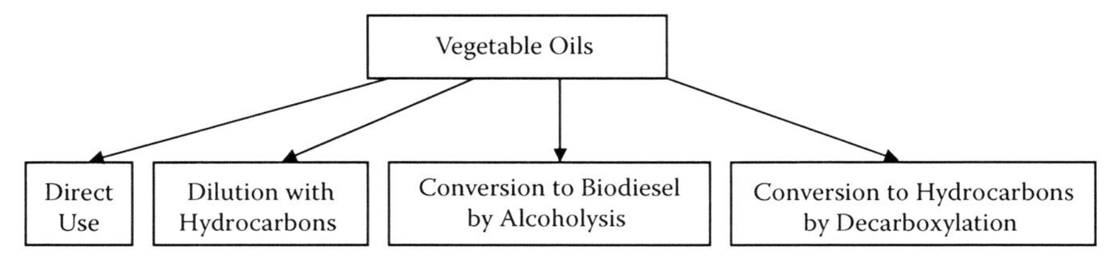

FIGURE 4.3 Use of vegetable oils as petroleum alternative fuels.

from agricultural products and is renewable and biologically less objectionable in the environment. However, methanol is preferred because of its low cost and its physical and chemical advantages (polar and shortest chain alcohol).

Methyl, ethyl, 2-propyl, and butyl esters have been prepared from canola and linseed oils through transesterification using KOH and/or sodium alkoxides as catalysts. In addition, (m)ethyl esters were prepared from rapeseed and sunflower oils using the same catalysts. Today, biodiesel is made from a variety of natural oils. Important among these are soybean oil and rapeseed oil. Rapeseed oil, a close cousin of canola oil, dominates the growing biodiesel industry in Europe. In the United States, biodiesel is being made from soybean oil because more soybean oil is produced than all other sources of fats and oil combined. There are many candidates for feedstocks, including recycled cooking oils, animal fats, and a variety of other oilseed crops.

For the preparation of biodiesel using a catalytic method, the catalyst (KOH) is dissolved into methanol by vigorous stirring in a small reactor. The oil is transferred into the biodiesel reactor and then the catalyst/alcohol mixture is pumped into the oil. The final mixture is stirred vigorously for 2 h at 340 K at ambient pressure. A successful transesterification reaction produces two liquid phases: ester and crude glycerin. The crude glycerin, the heavier liquid, settles at the bottom after a few hours of settling. The phase separation can be observed within 10 minutes and can be complete within 2 h of the settling. Complete settling can take as long as 20 h. After the settling is complete, water is added at the rate of 5.5%, v/v of the methyl ester of oil and then stirred for 5 minutes and the glycerin is allowed to settle again. Washing the ester is a two-step process, which is carried out with extreme care. A water wash solution at the rate of 28% by volume of oil and 1 g of tannic acid per liter of water is added to the ester and gently agitated. Air is carefully introduced into the aqueous layer while simultaneously stirring very gently. This process is continued until the ester layer becomes clear. After settling, the aqueous solution is drained and water alone is added at 28% by volume of the oil for the final washing (Demirbas 2002). Figure 4.4 shows the catalytic biodiesel production diagram.

The physical characteristics of the fatty acid (m)ethyl esters are very close to those of diesel fuel and the process is relatively simple. Furthermore, the (m)ethyl esters of the fatty acids can be burned directly in unmodified diesel engines, with very low deposit formation. The methyl and ethyl esters prepared from a particular vegetable oil have similar viscosities, cloud points, and pour points, whereas methyl, ethyl, 2-propyl, and butyl esters derived from a particular vegetable oil have similar gross heating values. However, their densities, which were 2 to 7% higher than those

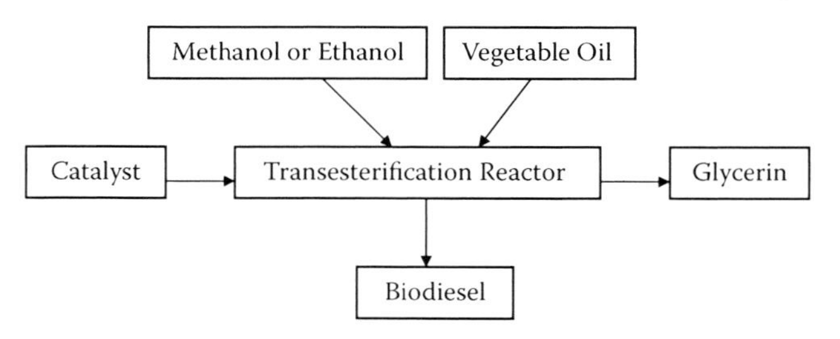

FIGURE 4.4 Catalytic biodiesel production.

of diesel fuels, statistically decreased in the order 2-propyl > ethyl > butyl esters. The higher heating values of the biodiesel fuels, on a mass basis, are 9 to 13% lower than No. 2 petroleum diesel. The viscosities of biodiesel fuels are twice that of No. 2 petroleum diesel. The cloud and pour points of No. 2 petroleum diesel are significantly lower than those of biodiesel fuels. The biodiesel fuels produced slightly lower power and torque and higher fuel consumption than No. 2 petroleum diesel.

Biodiesel is an efficient, clean, 100% natural energy alternative to the petroleum fuels. Among the many advantages of biodiesel fuel include the following: it is safe for use in all the conventional diesel engines; offers the same performance and engine durability as petroleum diesel fuel; is nonflammable and nontoxic; and reduces tailpipe emissions, visible smoke, and noxious fumes and odors. It is better than diesel fuel in terms of the sulfur content, flash point, aromatic content, and biodegradability (Demirbas 2003). If biodiesel is valorized efficiently for energy purposes, it would benefit the environment and the local population through job creation, provision of modern energy carriers to rural communities, avoiding urban migration and reduction of CO_2 and sulfur levels in the atmosphere.

4.4 NONCATALYTIC TRANSESTERIFICATION WITH SUPERCRITICAL ALCOHOL TRANSESTERIFICATION

In the conventional transesterification of animal fats and vegetable oils for biodiesel production, the free fatty acids and water always produce negative effects, since the presence of the free fatty acids and the water causes soap formation, consumes catalyst and reduces catalyst effectiveness, all of which result in a low conversion (Komers, Machek, and Stloukal 2001). The transesterification reaction may be carried out using either basic or acidic catalysts, but these processes are relatively time consuming and require the complicated separation of the product and the catalyst, which results in high production costs and energy consumption. In order to overcome these problems, Saka and Kusdiana (2001) and Demirbas (2002, 2003) proposed that biodiesel fuels may be prepared from vegetable oil via noncatalytic transesterification with supercritical methanol (SCM). The SCM is believed to solve the problems associated with the two-phase nature of normal methanol/oil mixtures by forming a single phase as a result of the lower value of the dielectric constant of the methanol in the supercritical state. As a result, the reaction is completed in a very short time (Han, Cao, and Zhang 2005). Compared with catalytic processes under barometric pressure, the SCM process is noncatalytic, purification of the products is much simpler, requires lower reaction time and lower energy, and is more environmentally friendly. However, the reaction requires temperatures of 525 to 675 K and pressures of 35 to 60 MPa (Demirbas 2003; Kusdiana and Saka 2001).

Supercritical transesterification is carried out in a high-pressure reactor (autoclave). In a typical run, the autoclave is charged with a given amount of the vegetable oil and liquid methanol with changed molar ratios. The autoclave is supplied with heat from an external heater, and the power is adjusted to give an approximate heating time of 15 min. The temperature of the reaction vessel can be measured with an iron-constantan thermocouple and controlled at ±5 K for 30 min. The transesterification occurs during the heating period. After each run, the gas is vented, and the autoclave is poured into a collecting vessel. All the contents are removed from the autoclave by washing with methanol. Table 4.1 shows critical temperatures and critical pressures of various alcohols. Table 4.2 shows comparisons between the catalytic methanol method and supercritical methanol method for the production of biodiesel from vegetable oils by transesterification.

4.5 TRANSESTERIFICATION REACTION MECHANISM OF VEGETABLE OIL

Transesterification consists of a number of consecutive, reversible reactions. The triglyceride is converted stepwise to diglyceride, monoglyceride, and finally glycerol. The formation of alkyl esters from monoglycerides is believed to be a step that determines the reaction rate, because monoglycerides are the most stable intermediate compound (Ma and Hanna 1999). Several aspects, including the type of catalyst (alkaline, acid, or enzyme), alcohol to vegetable oil molar ratio, temperature, water content, and free fatty acid content, have an influence on the course of the transesterification. In the transesterification of vegetable oils and fats for biodiesel

TABLE 4.1
Critical Temperatures and Critical Pressures of Various Alcohols

Alcohol	Critical Temperature (K)	Critical Pressure (MPa)
Methanol	512.2	8.1
Ethanol	516.2	6.4
1-Propanol	537.2	5.1
1-Butanol	560.2	4.9

TABLE 4.2
Comparisons between the Catalytic Methanol (MeOH) method and Supercritical Methanol (SCM) Method for the Production of Biodiesel from Vegetable Oils by Transesterification

Method	Catalytic MeOH Process	SCM Method
Methylating agent	Methanol	Methanol
Catalyst	Alkali (NaOH or KOH)	None
Reaction temperature (K)	303–338	523–573
Reaction pressure (MPa)	0.1	10–25
Reaction time (min)	60–360	7–15
Methyl ester yield (wt%)	96	98
Removal for purification	Methanol, catalyst, glycerol, soaps	Methanol
Free fatty acids	Saponified products	Methyl esters, water
Smelling from exhaust	Soap smelling	Sweet smelling

production, free fatty acids and water always produce negative effects, as the presence of free fatty acids and water causes soap formation, consumes catalyst, and reduces catalyst effectiveness, all of which result in a low conversion (Ali, Hanna, and Cuppett 1995). The transesterification is an equilibrium reaction and the transformation occurs essentially by mixing the reactants. In the transesterification of the vegetable oils, a triglyceride reacts with an alcohol in the presence of a strong acid or base, producing a mixture of fatty acids, alkyl esters, and glycerol. The stoichiometric reaction requires 1 mol of a triglyceride and 3 mol of the alcohol. However, an excess of the alcohol is used to increase the yields of the alkyl esters and to allow its phase separation from the glycerol formed (Bala 2005).

A reaction mechanism of the vegetable oil in the SCM was proposed based on the mechanism developed by Krammer and Vogel (2000) for the hydrolysis of the esters in sub- or supercritical water. The basic idea of supercritical treatment is the effect of the pressure and temperature on the thermophysical properties of the solvent, such as dielectric constant, viscosity, specific weight, and polarity (Kusdiana and Saka, 2001).

4.6 VARIABLES AFFECTING METHYL ESTER YIELD

The most important variables affecting the methyl ester yield during the transesterification reaction are molar ratio of the alcohol to vegetable oil and the reaction temperature. The viscosities of the methyl esters from the vegetable oils were slightly higher than that of No. 2 diesel fuel. Figure 4.5 shows a typical example of the relationship between the reaction time and the temperature (Demirbas 2002). A hazelnut kernel oil sample was used; the critical temperature and the critical pressure of methanol were 512.4 K and 8.0 MPa, respectively. The variables affecting the ester yield during the transesterification reaction were molar ratio of the alcohol to the vegetable

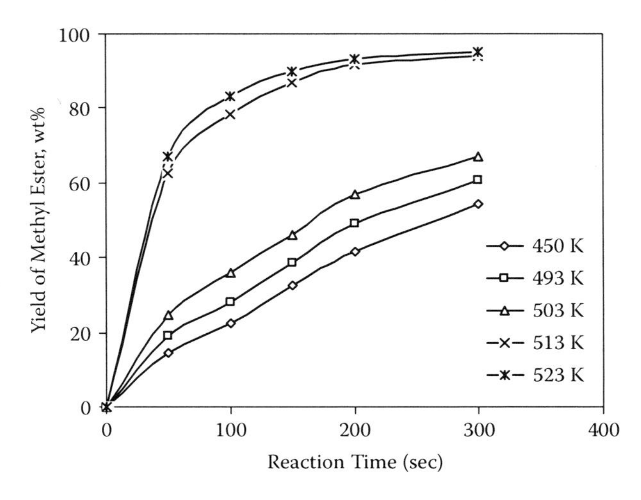

FIGURE 4.5 Changes in yield percentage of methyl esters as treated with subcritical and supercritical methanol at different temperatures as a function of reaction time. Molar ratio of vegetable oil to methyl alcohol: 1:41. Sample: hazelnut kernel oil.

oil, reaction temperature, reaction time, water content, and catalyst. It was observed that increase in the reaction temperature, especially to supercritical temperatures, had a favorable influence on the ester conversion (Demirbas 2002). The stoichiometric ratio for the transesterification reaction requires three moles of alcohol and one mole of triglyceride to yield three moles of fatty acid ester and one mole of glycerol. Higher molar ratios result in higher ester production in a shorter time.

As has been mentioned above, in the catalyzed methods, the presence of water has negative effects on yields of methyl esters. However, the presence of water positively affected the formation of methyl esters in our supercritical methanol method. Figure 4.6 shows the plots for yields of methyl esters as a function of water content in the transesterification of triglycerides.

Most diesel engines are designed to use highly lubricating, high sulfur content fuel. Recent environmental legislature has forced diesel fuel to contain only a minimum amount of the sulfur for lubricating purposes. Thus, the slightly higher viscosity of biodiesel is helpful and lubricating to most diesel motors. Compared to No. 2 diesel fuel, all of the vegetable oils are much more viscous, much more reactive to oxygen, and possess higher cloud point and pour point temperature. The flash point of all vegetable oils is far above that of diesel fuel, reflecting the nonvolatile nature of the vegetable oils. Vegetable oils are not directly volatile, but crack during distillation into a series of hydrocarbons or can be converted by transesterification into more volatile methyl esters.

4.7 COMPARISON OF FUEL PROPERTIES AND COMBUSTION CHARACTERISTICS OF METHYL AND ETHYL ALCOHOLS AND THEIR ESTERS

In general, the physical and chemical properties and the performance of the ethyl esters are comparable to those of the methyl esters. The methyl and ethyl esters have almost the same heat content. The viscosities of the ethyl esters are slightly higher and the cloud and pour points are slightly lower than those of the methyl esters.

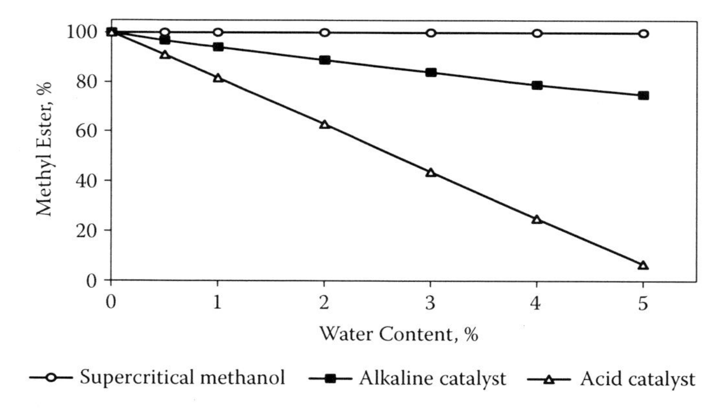

FIGURE 4.6 Plots for yields of methyl esters as a function of water content in transesterification of triglycerides.

Engine tests demonstrate that methyl esters produce slightly higher power and torque than the ethyl esters (Bala 2005). Some desirable attributes of the ethyl esters over methyl esters are: they have significantly lower smoke opacity, lower exhaust temperatures, and lower pour point. The ethyl esters tend to have more injector coking than the methyl esters.

4.8 BIODIESEL ECONOMY

The cost of the biodiesel fuels varies depending on the base stock, geographic area, variability in the crop production from season to season, the price of the crude petroleum, and other factors. Biodiesel has over double the price of diesel. The high price of biodiesel is largely due to the high price of the feedstock. However, biodiesel can be made from other feedstocks, including beef tallow, pork lard, and yellow grease, which are relative cheaper raw materials. Cooking oils can also be used as cheaper raw materials. The problem with processing these low-cost oils and fats is that they often contain large amounts of free fatty acids (FFA) that cannot be converted to biodiesel using an alkaline catalyst.

A review of twelve economic feasibility studies shows that the projected costs for biodiesel from oilseed or animal fats have a range of US$0.30 to $0.69/l, including the meal and glycerin credits and the assumption of reduced capital investment costs by having the crushing and/or esterification facility added onto an existing grain or tallow facility. Rough projections of the cost of biodiesel from vegetable oil and waste grease are US$0.54 to $0.62/l and US$0.34 to $0.42/l, respectively. With pretax diesel priced at US$0.18/l in the United States and US$0.20 to $0.24/l in some European countries, biodiesel is thus currently not economically feasible, and more research and technological development are needed.

4.9 CONCLUSIONS AND FUTURE PERSPECTIVES

The biodiesels are the mono-alkyl esters of long chain fatty acids derived from renewable feedstocks, such as vegetable oils or animal fats, for use in diesel engines. The purpose of the transesterification of vegetable oils to their methyl esters is to lower the viscosity of the oil. The main factors affecting transesterification are molar ratio of glycerides to alcohol, catalyst, reaction temperature and pressure, reaction time, and the contents of free fatty acids and water in oils.

Vegetable oils are a renewable and potentially inexhaustible source of energy, with an energy content close to that of diesel fuel. The vegetable oil fuels are not feasible because they are more expensive than petroleum fuels. However, with recent increases in petroleum prices and the uncertainties concerning petroleum availability, there is renewed interest in vegetable oil fuels for diesel engines.

REFERENCES

Ali, Y., M. A. Hanna, and S. L. Cuppett. 1995. Fuel properties of tallow and soybean oil esters. *JAOCS* 72:1557–1564.

Bala, B. K. 2005. Studies on biodiesels from transformation of vegetable oils for diesel engines. *Energy Edu. Sci. Technol.* 15:1–43.

Demirbas, A. 2002. Biodiesel from vegetable oils via transesterification in supercritical methanol. *Energy Convers. Mgmt.* 43:2349–2356.

Demirbas, A. 2003. Biodiesel fuels from vegetable oils via catalytic and non-catalytic supercritical alcohol transesterifications and other methods: A survey. *Energy Convers. Mgmt.* 44:2093–2109.

Demirbas, A. 2005. Biodiesel production from vegetable oils by supercritical methanol. *J. Sci. Ind. Res.* 64:858–865.

Demirbas, A. 2006. Global biofuel strategies. *Energy Edu. Sci. Technol.*17:32-63.

Han, H., W. Cao, and J. Zhang. 2005. Preparation of biodiesel from soybean oil using supercritical methanol and CO_2 as co-solvent. *Process Biochemistry* 40:3148–3151.

Komers, K., J. Machek, and R. Stloukal. 2001. Biodiesel from rapeseed oil and KOH. II. Composition of solution of KOH in methanol as reaction partner of oil. *Eur. J. Lipid Sci. Technol.* 103:359–362.

Krammer, P., and H. Vogel. 2000. Hydrolysis of esters in subcritical and supercritical water. *Supercrit. Fluids* 16:189–206.

Kusdiana, D., and S. Saka. 2001. Kinetics of transesterification in rapeseed oil to biodiesel fuels as treated in supercritical methanol. *Fuel* 80:693–698.

Ma, F., and M. A. Hanna. 1999. Biodiesel production: A review. *Biores. Technol.* 70:1–15.

Madras, G., C. Kolluru, and R. Kumar. 2004. Synthesis of biodiesel in supercritical fluids. *Fuel* 83:2029–2033.

Saka, S., and D. Kusdiana. 2001. Biodiesel fuel from rapeseed oil as prepared in supercritical methanol. *Fuel* 80:225–231.

Section II

Production of Bioethanol

5 Fuel Ethanol
Current Status and Outlook

Edgard Gnansounou

CONTENTS

ABSTRACT

An analysis of the current situation and perspective on biomass-to-ethanol is provided in this chapter. Various conversion pathways are compared from technical, economic, and environmental points of view. It is found that, due to a learning curve and other economic reasons, the United States and Brazil will maintain their comparative advantage in the next decades. However, the fast growth of the world fuel ethanol demand, as well as the perspectives of the oil market, may notably influence the international market price of ethanol and open opportunities for wide-scale production in other regions such as Europe and Asia. In the long term, lignocellulose-to-ethanol is the most viable pathway from a sustainability point of view. However, its production cost must be reduced significantly for this process to have a chance to drive forward the strategy of biomass-to-ethanol worldwide.

5.1 INTRODUCTION

Liquid biofuels are receiving increasing attention worldwide as a result of the growing concerns about oil security of supply and global climate change. In most developing countries, the emerging biofuels industry is perceived as an opportunity to enhance economic growth and create or maintain jobs, particularly in rural areas. The liquid biofuels market is shared mainly between bioethanol and biodiesel, with more than 85% market share for the former in 2005. The main advantage of bioethanol is the possibility to blend it in low proportions with gasoline (5 to 25% bioethanol by volume) for use, without any significant change, in internal combustion engines. That technology constitutes the highest proportion of the world's light duty vehicles fleet. Flexible fuelled vehicles (FFVs) are presently booming as well, particularly in Brazil and Sweden, creating a new opportunity for bioethanol to compete directly with gasoline.

The use of ethanol as a fuel has a long history, starting in 1826 when Samuel Morey used it with the first American prototype of the internal combustion engine. The renewal of interest in fuel ethanol started, however, from the 1973–74 world oil crisis when the Brazilian government launched its pro-alcohol strategic program to substitute a large share of imported oil. In the United States, the Energy Tax Act of 1978 exempted from excise tax the gasohol (10% of bioethanol blends with gasoline v/v). Later on, another U.S. federal program guaranteed loans for investment in ethanol plant construction. Brazil and the United States are still the two main producers and users of fuel ethanol worldwide.

Ethanol has good properties for internal combustion engines. Its average octane number of 99 is high compared to 88 for regular gasoline. However, the lower heating value (LHV) of ethanol (21 MJ/l) is 70% that of gasoline (about 30 MJ/l). Fuel ethanol is used in several manners in internal combustion engines: as 5% to 25% anhydrous ethanol blends with gasoline (5% maximum in Europe and India, 10% in the United States and China, 20 to 25% mandatory blends in Brazil), as pure fuel (100% of hydrated ethanol) in dedicated vehicles, or up to 85% in FFVs. When anhydrous bioethanol is blended with gasoline in small proportion (up to 15%), the influence of the lower heating value has no significant effect. For higher blend levels, the fuel economy is reduced compared to that with conventional gasoline.

Ethanol dedicated vehicles are optimized so that the engine efficiency is improved by running at higher compression ratios to take advantage of the better octane number of ethanol compared to gasoline. Therefore, for pure hydrated ethanol used in optimized vehicles, the ethanol can achieve about 75% or more of the range of gasoline on a volume basis. FFVs are equipped with line sensors that measure ethanol levels and adapt the air-fuel ratio to maintain good combustion conditions.

The use of bioethanol in internal combustion engines exhibits a few disadvantages: low levels of ethanol blended with gasoline increase vapor pressure and favor evaporative emissions that contribute to smog formation. For higher ethanol blend levels, the vapor pressure drops significantly, leading to more difficulty in cold weather conditions.

Due to its low cetane number, ethanol does not burn efficiently by compression ignition. Moreover, ethanol is not easily miscible with diesel fuel. Three methods are used to improve the use of ethanol in compression ignition vehicles. The first, used in direct blends of ethanol with diesel, involves addition of an emulsifier in order to improve ethanol-diesel miscibility. Other additives are used, such as ethylhexylnitrate or diterbutyl peroxide, to enhance the cetane number. Most blends of ethanol to diesel (E-diesel) have a limit of up to 15% ethanol and up to 5% emulsifiers (MBEP 2002). The second method is a dual fuel operation in which ethanol and diesel are introduced separately into the cylinder (SAE 2001). Finally, modification of diesel engines has been done to adapt their characteristics of auto-ignition and make them capable of using high blends such as 95% ethanol.

Even if bioethanol has a bright future, its environmental and economic performances vary significantly from one production pathway to the other. Its future development will depend mostly on the possibility to develop sustainable feedstocks, efficient technologies and to prevent potential risks such as local environmental hurdles and competition with food. In Section 5.2, the current status is analysed, including conversion chains and the situation in main producer countries. Section 5.3 presents the outlook to 2015. Finally, in Section 5.4, a few considerations are given on the necessity to define sustainability standards for biofuels in a neutral framework in order to promote best practices and sustainable pathways of bioethanol.

5.2 CURRENT STATUS

5.2.1 GENERIC CONVERSION SCHEME

Bioethanol can be produced from a large variety of carbohydrates: monosaccharides, disaccharides, and polysaccharides. The large-scale biomass-to-ethanol industry mostly uses the following feedstocks: sweet juice (e.g., sugarcane, sugar beet juice, or molasses) and starch (e.g., corn, wheat, barley, cassava). Ethanol is also commercially produced in the pulp and paper industry as a by-product of an acid-based conversion process. Modern lignocellulosic biomass-to-ethanol processes are envisaged to provide a significant percentage of bioethanol in the long term due to the expected low cost of the feedstock (agricultural and forestry residues) and to their high availability. The feedstock for bioethanol production is currently based mostly on agricultural crops, which can be devoted to both food and ethanol markets or dedicated solely to ethanol, that is, crops cultivated on fallow or undeveloped lands. In case of a high world production of bioethanol, the correlation between food and ethanol markets may generate a high volatility of agricultural crops with regard to fluctuations in energy prices.

Figure 5.1 outlines a generic biomass-to-ethanol process. One or more steps may be omitted and several may be combined, depending on the feedstock and the conversion technology. Once the biomass is delivered to the ethanol plant, it is stored in a warehouse and conditioned to prevent early fermentation and bacterial contamination. Through pretreatment, carbohydrates are extracted or made more accessible for further extraction. During this step, simple sugars may be made available in proportions depending on the biomass and the pretreatment process.

A large portion of fibers may remain for conversion to simple sugars through hydrolysis reactions or other techniques. In the fermentation step, batch operations

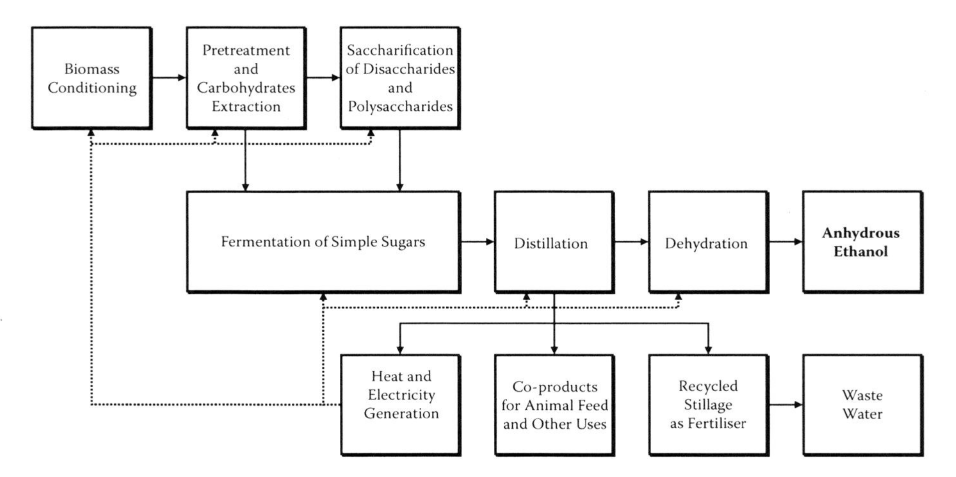

FIGURE 5.1 Schematic outline of the biomass-to-ethanol process. (From Gnansounou, E. and A. Dauriat. 2005. *Journal of Scientific and Industrial Research* 64:809–821. With permission.)

may be used in which the hydrolysate, the yeasts, nutriments, and other ingredients are added from the beginning of the step. In a fed batch process, one or more inputs are added as fermentation progresses. Continuous processes in which ingredients are constantly input and products removed from the fermentation vessels are also used (Wyman 2004). In efficient processes, the cell densities may be made high by recycling or immobilizing the yeasts in order to improve their activity and increase the fermentation productivity. The fermentation reactions occur at temperatures between 25 and 30°C and last between 6 and 72 h depending on the composition of the hydrolysate, and the type, density, and activity of the yeasts. The broth typically contains 8 to 14% of ethanol on a volume basis. Above this latter concentration, inhibition of yeasts may occur that reduces their activity. The distillation step yields an azeotropic mixture of 95.5% alcohol and 4.5% water that is the "hydrous" or "hydrated" ethanol which is then dehydrated to obtain an "anhydrous" ethanol with 99.6% alcohol and 0.4% water.

The remaining flow from the distillation column, known as vinasse, or stillage, can be valorized to produce co-products, which may include process steam and electricity, products for feeding animals, more or less concentrated stillage used as fertilizer, and other valuable by-products.

In 2005, around 36 billion liters of fuel bioethanol were produced in the world; Brazil and the United States provided 86% of the production.

5.2.2 Sucrose-to-Ethanol

The most common disaccharide used for bioethanol production is sucrose, which is composed of glucose and fructose. Sucrose represented 48% of the world's fuel ethanol production in 2006 (F. O. Licht 2006). Fermentation of sucrose is performed using commercial yeast such as *Saccharomyces cerevisiae*. The chemical reaction is composed of enzymatic hydrolysis followed by fermentation of simple sugars. First, invertase (an enzyme present in the yeast) catalyzes the hydrolysis of sucrose to convert it into glucose and fructose. Then, another enzyme (zymase), also present in the yeast, converts the glucose and the fructose into ethanol and CO_2. One tonne of hexose (glucose or fructose) theoretically yields 511 kg of ethanol. However, practical efficiency of fermentation is about 92% of this yield.

In the bioethanol industry, the sucrose feedstock is mainly sugarcane and sugar beet. It may also be sweet sorghum. A significant share of the fuel ethanol worldwide comes from sugarcane juice, Brazil being the main producer. In 2005, Brazil produced 16 billion liters of fuel ethanol, 2 billion of which were exported. Another potential large producer of sugarcane-to-ethanol is India, as this country is, with Brazil, the world leader of sugarcane production. However, Indian bioethanol production is currently low; around 300 million liters were produced in 2005 mainly from sugarcane molasses. The European Union (EU) is also a potentially large producer of ethanol based on sugar beet juice. Sugar beet currently plays a minor role in the production of ethanol in the EU compared to wheat but can increase significantly its market share in the future due to new incentives given by the EU for energy crops. In 2005, around 950 million liters of bioethanol were produced in the EU.

5.2.3 Starch-to-Ethanol

For converting starch to ethanol, the polymer of alpha-glucose is first broken through a hydrolysis reaction with glucoamylase enzyme. The resulting sugar is known as dextrose, or D-glucose that is an isomer of glucose. The enzymatic hydrolysis is then followed by fermentation, distillation, and dehydration to yield anhydrous ethanol.

In the fuel bioethanol industry, starch is mainly provided by the grains (corn, wheat, or barley). Corn, which is the dominant feedstock in the starch-to-ethanol industry worldwide, is composed of 60 to 70% starch. Conversion to ethanol is achieved in dry or wet mills. In the dry-milling process, the grain is ground to a powder, which is then hydrolyzed and the sugar contained in the hydrolysate is converted to ethanol while the remaining flow containing fiber, oil, and protein is converted into a co-product known as distillers grains (DG), or DGS when it is combined to process syrup. The co-product is either made available wet (WDGS), or more commonly dried (DDGS) and is sold as animal feed. WDGS is preferably reserved to local markets while the co-product is usually dried if the feed has to be shipped far away. Another co-product may be carbon dioxide, which can be sold for different applications (e.g., carbonated beverages or dry ice). Dry mills are dominant in the grain-to-ethanol industry. However, in a number of large facilities, the mills are kinds of biorefineries in which the grains are wet milled for first separating the different components, that is, starch, protein, fiber, and germ, before converting these intermediates into final co-products.

The United States is the leading grain-based ethanol producer in the world and the second producer, all feed-stocks inclusive. Its production of fuel ethanol increased rapidly recently, from 8 billion liters in 2002 to 15 billion liters in 2005. Corn-to-ethanol mills represented around 93% of the 18.5 billion liters of U.S. bioethanol capacity in 2006. The renaissance of fuel ethanol in the United States started from the world oil crises of 1973 and 1979 with the aim to improve the U.S. energy supply security. Later on, ethanol was used as a substitute to lead in gasoline. Finally, the Clean Air Act of the 1990s spurred on the use of bioethanol as an oxygenated compound in the reformulated gasoline, especially in areas where smog was an issue. As oxygenate, ethanol competes with methyl-tertiary-butyl-ether (MTBE). The ban of MTBE in several states launched the irresistible rise of ethanol in the U.S. oxygenates market. Besides these uses, fuel ethanol is also marketed as a gasoline extender and octane booster. Gasohol, a blend of 10% ethanol, 90% gasoline by volume, is used in conventional internal combustion engines. FFVs are currently emerging in the new car market. Other major grain-to-ethanol producers are the European Union, where wheat is the dominant feedstock. Canada and China are producers as well. South Africa has launched an ambitious corn-to-ethanol program.

5.2.4 Lignocellulosics-to-Ethanol

The main drawbacks of the current biomass-to-ethanol processes are as follows: the use of agricultural feedstock and the potential effects on food markets, the potential pressure on land use and natural resources such as water. The perspectives of bio-based fuels as options for partial fossil fuels substitution has encouraged research on the availability of biomass feedstock and development of efficient conversion processes. In the case of fuels for transport, bioconversion of lignocellulosic materials to ethanol has been recognized as one of the promising routes of producing competitive substitutes to gasoline. Lignocellulosics are the most abundant source of unutilized biomass. Their availability does not necessarily impact land use. Agricultural or forestry residues are available though their collection is costly. However, conversion of lignocellulosic materials to ethanol is more complex. Lignocellulose is composed mainly of cellulose, hemicelluloses, and lignin (see Figure 5.2).

Cellulose molecules consist of long chains of beta-glucose monomers gathered into microfibril bundles. The hemicelluloses can be xyloglucans or xylans depending on the type of plant. The backbone of the former consists of chains of beta-glucose monomers to which chains of xylose (a five-carbon sugar) are attached. Xylans are

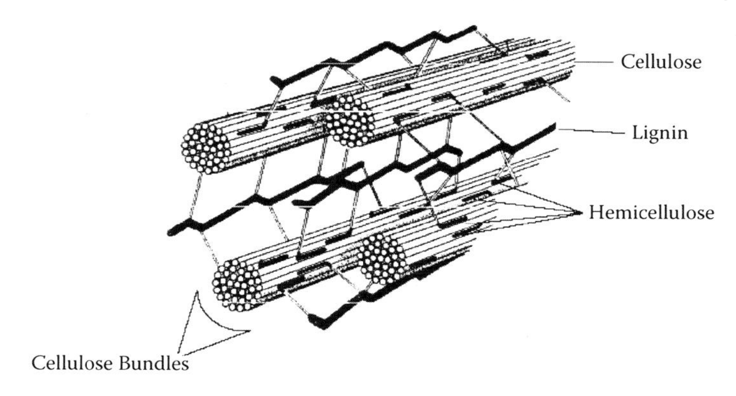

FIGURE 5.2 Structure of plant cell walls. (From Shleser, R. 1994. *Ethanol Production in Hawaii*. Honolulu: State of Hawaii, Energy Division, Department of Business, Economic Development and Tourism. With permission.)

composed mainly of xylose linked to arabinose or other compounds that vary from one biomass source to the other. The hemicellulose molecules are linked to the microfibrils by hydrogen bonds. Lignins are phenolic compounds formed by polymerization of three types of monomers (i.e., *p*-coumaryl, coniferyl, and synapyl alcohols), the proportion of which differs significantly depending whether the plant is from the family of gymnosperms, woody angiosperms, or grasses. Lignin adds to the cell wall a compressive strength and stiffness (Raven, Evert, and Eichhorn 1999).

Lignocellulose does not compete with food. Typical sources of lignocellulosic biomass are bagasse of sugarcane or sweet sorghum, corn stover, grasses, woody biomass, industrial wastes, and dedicated woody crops (e.g., poplar). Table 5.1 gives proportions of each component in a typical lignocellulosic biomass.

Once the lignocellulosic biomass is pretreated and hydrolyzed, the released sugars are fermented. The downstream process is similar to that used for sweet juice and starch. The aim of the pretreatment is the delignification of the feedstock in order to make cellulose more accessible in the hydrolysis step. Existing methods can be classified as physical, physicochemical, chemical, and biological treatment (Sun and Cheng 2002). In Table 5.2, the performance of a few methods is assessed with regard to the yield of fermentable sugars, inhibitors, the recycling of chemicals, the production of wastes, and the investments.

This comparison shows that *carbonic acid* and *alkaline extraction* have the best performance. However, the most common methods are *steam explosion* and *dilute acid prehydrolysis*, followed by *enzymatic hydrolysis*. In the steam explosion method, the lignocellulosic materials are treated with high-pressure saturated steam (0.69–4.83 MPa) at high temperature (160–260°C) for several seconds to a few minutes. Then the pressure is suddenly dropped to atmospheric pressure, causing the material to explode. Most of the hemicellulose is solubilized during the process, the efficiency of which depends on the temperature and residence time. It is reported that lower temperature and longer residence time give a higher efficiency (Wright 1998). Sulfuric acid or carbon dioxide is often added in order to reduce the production of inhibitors and improve the solubilization of hemicellulose (Morjanoff and Gray 1987). Steam explosion has a few limitations: the lignin-carbohydrate matrix is not completely broken down; degradation products are generated that reduce the efficiency of hydrolysis and fermentation steps; a portion of the xylan fraction is destroyed.

The use of dilute acid is the method prefered by the U.S. National Renewable Energy Laboratory (Wooley, Sheehan, and Ibsen 1999; Aden et al. 2002). In this method, the structure of the lignocellulosic materials is attacked with a solution of

TABLE 5.1
Typical Proportion of Cellulose, Hemicellulose, and Lignin in Lignocellulosic Biomass

Component	Percentage of Dry Weight
Cellulose	40–60
Hemicellulose	20–40
Lignin	10–25

TABLE 5.2
Advantages and Weaknesses of Selected Pretreatment Processes

Pretreatment Process	Yield of Fermentable Sugars	Inhibitors	Chemical Recycling	Wastes	Investment
Physical					
- Mechanical	-	++	++	++	+
Physicochemical					
- Steam explosion	+	--	++	+	--
- Ammonia fiber explosion (AFEX)	+/-	++	--	+	–
- Carbonic acid	++	++	++	++	+
Chemical					
- Dilute acid	++	--	--	–	+/-
- Concentrated acid	++	--	--	_	–
- Alkaline extraction	++/+	++	--	_	++
- Wet oxidation	+/-	+	++	+	+
- Organosolv	++	++	–	+	--

++: very good with regard to; +: good with regard to; -: bad with regard to; --: very bad with regard to

Based on de Bont, J. A. M. and J. H. Reith, personal communication.

0.5 to 1.0% sulfuric acid at about 160 to 190°C for approximately 10 minutes. During this reaction, the hemicellulose is largely hydrolyzed, releasing different simple sugars (e.g., xylose, arabinose, mannose, and galactose) but also other compounds of the cellulosic matrix, a few of which can inhibit the enzymatic hydrolysis and fermentation. The stream is then cooled. Part of the acetic acid, much of the sulfuric acid and other inhibitors produced during the degradation of the materials are removed. Finally neutralization is performed and pH is set to 10 before hydrolysis and fermentation.

Enzymatic hydrolysis of cellulose is achieved using cellulases, which are usually a mixture of groups of enzymes such as endoglucanases, exoglucanases, and beta-glucosidases acting in synergy to attack the crystalline structure of the cellulose, removing cellobiose from the free chain ends and hydrolyzing cellobiose to produce glucose. Cellulases are produced by fungi such as *Trichoderma reesei*, the most common fungus used for this purpose. Other fungi are species of *Aspergillus*, *Schisophyllum,* and *Penicillium.* Efficiency of cellulose enzymatic hydrolysis has been reported to be affected by the substrate to enzyme ratio, cellulase dosage, and the presence of inhibitors. Cellulase loading may vary from 7 to 33 FPU/g (substrate) depending on the substrate structure and concentration (Sun and Cheng 2002). High concentration of cellobiose and glucose inhibits the activity of cellulase enzymes and reduces the efficiency of the saccharification. One of the methods used to decrease this inhibition is to ferment the reduced sugars along their release. This is achieved by *simultaneous saccharification and fermentation* (SSF) in which fermentation using yeasts such as *Saccharomyces cerevisiae* and enzymatic hydrolysis are achieved simultaneously in the same reactor. The fermentation of the xylose released from the prehydrolysis process can be carried out in a separate vessel or in the SSF reactor using a genetically modified strain from the bacterium *Zymomonas mobilis* that can convert both glucose and xylose. The latter method is named *simultaneous saccharification and co-fermentation* (SSCF).

Compared to the sequential saccharification and fermentation process, the SSCF exhibits several advantages, including lower requirement of enzyme, shorter process time, and cost reduction due to economy in fermentation reactors (only one reactor compared to three sets). However, a few disadvantages need to be taken into consideration, including the difference between the optimal temperatures for saccharification (50–60°C) and fermentation (30°C), the inhibition of enzymes and yeast to ethanol, and the insufficient robustness of the yeast in co-fermenting C5 and C6 sugars.

The main co-product of lignocellulose conversion to ethanol is energy. The effluent from the distillation column that contains most of the lignin and other nonfermentable products is sent to a combined heat and power (CHP) plant to produce process steam and electricity required by the ethanol plant. Depending on the proportion of lignin in the feedstock, excess electricity may be available for export sale.

Contrary to the conversion of sweet juice and that of starch to ethanol, which are mature technologies, the modern lignocellulose-to-ethanol process is still in the pilot and demonstration stages. A few facilities exist: the U.S. National Renewable Energy Laboratory has built a pilot plant based on the SSCF method capable of processing one ton of dry material per day (DOE 2000); Iogen Corporation (Canada) in 2003 built a demonstration plant with an annual production of 320,000 liters of ethanol, using wheat straw as feedstock and a sequential steam explosion prehydrolysis (cellulose production), enzymatic hydrolysis of cellulose and co-fermentation of xylose and glucose; in 2004, a Swedish company ETEK developed a pilot plant capable of producing 150,000 liters of ethanol per year using soft wood as feedstock (Lindstedt 2003).

5.3 OUTLOOK FOR BIOETHANOL DEVELOPMENT

5.3.1 Drivers for Fuel Ethanol Development

The following key factors can influence the future development of fuel ethanol worldwide: security of the energy supply, economic drivers, environmental drivers, and technological development.

5.3.1.1 Security of the Energy Supply

The prospective of fossil sources depletion in the long term, particularly the pressure on world oil reserves, is the subject of growing concerns in net oil import countries. Geopolitical instability in several oil producing countries and the rising oil demand in emerging Asian economies such as China and India add to the threat of oil supply insecurity in the medium to long term. Development of biofuels is considered a viable option for energy supply diversification. Furthermore, potential biofuel-producing countries are more diverse geographically than oil-producing countries. However, due to several factors, such as land use, risk of competition with food, and ecological risks, biofuels can only substitute for a small part of world road-transport fuel demand, for example, 4 to 7% in 2030 compared to 1% in 2005 (IEA 2006).

5.3.1.2 Economic Drivers

The cost of bioethanol to end users is one of the most important drivers of fuel ethanol development. That cost is composed of the price of bioethanol, investment and operating costs of vehicles using bioethanol. In several countries, the production cost of bioethanol is higher than that of gasoline at the current price of oil, requiring governmental incentives such as partial or total tax exemption to make fuel ethanol competitive.

The production cost of bioethanol fuel depends on many factors, including the conversion pathway, plant size and location, feedstock and co-products markets, which may vary from one country to the other and within the same country projects may have different production costs (see Table 5.3). The ethanol derived from sugarcane juice is commonly cheaper than the others; production in North America (Brazil and the United States) is less expensive than that in Europe due to a learning curve, low cost of feedstock, and other differences in expenditures. The possibility to valorize co-products contributes to reducing the production cost of bioethanol. Finally, lignocellulose-to-ethanol is expected to be, in the long term, more competi-

TABLE 5.3
Typical Bioethanol Fuel Production Costs

Reference	Feedstock	Country or Region	Range of Sizes (Million Liters per Year)	Production Cost (US$/Liter)
	Sweet Juice			
Walter[a]	Sugarcane	Brazil	–	0.17–0.19
Gnansounou et al. 2005	Molasses	China	125	0.30
Gnansounou et al. 2005	Sweet sorghum	China	125	0.27
F. O. Licht 2006	Sugar beet	Germany	200	0.48
F. O. Licht 2006	Sugar beet	Germany	50	0.55
	Starch			
F. O. Licht 2006	Corn	U.S.	–	0.25
Gnansounou et al. 2005	Corn	China	125	0.31
Gnansounou et al. 2005	Cassava	China	125	0.23
F. O. Licht 2006	Wheat	Germany	50	0.51
F. O. Licht 2006	Wheat	Germany	200	0.44
	Lignocellulose			
Wooley et al. 1999	Yellow poplar	U.S.	197	0.38
Aden et al. 2002	Corn stover	U.S.	262	0.28
Gnansounou et al. 2005	Bagasse of sweet sorghum	China	125	0.30

[a] Walter, A. Experience with large-scale production of sugar cane and plantation wood for the export market in Brazil; impacts and lessons learned (Based on Walter, A., personal communication) March, 2005.

tive than ethanol from corn although its reported production cost is currently based on engineering estimates as no commercial plant exists.

Assuming that the production cost of gasoline in 2015 will be between 0.45 and 0.55 US$ (2000) per liter, on a volume basis, bioethanol at its current production cost will be competitive. The situation is different if the comparison is made on an energy basis as the LHV of ethanol is 30% lower than that of gasoline. Therefore, in several countries, subsidies and tax reductions by the government will still be required for sustaining the penetration of bioethanol. However, this conclusion will depend a lot on the market price of gasoline. Another way to promote bioethanol introduction in the market is to cross-subsidize ethanol by fossil fuel. This approach increases the price of fuel for consumers and is neutral from a taxation point of view. When the difference between the production cost of ethanol and fossil fuels is low and the blend level is about 5%, the increase in price is not significant as the oil price is very volatile. In the case of high ethanol production cost as in Europe, direct subsidies are required in order to make ethanol introduction affordable to most of consumers. Finally, the costs borne by the end-users can be lower if international bioethanol trade is encouraged. At present, various barriers limit that trade to a low percentage of the demand. One such barrier is the lack of international quality and sustainability standards.

5.3.1.3 Environmental Drivers

The main environmental drivers of bioethanol supply chains are as follows: net energy balance, greenhouse gas (GHG) emissions balance, and local environmental effects. The net energy balance of biomass-to-ethanol measures, from a life cycle assessment (LCA) viewpoint, is the ratio of the energy content of bioethanol to the net nonrenewable primary energy (allocated to ethanol) consumed in the whole production process, from biomass production to its conversion into ethanol. On average, the ratio (output/input) between the produced ethanol and the input of nonrenewable energy varies from 1.0 to 5.0 or more. These values depend on the following factors: allocation between ethanol and co-products; the use of renewable energy for fuelling the process, the agricultural practices for producing the feedstock, the energy integration within the production plant, the size of the plant, and transport distances between the plant and the area of biomass collection. Intensive agriculture needs more fertilizers and leads to a larger grey energy input. Recycling the residues to produce process steam and electricity, as is often the case for sugarcane, improves the net energy balance.

5.3.1.4 Greenhouse Gas Balance

The net GHG balance is a key driver of bioethanol development, as in several countries reduction of GHG emissions is one of the main objectives of the promotion of bioethanol. Particularly in Kyoto Protocol Annex I countries, development of biofuels consumption is expected to contribute significantly to the achievement of GHG emissions reduction. However, as is the case for net energy balance, the performance of bioethanol with regard to GHG emissions varies from one supply chain to the other. It also depends closely on the allocation method and the reference system adopted for the LCA. Based on several assessments undertaken by the Laboratory of Energy Systems (LASEN), it is found that with an incorporation rate of 5% anhydrous ethanol within gasoline and an equal performance with respect to conventional gasoline, the net savings of GHG emissions vary between 1.5 kg (low performance agricultural feedstock) and 2.5 kg (waste lignocellulosic biomass) of CO_2 equivalent per liter of ethanol incorporated to gasoline. In these evaluations, the life cycle inventory was described in the context of Switzerland with economic allocation, and the reference vehicle was a recent standard 1.6 l light passenger vehicle (Gnansounou and Dauriat 2004).

5.3.1.5 Other Environmental Effects

As ethanol contains more oxygen than gasoline, its use favors more complete combustion and reduces the emission of particulate matter (PM) and hydrocarbons (HC) which result from incomplete combustion of gasoline. Tailpipe emissions of carbon monoxide (CO) and sulphur dioxide (SO_2) are also improved. However, low-level blends of ethanol with gasoline can increase the emissions of volatile compounds (VOCs) and oxide of nitrogen (NO_x). These emissions favor ozone formation. Emissions of aldehydes (mostly acetaldehydes) and peroxyacetyl nitrate (PAN) also increase, to an extent that depends on weather conditions. The use of catalytic converters reduces the emissions of aldehydes. VOCs emissions can be prevented by reducing in refinery the vapor pressure of gasoline that is blended with ethanol. Experiments about different percentages of ethanol-diesel blends show significant advantages concerning PM, NO_x, and CO. However, no evidence is given for improvement of HC emissions (Ahmed 2001). Furthermore, ethanol is more corrosive than gasoline and diesel and at high concentration can damage fuel system components. For low-level blends, these concerns are limited and E5 or E10 can be used in existing vehicles without violating most manufacturers' warranties. For high concentrations of ethanol, compatible materials are used in adapted or dedicated designed vehicles. Finally, biomass-to-ethanol impacts land use unless the feedstock is an agricultural or forestry waste that is not required for soil fertilization.

5.3.1.6 Technological Development

The goal of technological advances is to achieve reduction of GHG all along the supply chain, from good practices in agriculture to the valorization of the whole biomass through the biorefinery concept; reduction of production costs through process and value chain optimization; development of low-cost lignocellulose-to-ethanol. The overall goal of this progress will be to decrease significantly the cost of GHG reduction.

5.3.2 Future Demand and the Production of Bioethanol

World demand for fuel ethanol in 2015 is estimated to range between 65 and 90 billion liters. Brazil and the United States will remain the leading consumers followed by the European Union. Several other countries will emerge, especially in Asia. In Brazil, the evolution of the FFVs market share is a key driver of future fuel ethanol demand; the market share of FFVs in new gasoline used cars was more than 80% in 2006. The rush on that technology has created a new situation from which bioethanol becomes a direct competitor of gasoline. Consumers will maintain their preference for high-level ethanol blend to gasoline as long as the price of bioethanol in Brazil reflects the production cost and the price of gasoline is higher. However, another scenario is also possible. The role of Brazil as an exporter of bioethanol may be enhanced in the future as international demand for bioethanol may increase owing to the growth of the carbon market. The price of bioethanol in Brazil will be influenced both by the international price of gasoline and by the prices of ethanol and sugar in international markets. These interrelations may enhance the volatility of the local price of ethanol and contribute to escalating the local price. This scenario would result in a low growth rate of the internal demand of fuel ethanol. The production of fuel ethanol in Brazil in 2015 is estimated to be in the range of 28 to 35 billion liters, with export volume of 4 to 8 billion liters.

In the United States, bioethanol demand will continue to grow, boosting by the ban of MTBE and the Energy Policy Act (EPACT) of 2005 that sets up a national Renewable Fuels Standards (RFS). This new legislation establishes a baseline for use of renewable fuels, starting from 4 billion American gallons (15.14 billion liters) in 2006 to 7.5 billion American gallons (28.39 billion liters) in 2012. The Renewable Fuels Association (2006) foresees that a large share of the renewable fuel will be bioethanol. The estimated demand for fuel ethanol in the United States ranges from 25 to 30 billion liters in 2015 with a maximum net import of 2 billion liters.

In 2003, the European Union adopted an alternative motor fuels directive which sets up indicative biofuel market share targets of 2% and 5.75% (in energy content) for 2005 and 2010, respectively. In a progress report issued on the January 10, 2007, the European Commission (Commission of the European Communities 2007) estimated the market share in 2005 to be 1% and then envisaged from 2020 a mandatory minimum target of 10% biofuel in 2020. Assuming a 128 billion liter gasoline demand in 2015 and applying a mandatory target of 8% results in a fuel ethanol demand of 13 billion liters. The European fuel ethanol demand in 2015 is estimated to range between 10 and 15 billion liters. Asia is another region where the fuel bioethanol market is increasing very rapidly. China's program for bioethanol fuel is promising in the short term; however, in the long term, it is forecast that China will become a net importer of corn. The success of Indian production of sugarcane-to- ethanol will mainly depend in technology progress and improvements in agricultural practices. The low availability of water is the main bottleneck. Thailand is developing a bioethanol program on a wide scale with the aim to diversify feedstocks between sugarcane and cassava. Japan is also on track, with a program of production in Brazil for import. It is expected that the South African bioethanol development program will expand at a regional level: the Southern Africa Development Community (SADC) is prepared to adopt such a vision. Fuel ethanol demand in 2015 for the rest of the world is estimated between 10 to 15 billion liters.

5.4 CONCLUSIONS

From 2005 to 2015, world demand for fuel ethanol will more than double. Assuring this growth without a significant environmental footprint and avoiding social and economic hurdles are challenging. The idea is progressing in several countries that biofuels should be developed in a regulated framework. Efforts to set up standards for sustainability of biofuels are in progress, especially in the United Kingdom (Tipper et al. 2006), Germany (Öko-Institut 2006), and the Netherlands (Cramer Commission 2007). The following themes are often included in draft standards: greenhouse gas balance; local environment (air, water, soil) and biodiversity; social well-being, that is, competition with food, local energy supply, medicine, and building materials; economic prosperity.

Having developed long experience with low-level ethanol blends (E10 to E25), as well as with nearly pure ethanol (E85), Brazil and the United States benefit from the learning curve and particularly favorable conditions with regard to agricultural feedstocks, that is, sugarcane for Brazil and corn for the United States. Especially, Brazil exhibits the lowest production costs of fuel ethanol worldwide and is in position to capture a large share of the international market in the future. However, the market price of fuel ethanol will fluctuate as a result of the balance between demand and supply of bioethanol, oil, and sugar. It is likely that the trend will be for an increase due to rapid growth in the world ethanol demand in the future. In industrialized countries, the economy of fuel ethanol development will depend a lot on possibilities to alleviate market barriers that limit international trade. It is expected that future negotiations in the framework of the World Trade Organization will help find a good balance between the desire of bioethanol producers from developing countries to export and the desire of the industrialized countries to protect their local ethanol industry. In this respect, it is important to prevent using standards of biofuel sustainability as a new instrument of protectionism. Lignocellulosics-to-ethanol is expected to equalize comparative advantages among most of the countries while enhancing the GHG balance. Research to decrease production cost is a key driver of the smooth development of the fuel ethanol market in the future. However, competition with other energy uses of lignocellulosics has to be considered, such as biocombustibles, biomass-to-liquids technologies (BTL) such as the Fischer-Tropsch process and biomass-to-gas (BTG). There is a need for a neutral framework for defining internationally acceptable standards for bioenergies that will enable the promotion of the most viable pathways of biomass-to-energy. The initiative (Frei, Gnansounou, and Püttgen 2006) launched in November 2006 by the Energy Center of the Swiss Federal Institute of Technology, aimed at defining such standards for biofuels, is in line with that requirement.

REFERENCES

Aden, A., M. Ruth, K. Ibsen, and J. Jechura. 2002. Lignocellulosic Biomass to Ethanol Process Design and Economics Utilizing Co-Current Dilute Acid Prehydrolysis and Enzymatic Hydrolysis for Corn Stover. Report T-P510-32438. Golden, CO: National Renewable Energy Laboratory.

Ahmed, I. 2001. Oxygenated diesel: emissions and performance characteristics of ethanol-diesel blends in CI engines. SAE report No 2001-01-2475.

Commission of the European Communities. 2007. Biofuels Progress Report. COM(2006) 845 final, Brussels.

Cramer Commission. 2007. Testing framwork for sustainable biomass. Final report from the project group 'Sustainable production of biomass.' IPM, March 2007, The Netherlands. www.nl/biobrandstoffen/download/070427/cramer/finalreportEN.pdf.

DOE (Department of Energy). 2000. The DOE Bioethanol Pilot Plant. DOE leaflet GO-10200-1114. Washington, DC: DOE.

F. O. Licht. 2006. World Ethanol Markets. The Outlook to 2015. An F.O. Licht Special Report No. 138.

Frei, C., E. Gnansounou, and H. B. Püttgen. 2006. Sustainable Biofuels Standards Initiative. Lausanne: EPFL-Energy Center.

Gnansounou, E. and A. Dauriat. 2004. Comparative Study of Fuels by Analysis of Their Life Cycle. Report prepared for Alcosuisse, Laboratory of Energy Systems (LASEN), EPFL, Switzerland (in French).

Gnansounou, E. and A. Dauriat. 2005. Ethanol fuel from biomass: A review. Journal of Scientific & Industrial Research 64: 809–821.

Gnansounou, E., A. Dauriat, and M. Amiguet. 2005. Economic and Social Profitability. Deliverable of WP6 of the EU project ASIATIC, Laboratory of Energy Systems (LASEN), Lausanne, Switzerland.

IEA (International Energy Agency). 2006. World Energy Outlook 2006. Paris: Organisation for Economic Co-operation and Development.

Lindstedt, J. 2003. Alcohol production from lignocellulosic feedstock. In FVS Fachtagung, 228–237. Network Regenerative Kraftstoffe (RefuelNet) Stuttgart, Germany.

MBEP (Michigan Biomass Energy Program). 2002. Fact Sheet. http://www.michiganbioenergy.org/ethanol/ediesel facts.htm

Morjanoff, P. J. and P. P. Gray. 1987. Optimization of steam explosion as a method for increasing susceptibility of sugarcane bagasse to enzymatic saccharification. Biotechnology and Bioengineering 29: 733–741.

Öko-Institut. 2006. Sustainability Standards for Bioenergy. WWF, Germany.

Raven, P. H., R. F. Evert, and S. E. Eichhorn. 1999. *Biology of Plants*, 6th ed. New York: Freeman and Company/Worth Publishers.

Renewable Fuels Association. 2006. From Niche to Nation, Ethanol Industry Outlook 2006. http://www.ethanolRFA.org

SAE. 2001. National Ethanol Vehicle Challenge Design Competition. www.saeindia.org

Shleser, R. 1994. Ethanol Production in Hawaii. State of Hawaii, Energy Division, Department of Business, Economic Development and Tourism, Honolulu.

Sun, Y. and J. Cheng. 2002. Hydrolysis of lignocellulosic materials for ethanol production: A review. *Bioresource Technology* 83: 1–11.

Tipper, R., J. Gărstang, W. Vorley, and J. Woods. 2006. Draft Environmental Standards for Biofuels. London LOWCVP. 2006. http://www.lowcvp.org.uk/resources/reportsstudies/

Wooley, R. R., J. Sheehan, and K. Ibsen. 1999. Lignocellulosic Biomass to Ethanol Process Design and Economics Utilizing Co-Current Dilute Acid Prehydrolysis and Enzymatic Hydrolysis: Current and Futuristic Scenarios. Report TP580-26157. Golden, CO: National Renewable Energy Laboratory.

Wright, J. D. 1998. Ethanol from biomass by enzymatic hydrolysis. Chemical Engineering Progress 84 (8): 62–74.

Wyman, C. E. 2004. Ethanol fuel. In *Encyclopaedia of Energy*, ed. C. J. Cleveland, 541-555. New York: Elsevier.

6 Bioethanol from Biomass

Production of Ethanol from Molasses

Velusamy Senthilkumar and
Paramasamy Gunasekaran

CONTENTS

ABSTRACT

In recent years, much attention has been paid to the conversion of biomass into fuel ethanol, apparently the cleanest liquid fuel alternative to the fossil fuels. Agronomic residues such as corn stover (corn cobs and stalks), sugarcane waste, wheat, or rice straw, forestry and paper mill wastes, and dedicated energy crops are the major biomass resources considered for the production of fuel ethanol. Molasses, one of the renewable biomass resources, a main by-product of the sugar industry, represents a major fermentation feedstock for commercial ethanol production. Significant advances have been made in the last two decades in developing the technology for ethanol fermentation from molasses. This chapter gives an overview of the status of ethanol fermentation from molasses and processes applied for the improvement of ethanol production by ethanologenic microorganisms such as the yeasts *Saccharomyces* and *Kluyveromyces* and the bacterium *Zymomonas mobilis*.

6.1 INTRODUCTION

Much biofuel research is presently directed towards the improvement of the bioconversion strategies, exploring the technical and economic potential and possible environmental impacts of such processes. In particular, for several years the production of ethanol from molasses has been the subject of research. Two aspects of investigation have been mostly carried out, the supplementation of molasses and the use of thermotolerant strains for improving both the rate of alcohol production and the final ethanol concentration (Damiano and Wang 1985).

Cane molasses is the final run-off syrup from sugar manufacture and is an important by-product. It is a dark brown, viscous liquid obtained as a residue. Total residual sugars in molasses can amount to 50–60% (w/v), of which about 60% is sucrose, which makes this a suitable substrate for industrial-scale ethanol production. The commercial production of ethanol is carried out by the fermentation of molasses with yeast. The majority of distilleries in India practice a batch process with open fermentation system for ethanol production from diluted cane molasses. In spite of the fact that India is the world's largest producer of sugar and sugarcane, ethanol yield has not exceeded more than 1.5 billion liters per year—a capacity utilization of about 60%. This could, among many other factors, be due to the fact that most of the distilleries situated in the tropical regions of India carry out fermentation at temperatures not controlled and even range above 40°C during the summer season. Such high temperatures adversely affect the activity of the fermenting organisms and increase the toxic effect of ethanol (Jones, Pamment, and Greenfield 1981), leading to decreased fermentation efficiency and premature termination of the fermentation.

6.2 TYPES OF MOLASSES

The Association of American Feed Control Officials (AAFCO, 1982) has described the types of molasses and their composition (Table 6.1). *Cane molasses* is a by-product of the manufacture or refining of sucrose from sugarcane. It contains total sugars not less than 46%. *Beet molasses* contains total sugars not less than 48% and its density is about 79.5° Brix. *Citrus molasses* is the partially dehydrated juice obtained from the manufacture of dried citrus pulp, with total sugars not less than 45% and its density is about 71.0° Brix. *Hemicellulose extract* is a by-product of the manufacture of pressed wood. It is the concentrated soluble material obtained from the treatment of wood at elevated temperature and pressure without the use of acids, alkalis, or salts. It contains pentose and hexose sugars, and has total carbohydrate content not less than 55%. *Starch molasses* is a by-product of dextrose manufacture from starch derived from corn or grain sorghum where the starch is hydrolyzed by enzymes or acid. It contains about 43% reducing sugars and 73% total solids. The estimates for the production of various types of molasses show that of the total U.S. supply, 60%

TABLE 6.1
Composition of Different Types of Molasses

	Type of molasses				
Item	**Cane**	**Beet**	**Citrus**	**Extract**	**Starch**
Brix	79.5	79.5	71.0	65.0	78.0
Total Solids (%)	75.0	77.0	65.0	65.0	73.0
Specific Gravity	1.41	0.41	1.36	1.32	1.40
Total Sugars (%)	46.0	48.0	45.0	55.0	50.0
Crude Protein (%)	3.0	6.0	4.0	0.5	0.4
Nitrogen Free Extract (%)	63.0	62.0	55.0	55.0	65.0
Total Fat (%)	0.0	0.0	0.2	0.5	0.0
Total Fiber (%)	0.0	0.0	0.0	0.5	0.0
Ash (%)	8.1	8.7	6.0	5.0	6.0
Calcium, (%)	0.8	0.2	1.3	0.8	0.1
Phosphorus, (%)	0.08	0.03	0.15	0.05	0.2
Potassium, (%)	2.4	4.7	0.1	0.04	0.02
Sodium, (%)	0.2	1.0	0.3	---	2.5
Chlorine, (%)	1.4	0.9	0.07	---	3.0
Sulfur, (%)	0.5	0.5	0.17	---	0.05
Swine (ME)	2343	2320	2264	2231	---

is cane molasses, 32% is beet molasses, 7% is starch molasses, and 1% citrus molasses. The production of citrus molasses, starch molasses, and hemicellulose extract is quite limited.

6.3 GENERAL PROCESS FOR THE PRODUCTION OF ETHANOL FROM MOLASSES

Ethanol manufacture in distilleries involves three main steps, namely feed preparation, fermentation, and distillation (Figure 6.1). Molasses is diluted with water to obtain a feed containing suitable concentration of the sugars. The pH is adjusted, if required, by the addition of sulfuric acid. The diluted molasses solution is transferred to the fermentation tank, where it is inoculated with typically 10% seed culture of the yeast. The mixture is then allowed to ferment without aeration under controlled conditions of temperature and pH. Because the reaction is exothermic, the fermenter is cooled to maintain a reaction temperature of 25°C. Fermentation typically takes 48 to 80 h for completion and the resulting broth contains 6 to 8% ethanol. Once fermentation is complete, yeast is separated by settling and the cell-free broth is taken for distillation. Indian distilleries typically employ six to nine fermenters for ensuring continuous feed to the alcohol distillation system. Fermentation is carried out under batch or continuous mode. Because of higher efficiency (89 to 90% compared to 80 to 84% in the batch mode), ease of operation, and substantial saving in water consumption, distilleries employ continuous fermentation. The cell-free fermented

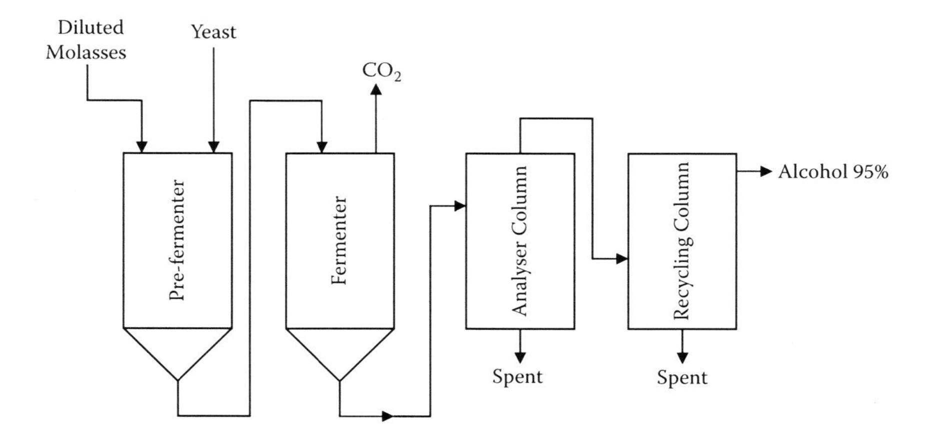

FIGURE 6.1 Scheme of the ethanol manufacturing process from molasses.

broth is preheated to about 90°C and is sent to the degasifying section of the analyzer column. The bubble cap fractionating column removes any trapped gases (CO_2, etc.) from the liquor, which is then steam heated and fractionated to give 40% alcohol. The bottom discharge from the analyzer column is the effluent (spent wash). The alcohol vapors from the analyzer column are further taken to the rectifying column where by reflux action, 95 to 99% rectified alcohol is collected.

6.4 FERMENTATION OF MOLASSES BY *SACCHAROMYCES* SPP.

The production of ethanol from cane molasses mostly utilizes the yeast strains belonging to *Saccharomyces* spp. A prerequisite for an efficient process is the availability of yeast strains with high specific ethanol productivity and adequate tolerance towards the substrate and product concentrations at the ambient temperatures prevailing in the regions. Osmotolerant yeast is particularly important when high-salt-containing cane and other blackstrap molasses are used as the raw material. Flocculation is also another desirable feature, which enhances the ease of cell recovery in the batch fermentation and permits the retention of yeast cells in tower reactors in continuous fermentation (Royston, 1966). Several yeast strains have been tested for their performance for ethanol fermentation and few of them have been used for industrial-scale ethanol production (Table 6.2). There are relatively few data on the comparative performance of different yeasts on high-salt molasses. Ragav et al. (1989) studied the performance of an adapted culture of the flocculent *Saccharomyces uvarum* strain 17 in batch fermentation of sugarcane molasses and compared it with a standard brewing strain, *S. uvarum* ATCC 26602 and of a substrate- and ethanol-tolerant strain, *S. cerevisiae* Y-10. *S. uvarum* strain 17 has been used by Comberbach and Bu'Lock (1984) for rapid and efficient continuous fermentation of glucose to ethanol.

S. cerevisiae strains isolated from the molasses or jaggery were examined for their ethanol production ability in molasses with high sugar concentrations and other

TABLE 6.2
Yeast Strains Used for Commercial Production of Ethanol and Their Relative Efficiency

Yeast strain	Fermentation efficiency (%)	Ethanol/ton of molasses (gallons)
ATCC 4132	93	73
CBS 237	90	70
Y 7494	86	67
UCD 505	83	65
UCD 595	81	63
ATCC 26603	81	63
DADY	77	60
BAKER	77	60
ATCC 26602	62	48
NCYC 90	57	44
Y 2034	55	43
CBS 1235	35	27

desirable fermentation characteristics. Four strains, isolate 3B, *S. cerevisiae* HAU-11, *S. cerevisiae* MTCC 174, and *S. cerevisiae* MTCC 172, gave high efficiency of ethanol production, that is, 71.0, 67.0, 66.7, and 61.5%, respectively, in the concentrated molasses (40% sugars). Viability of the yeast strains was quite high in the diluted molasses but decreased drastically with increase in the concentration of the sugars in the medium and also with prolonged incubation. The four superior strains (3B, *S. cerevisiae* MTCC 172, *S. cerevisiae* MTCC 174, and *S. cerevisiae* HAU-11) showed cell viability between 57 and 71% in molasses with sugar concentration of 35 to 40% (Bajaj et al. 2003). Thermotolerant *S. cerevisiae* MT15 was isolated after ultraviolet treatment, extensive screening, and optimization of fermentation in molasses medium (Rajoka et al. 2005). The mutation altered the culture's behavior and its potential to form metabolites. This mutant, when grown on molasses (containing 15% sugars, w/v), produced the highest volumetric alcohol yield of 72 g/l at 40°C, which was higher than those reported on well-documented *Kluyveromyces marxianus* IMB-3 on molasses or glucose. The organism was capable of rapid fermentation at a temperature of up to 40°C with significantly ($P \leq 0.05$) higher substrate consumption parameters (Table 6.3), better than its wild strain and five other strains of *K. marxianus* (Banat and Marchant 1995; Banat et al. 1998). The mutant showed 1.45-fold improvement over its wild parent with respect to ethanol productivity (7.2 g/l/h), product yield (0.44 g ethanol/g substrate utilized), and specific ethanol yield (19.0 g ethanol/g cells). The improved ethanol productivity was directly correlated with the titers of intracellular and extracellular invertase activities. The mutant supported higher volumetric and product yield of ethanol, significantly ($P \leq 0.05$) higher than the parental and other strains. Thermodynamic studies revealed that the cell system exerted protection against thermal inactivation during formation of ethanol (Rajoka et al. 2005).

TABLE 6.3
Different Strategies Employed for the Maximum Production of Ethanol from Molasses by *K. marxianus* Strains

Substrate (g/l of sugar)	Ethanol productivity (g/l)	Specific ethanol yield (g/g)	Fermentation efficiency (%)	Reactor type	Strategy for the improvement	Reference
Diluted molasses (23%)	74.0	-	94.9	Shake flask	Nelder and Mead optimization strategy	Gough et al., 1998
Diluted molasses (140)	57.0	-	74	Shake flask	Calcium alginate immobilization	Gough et al., 1998
Molasses (100 glucose+110)	55.9	0.47	78.64	Continuous	Immobilization on mineral Kissiris	Nigam et al., 1996
Diluted molasses (140)	58	-	71	Shake flask	Amberlite IRN 150 pretreatment of molasses	Gough et al., 1998
Diluted molasses (140)	60	-	84	Continuous	Alginate-immobilization	Gough et al., 1998

6.4.1 Ethanol Fermentation by the Cell Recycle System

The continuous cell recycle fermentation of *S. cerevisiae* showed that the productivity was affected by the recycling ratio and dilution rate (Sittikat and Jiraarun 2005). It was found that ethanol productivity increased with increasing dilution rate from 0.2/h to 0.3/h but decreased when the dilution rate increased more than this value. This was probably due to cell wash out from the system at higher dilution rates. The maximum productivity of the pilot recycling circulating culture, 20.61 ml/l/h, was obtained at the dilution rate of 0.3/h and the recycling ratio of 9. As dilution rate increased, the concentration of cells in the fermenter decreased. The increase of dilution rate above 0.3/h caused an increase in the up-flow rate in the sedimentation vessel, resulting in a low concentration of cells. On the other hand, increasing the recycling ratio caused an increase in the concentration of cells in the fermenter. Some unused medium was fed back to the main fermenter for fermenting again. At a circulating ratio higher than 9.0, the concentration was almost uniform in that cell concentrations in the fermenter and separation vessel were the same. The feed rate and circulating ratio affect the flow condition in the fermenter and the separation vessel. High growth rate and good separation at high ethanol concentrations are the criteria required for the selection of strains for ethanol fermentation (Sittikat and Jiraarun 2005).

6.5 FERMENTATION OF MOLASSES USING THE THERMOTOLERANT YEAST *K. MARXIANUS*

During molasses fermentation, the generation of heat is one of the main disadvantages of fermentation. Several strains of the thermotolerant yeast *K. marxianus* have been shown to address this problem (Table 6.4). It has been demonstrated that the thermotolerant, ethanol-producing yeast strain *K. marxianus* is capable of converting a number of simple and complex carbohydrate substrates to ethanol at relatively elevated temperatures, up to 45°C (Barron et al. 1995). It has also been demonstrated that the yeast is capable of producing ethanol from diluted, unsupplemented molasses (Gough et al. 1998). An immobilized yeast cell preparation can also be used as the biocatalyst in a variety of fermentations (Gough et al. 1998). Ethanol production by *K. marxianus* IMB3 was maximum at 23% (v/v) molasses. At this concentration, 7.4% (v/v) ethanol was produced, representing 84% of the apparent theoretical maximum yield. The rate of ethanol production was 1 g/l/h. Above 23% (v/v) molasses concentration, the maximum ethanol concentration and the biomass concentration decreased. At 44% (v/v) of the molasses, no ethanol was produced. On addition of increasing amounts of sucrose from 140 to 180 g/l, to correspond with the total sugar concentration in the molasses dilution experiments, a decrease in the concentration of ethanol was noted and was comparable to that achieved in the molasses dilution

TABLE 6.4

Comparative Growth Kinetics of *S. cerevisiae* and Its Thermotolerant Mutant MT15 Grown on Molasses (15% sugars), Different Temperatures in 15 l Fermentation Medium in a Fully Controlled Bioreactor

Strain	μ/h	Qs (g/l/h)	Qx (g/l/h)	qS (g/g/h)
30°C				
Parent	0.20	2.6	0.65	78.
MT15	0.24	3.6	0.70	7.9
35°C				
Parent	0.23	2.5	0.70	8.6
MT15	0.26	3.7	0.75	8.8
38°C				
Parent	0.20	2.0	0.65	7.8
MT15	0.23	3.4	0.70	8.0
40°C				
Parent	0.18	1.7	0.55	6.8
MT15	0.20	2.9	0.65	7.8

Each value is a mean of three independent fermenter runs. Values followed by different letters differ significantly at $P \leq 0.05$. μ, specific growth rate; Qx, grams cells synthesized per liter per hour; Qs, grams substrate consumed per liter per hour; qS is specific rate of substrate uptake that was a result of division of μ.

From *Rajoka et al. 2005. Lett. Appl. Microbiol. 40:* 316–321. With permission.

experiments. A study on the effects of the four supplements, magnesium, nitrogen, potassium, and linseed oil, on the fermentation rate and final ethanol concentration showed a significant increase in both the ethanol production rate (4.8 g/l/h) and ethanol concentration (8.5% v/v) (Gough et al. 1998). As the biomass concentration was not determined, it was not possible to differentiate the effects on the biomass concentration and specific ethanol production. Magnesium sulfate and linseed oil have been reported to exert a positive effect on ethanol production rate (Karunakaran and Gunasekaran 1986).

6.5.1 Strategies for the Improvement of the Production of Ethanol by *K. marxianus*

A thermotolerant strain of *K. marxianus* IMB3 was immobilized in calcium alginate matrices. The ability of the biocatalyst to produce ethanol from cane molasses originating in Guatemala, Honduras, Senegal, Guyana, and the Philippines was examined (Gough et al. 1998). In each case, the molasses was diluted to yield a sugar concentration of 140 g/l and fermentations were carried out in batch-fed mode at 45°C. During the first 24 h, the maximum ethanol concentrations obtained ranged from 43 to 57 g/l, with the optimum production on the molasses from Honduras. Ethanol production during the subsequent refeeding of the fermentations at 24 h intervals over a 120-h period decreased steadily to concentrations ranging from 20 to 36 g/l; the ethanol productivity remained highest in fermentations containing the molasses from Guyana. When each set of fermentation was refed at 120 h and allowed to continue for 48 h, ethanol production again increased to a maximum, with concentrations ranging from 25 to 52 g/l. However, increasing the time between the refeeding at this stage in fermentation had a detrimental effect on the functionality of the biocatalyst (Gough et al. 1998).

Tamarind wastes, such as tamarind husk, pulp, seeds, fruit, and the effluent generated during the tartaric acid extraction, were used as supplements to evaluate their effects on alcohol production from cane molasses (Patil et al. 1998). Small amounts of these additives enhanced the rate of ethanol production in batch fermentations. Tamarind fruit increased ethanol production 6.5 to 9.7% (w/v) from the 22.5% reducing sugars of the molasses. In general, the addition of tamarind to the fermentation medium showed more than 40% improvement in the production of ethanol using higher cane molasses sugar concentrations. The direct fermentation of the aqueous tamarind effluent also yielded 3.25% (w/v) ethanol, suggesting its possible use as a diluent in the molasses fermentations (Patil et al. 1998). Fresh, defrosted, and delignified brewer's spent grains (BSG) were used to improve the alcoholic fermentation of molasses by yeast (Kopsahelis et al. 2007). Glucose solution (12% w/v) with and without nutrients was used for cell immobilization on fresh BSG, without nutrients for cell immobilization on defrosted and with nutrients for cell immobilization on delignified BSG. Repeated fermentation batches were performed by the immobilized biocatalysts in molasses of 7, 10, and 12 initial Baume density without additional nutrients at 30 and 20°C. The defrosted BSG immobilized biocatalyst was used only for repeated batches of 7 initial Baume density of molasses without nutrients at 30 and 20°C. After the immobilization, the immobilized microorganism population was at 10^9 cells/g support for all the immobilized biocatalysts. The fresh BSG immobilized biocatalyst without additional nutrients for the yeast immobilization resulted in higher fermentation rates, lower final Baume densities, and higher ethanol productivities in the molasses fermentation at 7, 10, and 12 initial degrees Baume densities than the other biocatalysts. Adaptation of the defrosted BSG immobilized biocatalyst in the molasses fermentation system was observed from batch to batch approaching kinetic parameters reported in the fresh BSG immobilized biocatalyst. Therefore, the fresh or defrosted BSG as yeast supports could be promising for the scale-up operation (Kopsahelis et al. 2007). *S. cerevisiae* immobilized on orange peel pieces was examined for alcoholic fermentation of molasses at 30 to 15°C. The fermentation times in all the cases were low (5–15 h) and ethanol productivities were high (150.6 g/l/d), showing good operational stability of the biocatalyst and suitability for commercial applications. Reasonable amounts of volatile by-products were produced at all the temperatures studied, revealing potential application of the proposed biocatalyst in fermented food applications to improve productivity and quality (Plessas et al. 2007).

With respect to the use of alginate as the immobilizing matrix, it was found that the integrity of the matrix becomes compromised over prolonged operating times and it becomes necessary to supplement the media/reactor feeds with calcium. As an alternative immobilization matrix to alginate for the immobilized cells in continuous or semicontinuous processes, poly vinyl alcohol cryogel (PVAC) beads were attempted (Gough et al. 1998). In a fed-batch mode, the alginate-immobilized biocatalyst produced ethanol concentrations of up to a maximum of 57 g/l within 48 h from 140 g/l sugar concentration (80% theoretical yield). When the fermentations containing the alginate-based biocatalyst were refed for a further 425 h the ethanol concentration decreased dramatically to 20 g/l. Over the extended period of time from 60 to 500 h, the concentration of ethanol remained low. The average concentration of ethanol produced during the 500 h period was calculated to be 21 g/l and this represented 29% of the maximum theoretical yield. The PVAC-immobilized biocatalyst was used to convert molasses to ethanol at 72 h to maximum concentrations of 52 to 53 g/l (73% theoretical yield) (Gough et al. 1998). The average concentration of ethanol produced over a 600 h period was calculated to be 45 g/l (63% theoretical yield). Reasons for this dramatic difference in productivity, particularly at prolonged running times, are as yet unknown, although preliminary results suggest that the PVAC-immobilized biocatalyst remains viable for a longer period of time when compared with the immobilized alginate-based system (Gough et al. 1998).

The effect of molasses sugar concentration on the production of ethanol by alginate-immobilized *K. marxianus* in a continuous flow bioreactor was examined (Gough et al. 1998). Maximum ethanol concentrations were obtained using sugar concentrations of 140 g/l at 10 h. Ethanol concentrations subsequently decreased to lower levels over a 48 h period. Yeast cell number within the immobilization matrix was dramatically reduced over this time. At lower molasses concentrations, ethanol production remained relatively constant. The effect of residence time on ethanol production in a continuous flow bioreactor was examined. At a fixed molasses sugar concentration (120 g/l) a residence time of 0.66 h was found to be optimal on the basis of volumetric productivity.

6.6 POTENTIAL OF *ZYMOMONAS MOBILIS* FOR THE PRODUCTION OF ETHANOL FROM MOLASSES

Higher demands for alcohol have resulted in several approaches for improving the ethanol fermentation process. In the search for an efficient ethanol-producing organism, the bacterium *Z. mobilis* has been found to have several advantages over yeast fermentation. These include (1) higher sugar uptake and ethanol yield, (2) lower biomass production, (3) higher ethanol tolerance, (4) no need for controlled addition of oxygen during the fermentation, and (5) amenability to genetic manipulations. The strains of *Z. mobilis* can use only glucose, fructose, and sucrose with high fermentation efficiency. However, the yields in sucrose are comparatively low due to the formation of by-products such as levan and sorbitol (Viikari 1984). Attempts have been made at ethanol fermentation using commercial substrates such as cane and beet molasses. However, the ethanol yields from molasses are low due to the presence of inorganic ions and also due to the formation of by-products (Gunasekaran et al. 1986). Reports indicated the selection of mutant strains to ferment cane and hydrolyzed beet molasses with high efficiency (Park and Baratti 1991).

6.6.1 Adaptation of *Z. mobilis* for Fermentation of Cane Molasses

The parameters for the fermentation of molasses (20% w/v) at 30°C by *Z. mobilis* ZM4A are shown in Table 6.5. The maximum ethanol yield was reached to 0.47 g/g with 91.2% substrate consumption (Jain and Singh 1994). Fermentation of molasses with the partial supplementation of mineral salts, or with the yeast extract by *Z. mobilis* has been reported (Gunasekaran et al. 1986). Maximum final ethanol concentration of 39.4 g/l was observed with a substrate utilization of 91.3 g/l at 24 h in the fermentation without mineral supplementation (Jain and Singh 1994). Therefore, the molasses medium did not require any addition of supplements and it also provided some buffering capacity as the pH was not changed. An ethanol yield of

TABLE 6.5
Ethanol Production by *Z. mobilis* from Molasses Medium

Overall parameters	
Initial sugar (g/l)	110.0
Residual sugar (g/l)	9.6
Biomass (g/l)	1.6
Ethanol (g/l)	47.0
Substrate utilized (g/l)	91.2
Ethanol yield (g/l)	0.47
Biomass yield (g/l)	0.016
Fermentation efficiency (%)	92.0
Fermentation time (h)	24.0

From Jain, V. K. and A. Singh. 1995. Fermentation of sucrose and cane molasses to ethanol by immobilized cells of *Zymomonas mobilis*. Vol. 10. *Journal of Microbial Biotechnology*. With permission.

0.48 g/g was obtained from the molasses (90 g/l sugar concentration) without the supplements. Park and Baratti (1991) had reported that the addition of 0.5 g/l of magnesium sulfate to the sugar beet molasses medium enhanced ethanol production by *Z. mobilis*.

6.6.2 Fermentation Kinetics of *Z. mobilis* at High Concentration of the Molasses

Since *Z. mobilis* was efficient in fermentation of 20% (w/v) of molasses, batch fermentation kinetics were carried out at higher molasses concentrations (Jain and Singh 1994). The batch fermentation with molasses (110 g/l sugar concentration) gave maximum ethanol productivity (47 g/l) with a maximum substrate consumption (91.0 % w/v). Other parameters, such as the specific growth rate (0.128 to 0.137 μ/h), specific ethanol productivity (3.12 to 3.56 g/g/h), specific substrate uptake (7.50 to7.74 g/g/h), and the fermentation efficiency (76.3 to 92.0) were higher than that of the 40% molasses medium. Molasses concentration of 40% (200 g/l sugar) inhibited cell growth. This can be explained by the combined effect of the inhibition by ethanol and the influence of high osmotic pressure with the increasing concentration of molasses (Park and Baratti 1991). Comparative studies of the *Z. mobilis* ZM4 on sucrose and molasses showed that the sucrose was more efficiently fermented to ethanol at high concentrations (200 g/l), yielding 88.0 g/l of ethanol, whereas the inhibitory effect of inorganic ions is significant for molasses medium with 200 g/l sugars (Table 6.6).

6.6.3 Continuous Fermentation of Diluted Molasses by *Z. mobilis*

Savvides et al. (2000) developed a series of *Z. mobilis* CP4 mutants and *inaZ* recombinant *Z. mobilis* strains for the production of ethanol from molasses. In complete sucrose medium, ethanol production followed the steady-state biomass. The wild-type strain and the strains suc40 and suc40/pDS3154-*inaZ* displayed almost constant ethanol production (43 g/l). Thereafter, an 80% decrease in the ethanol production occurred. When sugar beet molasses was used as the growth medium, both the strains (suc40 and suc40/pDS3154-inaZ) produced exactly the same amount of ethanol. The hypertolerant mutant exhibited fastest growth and high stability in medium containing 20% sugar beet molasses. Fatty acid analysis of the strains showed that the presence of high levels of long chain unsaturated fatty acids (vaccenic acid, 18:1), which was even greater in the mutant strain (about 80%). Carey and Ingram (1983) suggested that the presence of vaccenic acid, in particular, could explain the ability of this organism to grow in high ethanol concentrations, due to the ethanol destabilizing effect on the membrane structure being compensated by the presence of long chain unsaturated fatty acids (Savvides et al. 2000).

The effect of pH on ethanol fermentation by several *Z. mobilis* isolates in molasses showed maximum ethanol production between pH 5.0 and 5.6. A comparative study on ethanol production by *Z. mobilis* 10988 and these isolates revealed that the isolates produced considerably lower levels of ethanol. Fermentation at 32°C had a positive effect on ethanol production from 46 to 50 g/l and temperature above 34°C

TABLE. 6.6
Comparison of Batch Fermentation of Molasses with *Z. mobilis*

Substrate	Sugar concentration (g/l)	Conversion (%)	Final ethanol concentration (g/l)	Fermentation efficiency (%)	Productivity (g/l/h)	Reference
Cane molasses desalted	200	—	—	60.7	—	Gunasekaran et al., 1986
Cane molasses programmed feeding	200	—	82.0	80–85	—	Karunakaran and Gunasekaran, 1986
Cane molasses	200	42.0	26.8	34.9	—	Gunasekaran et al., 1986
Cane molasses	200	93.6	64.6	85.0	3.0	Gunasekaran et al., 1986
Hydrolysed beet	152	88.5	56.3	86.2	2.4	Park and Baratti, 1991
molasses	100	91.0	47.0	92.0	1.96	Jain and Singh , 1994
Cane molasses	150	83.0	52.7	82.2	2.20	Jain and Singh, 1994

severely inhibited ethanol production as well as the biomass. At higher concentrations of molasses (25 g/l sugar concentration), the yeast strain produced more ethanol and in lower concentrations (23 g/l sugar concentration) *Z. mobilis* produced high ethanol with a maximum theoretical yield. It is known that the strains of *Z. mobilis* produce ethanol at high sugar concentrations in synthetic media (Rogers et al. 1982). This low efficacy could be due to the presence of inhibitors in the molasses, which inhibit growth and ethanol production.

6.7 CONCLUSIONS

The utilization of bioethanol for transportation has the potential to contribute to a cleaner environment. It is expected that the bioethanol industry will benefit from the efficient exploitation of renewable resources such as sugarcane molasses. Process development for ethanol production with various microorganisms has an optimistic outlook. However, toxic compounds present in the molasses, which are formed during the sugar separation process, inhibit the fermentative microorganisms. To conquer these, genetic engineering approaches are being investigated for manipulating resistance traits such as tolerance to ethanol and inhibitors, thermotolerance, reduced need for nutrient supplementation, and improvement of sugar transport. Yeast strains such as *K. marxianus* and *S. cerevisiae* have several advantages for molasses fermentation at high temperature (40 to 45°C), such as reduced risk of contamination, faster recovery of ethanol, and considerable savings on capital and running costs of refrigerated temperature control in temperate countries. Global efforts are continuing to develop a thermo- and osmotolerant yeast strain, which could be effectively used for the production of molasses for ethanol.

ACKNOWLEDGMENTS

The authors gratefully acknowledge the Department of Biotechnology (DBT) New Delhi, India, for providing financial support through the project BT/PR3445/AGR/16/283/2002-III.

REFERENCES

Bajaj, B. K., T. Vikas, and R. L. Thakur. 2003. Characterization of yeasts for ethanol fermentation of molasses with high sugar concentrations. *J. Sci. Ind. Res.* 62: 1079–1085.

Banat, I. M. and R. Marchant. 1995. Characterization and potential industrial applications of five novel thermotolerant fermentative yeast strains. *World J. Microbiol. Biotechnol.* 11: 304–306.

Banat, I. M., P. Nigam, D. Singh, R. Merchant, and A. P. McHale. 1998. Ethanol production at elevated temperatures and alcohol concentrations: A review. I. Yeast in general. *World J. Microbiol. Biotechnol.* 14: 809–821.

Barron, N., R. Marchant, L. McHale, and A. P. McHale. 1995. Studies on the use of a thermotolerant strain of *Kluyveromyces marxianus* in simultaneous saccharification and ethanol formation from cellulose. *Appl. Microbiol. Biotechnol.* 43: 518–520.

Carey, V. C. and L. O. Ingram. 1983. Lipid composition of *Zymomonas mobilis*: Effects of ethanol and glucose. *J. Bacteriol.* 154: 1291–1300.

Comberbach, D. M. and J. D. Bu'Lock. 1984. Continuous ethanol production in the gas lift tower fermenter. *Biotechnol. Lett.* 6: 129–131.

Damiano, D. and S. S. Wang. 1985. Improvements in ethanol concentration and fermentor ethanol productivity in yeast fermentations using whole soy flour in batch and continuous recycle systems. *Biotechnol. Lett.* 7: 135–140.

Gough, S., N. Barron, A. L. Zubov, V. I. Lozinsky, and A. P. McHale. 1998. Production of ethanol from molasses at 45°C using *Kluyveromyces marxianus* IMB3 immobilized in calcium alginate gels and poly (vinyl alcohol) cryogel. *Bioproc. Eng.* 19: 87–90.

Gunasekaran, P., T. Karunakaran, and M. Kasthuribai. 1986. Fermentation pattern of *Zymomonas mobilis* strains on different substrates: A comparative study. *J. Biosci.* 1: 181–186.

Jain, V. K. and A. Singh. 1994. Ethanol production from cane molasses using free and immobilized cells of *Zymomonas mobilis*. Proceedings of national seminar held at HBTI, Kanpur, edited by V. K. Jain. New Delhi, India: Wisdom Publishing House.

Jones, R. P., N. Pamment, and P. F. Greenfield. 1981. Alcohol fermentation by yeasts. The effect of environmental and other variables. *Process Biochem.* 16: 42–45.

Karunakaran, T. and P. Gunasekaran. 1986. Mg^{++} positively regulates the fermentation efficiency in *Zymomonas mobilis*. *Curr. Sci.* 55: 857–859.

Kopsahelis, N., N. Agouridis, A. Bekatorou, and M. Kanellaki. 2007. Comparative study of spent grains and delignified spent grains as yeast supports for alcohol production from molasses. *Bioresour. Technol.* 98: 1440–1447.

Nigam, P., I. M. Banat, D. Singh, A P. McHale, and R. Marchant. 1996. Continuous ethanol production by thermotolerant Km IMB3 yeast immobilized on mineral kissiris at 45°C. *World J. Microbiol. Biotechnol.* 13: 283–288.

Park, S. C. and J. Baratti. 1991. Batch fermentation kinetics of sugar beet molasses by *Zymomonas mobilis*. *Biotech. Bieng.* 38: 304–313.

Patil, B. G., D. V. Gokhale, K. B. Bastawde, U. S. Puntambekar, and S. G. Patil. 1998. The use of tamarind waste to improve ethanol production from cane molasses. *J. Ind. Microbiol. Biotechnol.* 21: 307–310.

Plessas, S., A. Bekatorou, A. A. Koutinas, M. Soupioni, I. M. Banat, and R. Marchant. 2007. Use of *Saccharomyces cerevisiae* cells immobilized on orange peel as biocatalyst for alcoholic fermentation. *Bioresour. Technol.* 98: 860–865.

Ragav, R., V. Sivaraman, D. V. Gokhale, and B. Setharamaroa. 1989. Ethanol fermentation of cane molasses by highly flocculent yeast. *Biotech. Lett.* 10: 739–744.

Rajoka, M. I., M. Ferhan, and A. M. Khalid. 2005. Kinetics and thermodynamics of ethanol production by a thermotolerant mutant of *Saccharomyces cerevisiae* in a microprocessor controlled bioreactor. *Lett. Appl. Microbiol.* 40: 316–321.

Rogers, P. L., M. L. Skotnicki, and D.E. Tribe. 1982. Ethanol production by *Zymomonas mobilis*. *Adv. Biochen Eng. Biotechnol.* 23: 37–84.

Royston, M. G. 1966. Tower fermentation of beer. *Process Biochem.* 1: 215–221.

Savvides, A. L., A. Kallimanis, A. Varsaki, A. I. Koukkou, C. Drainas, M. A. Typas, and A. D. Karagouni. 2000. Simultaneous ethanol and bacterial ice nuclei production from sugar beet molasses by a *Zymomonas mobilis* CP4 mutant expressing the *inaZ* gene of *Pseudomonas syringae* in continuous culture. *J. Appl. Microbiol.* 89: 1002–1008.

Sittikat, J. and N. Jiraarun. 2005. Fermentation of molasses for producing ethanol by continuous circulating system. 31st Congress on Science and Technology of Thailand at Suranaree University of Technology, October 18–20.

Viikari, L. 1984. Formation of levan and sorbitol from sucrose by *Zymomonas mobilis*. *Appl. Microbiol. Biotechnol.* 19: 252–255.

7 Bioethanol from Starchy Biomass

Part I Production of Starch Saccharifying Enzymes

Subhash U Nair, Sumitra Ramachandran, and Ashok Pandey

CONTENTS

ABSTRACT

There has been a substantial shift in global perception of the production of ethanol from plant-based biomass. These could be of starchy or cellulosic (or lignocellulosic) nature, which requires a step of hydrolysis to produce fermentable sugars which then can be fermented to ethanol. The hydrolysis of the starchy raw materials can be done by acids or enzymes, but due to several advantages enzymatic hydrolysis offers, it is the preferred choice for industrial applications. The enzymes involved in hydrolysis include α-amylase, β-amylase, glucoamylase, and pullulanase. These enzymes can be obtained from plant and microbial sources but industrial demand is met through the latter. This chapter presents a brief description of the sources, applications, and production of these enzymes.

7.1 INTRODUCTION

The amylases (a term that refers to α-amylase and β-amylase here) can be generally defined as the enzymes that hydrolyze the *O*-glycosyl linkage of starch. α-Amylases are one of the most popular and important forms of the industrial amylases and have different reaction and product specificities, which include exo- and endo-specificity, preference for the hydrolysis, or the transglycosylation, α-(1,1), α-(1,4) or α-(1,6)-glycosidic bond specificity and glucan synthesizing activity. Depending on the type of cleavage caused on starch, the enzymes belonging to the amylase family are of the following types (see Figure 7.1; Kuriki and Umanaka 1999).

7.2 ENZYMES HYDROLYZING α-1,4-GLUCOSIDIC LINKAGES

7.2.1 Endo-amylases (EC 3.2.1.1)

This group of enzymes known as α-amylases cleave the α-1,4-glucosidic linkages. These enzymes generally do not cleave the α-1,6-glucosidic linkages. It is the most widespread enzyme among aerobes and anaerobic microbes. α-Amylases from different sources have been purified and many have been crystallized. These enzymes are chiefly required in the thinning of starch in the liquefaction process in the sugar, alcohol, and brewing industries.

7.2.2 Exo-amylases (β-amylases EC 3.2.1.2)

The β-amylases generally occur in plants such as malt, sweet potato, soy bean, etc., and also as extracellular enzymes in some aerobic and anaerobic microorganisms. This enzyme degrades the amylose, amylopectin, and glycogen in an exo-fashion from the nonreducing ends, by hydrolyzing the alternate glucosidic linkages. It is incapable of bypassing the α-1,6 linkages and, hence, causes the incomplete hydrolysis to form limit dextrins.

7.3 ENZYMES HYDROLYZING α-1,6-GLUCOSIDIC LINKAGES

The debranching enzymes hydrolyze the α-1,6-glycosidic linkages in branched polymers. Two types of direct debranching enzymes, namely pullulanase and isoamylase are known.

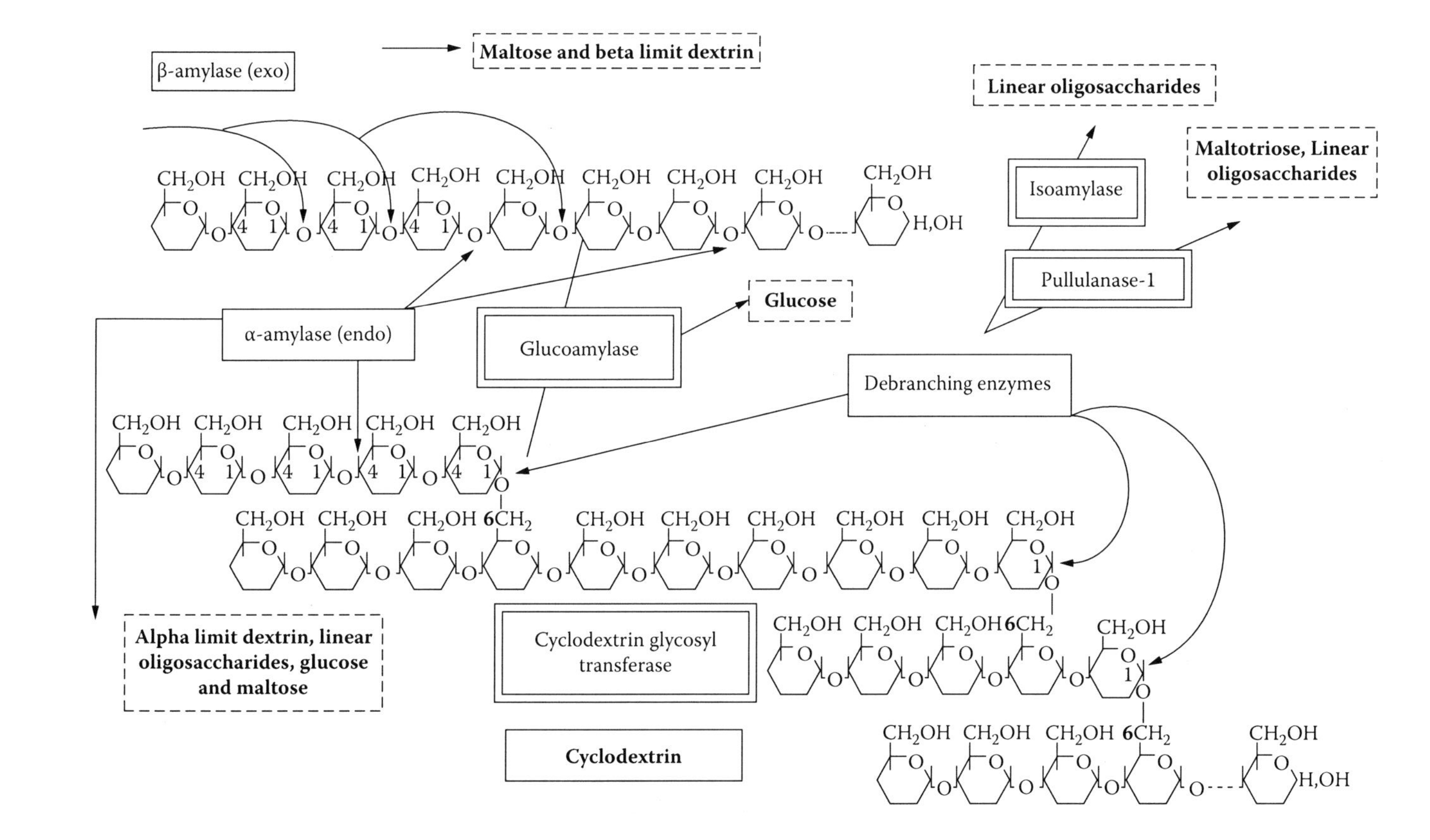

FIGURE 7.1 Schematic representation of different enzymes belonging to the amylase family.

7.3.1 Pullulanases (EC 3.2.1.41)

The enzymes that hydrolyze certain α-1,6-linkages are classified as pullulanases. These enzymes are capable of hydrolyzing the α-1,6-linkages in polysaccharides such as starch, amylopectin, and especially in pullulan to give the trisaccharide maltotriose. True pullulanases do not hydrolyze the α-1,4-linkages in pullulan. The pullulanases are different from isoamylase for their ability to degrade the branches of pullulan.

7.3.2 Isoamylases (EC 3.2.1.68)

Isoamylase generally occurs in higher plants (the enzyme from such sources is referred to as R-enzyme) but has also been found in yeast and bacteria. These enzymes have little or no effect on pullulan, but cleave all the α-1,6-linkages in amylopectin and glycogen.

7.4 Transglycosylation to Form α-1,4 and/or α-1,6-Glucosidic Linkages

7.4.1 Cyclodextrin Glucotransferase (EC 2.4.1.19; CGTases)

This is a unique enzyme capable of converting starch and related substances to cyclodextrin (CD). It is a multifunctional enzyme and catalyzes the conversion of the starch into CD by intramolecular transglycosylation (cyclization) and in the presence of acceptors it also catalyzes intermolecular transglycosylation (coupling and disproportionation).

7.5 Hydrolysis of Both α-1,4- and α-1,6-Glucosidic Linkages

7.5.1 Glucoamylases (EC. 3.2.1.3)

This class of enzymes, also referred to as amylogucosidases or saccharogenic amylase (1,4-α-D-glucan glucohydrolase), are capable of cleaving both α-1,6- and α-1,4-glucosidic linkages. These enzymes remove one glucose unit at a time from the nonreducing end of the large carbohydrate molecule. Although they are capable of hydrolyzing certain α-1,6-glucosidic linkages, they hydrolyze the α-1,4-glucosidic linkages much more rapidly. The microbial glucoamylase is important in the starch bioprocessing and brewing industry where the hydrolysis of starch is necessary. These enzymes can convert starch to glucose even in the absence of other enzymes.

There are a few deviations from the above classification, since many enzymes are reported to show mixed types of the specificity. For example, the α-amylases weakly catalyze the α-1,4 transglycosylation; the CGTases feebly catalyze the α-1,4 hydrolysis in addition to the main reaction, the α-1,4 transglycosylation. There are a few α-amylases that catalyze the α-1,6 hydrolysis. Some pullulanases from thermophilic microorganisms have recently been reported to hydrolyze not only α-1,6- but also α-1,4-glucosidic linkages (Guzman-Maldonado and Paredas-Lopez 1995; Kuriki and Umanaka 1999).

Most starch-converting enzymes belong to the GH-13 family. The α-amylase family or GH-13 family, is a large enzyme family that constitutes about 20 enzymes. This family can be further classified into clans, based on the three-dimensional structure of their catalytic components. Each of the clans may consist of two or more families with similar three-dimensional structure of their catalytic domain but with few sequence similarities, since protein structure is conserved to a large extent by the evolution as compared to the amino acid sequence. The α-amylase family (GH-13) belongs to the eighth clan among the fourteen clans described (Reddy, Nimmagadda, and Rao 2003). An enzyme should have the following characteristics to be included in this family: it should have specificity for α-1,4- or 1,6-glucosidic linkages and be capable of hydrolyzing them to produce mono- or oligosaccharides, or capable of transglycosylations to form α-glucosidic linkages. They should have Asp, Glu, and Asp residues as the catalytic sites, corresponding to the Asp 206, Glu 230, and Asp 297 of Taka amylase A and should possess four highly conserved regions in their primary structures consisting of catalytic and important substrate-binding sites. $\alpha(\beta/\alpha)_8$; the TIM barrel catalytic domain is also a required feature (Kuriki and Umanaka 1999, Henrissat 1991).

7.6 SOURCES

The starch saccharifying enzymes play a dominant role in carbohydrate metabolism and are produced by many plants and animals. They are also produced by several microorganisms, which are the preferred sources for commercial production. These enzymes are either cell bound, intracellular, or extracellular. From a commercial point of view, extracellular enzymes are preferred and since the applications of these enzymes are usually at temperatures higher than 50°C, it is always advantageous to isolate enzymes that are thermostable. The major microbial sources of the thermostable α-amylases are *Bacillus* sp, especially *B. subtilis, B. stearothermophilus, B. licheniformis*, and *B. amyloliquefaciens*, and are widely used for the commercial production of the enzyme for various applications (Bertoldo and Antranikian 2001). However, for industrial applications, *B. licheniformis* (Thermamyl) and *B. amyloliquefaciens* are commonly employed. Equally important are the filamentous fungi, predominantly from the genus *Aspergillus*, which are widely used for the production of these enzymes, especially glucoamylase (Pandey et al. 2000a; Gupta et al. 2003; Sivaramakrishnan et al. 2006).

However, in comparison to the ubiquitous presence of the α-amylases, β-amylases are produced by relatively fewer organisms. As regards glucoamylase, the bacterial sources do not provide sufficient glucoamylase for commercial needs. Currently, glucoamylase is produced mainly by the solid-state or submerged fermentation of many molds and bacteria, mainly from *Aspergillus niger, A. oryzae, A. phoenicis, A. candidus, A. awamori, Rhizopus delemar, Mucor rouxianus,* etc.

The debranching enzymes R-enzymes or pullulanases are reported from several plant, animal, and microbial sources. Pullulanase was first discovered in *Klebsiella* species. In the industry, the thermostable and acidophilic pullulanases are produced from *Bacillus acidopullulyticus*. The pullulanases from *Thermococcus* sp., *T. litoralis, Pyrococcus furiosus, and P. woesei* have been described as type II as they cleave both the α-1,4- and α-1,6-glucosidic bonds in starch and related polysaccharides. Neopullulanase, which hydrolyzes pullulan to panose (6-α-D-glucosylmaltose) by hydrolyzing its α-(1,4)-glucosidic linkages, are found in *Bacillus, Micrococcus, Thermoactinomycetes,* and *Listeria* species. The isopullulanases (IPU) hydrolyze α-(1,4) glucosidic linkages of pullulan to produce isopanose (6-*O*-α-maltosyl-glucose). Few strains of fungi produce these types of pullulanases, for example, *Aspergillus niger* and *Arthrobacte*r sp. The IPU does not attack the starch or dextran. Table 7.1 summarizes different bacterial and fungal sources of amylase, glucoamylase, and pullulanase (Doman-Pytka and Bardowski 2004).

TABLE 7.1
Microbial Sources of Starch Saccharifying Enzymes

α-Amylase	β-Amylase	Glucoamylase	Pullulanase
Aeromonas caviae	*Aspergillus terreus*	*Acremonium zonatum*	*Aerobacter aerogenes*
Acinetobacter	*Bacillus cereus*	*Amylomyces rouxii*	*Anaerobranca gottschalkii*
Alicyclobacillus acidocaldarius	*B. circulans*	*Arxula adeninivorans*	*Bacillus acidopullulyticus*
Alteromonas haloplanetis	*B. megatarium*	*Aspergillus* sp.	*Bacillus stearothermophilus*
Archaeobacterium pyrococcus woesei	*B. polymyxa*	*A. awamori*	*Bacillus cereus*
Aspergillus sp.	*Brettanomyces naardensis*	*A. candidus*	*Bacillus circulans*
A. awamori	*C. thermocellum*	*A. foetidus*	*Bacillus flavocaldarius*
A. flavus	*Clostridium thermosulfurogenes*	*A. niger*	*Bacillus macerans*
A. fumigatus	*Corynascus sepedonium*	*A. oryzae*	*Bacillus naganoensis*
A. kawachi	*Debaromyces* sp.	*A. phoenicus*	*Bacteroides thetaiotaomicron*
A. niger	*Emericella nidulans* 45	*A. saitri*	*Clostridium thermohydrosulfuricum*
A. oryzae	*E. nidulans* MNU 82	*A. terreus*	*Clostridium thermosaccharolyticum*
A. usanii	*Malbranchea sulfurea*	*Bacillus firmus/lentus*	*Desulfurococcus mucosus*
Bacillus sp.	*Paffia rhodozyma*	*B. stearothermophillus*	*Fervidobacterium pennavorans*
B. acidocoldarius	*Pichia anomala*	*Candida famata*	*Klebsiella aerogenes*
B. amyloliquefaciens	*P. holestii*	*Cephalosporium charticola*	*Klebsiella pneumoniae*
B. brevis	*Pseudomonas* sp.	*Chalara paradoxa*	*Klebsiella planticola*
B. circulans	*Rhizopus japonicus*	*Clostridium* sp.	*Klebsiella oxytica*
B. coagulans	*Saccharomyces* sp.	*C. acetobutylicum*	*Micrococcus* sp.

TABLE 7.1
Microbial Sources of Starch Saccharifying Enzymes

α-Amylase	β-Amylase	Glucoamylase	Pullulanase
Bacillus flavothermus	*Sacchoromyces cerevisiae*	*Clostridium thermohydrosulfuricum*	*Pyrococcus woesei*
B. globisporus	*Syncephalastrum racemosum* RR96	*C. thermosaccharolyticum*	*Rhodothermus marinus*
B. licheniformis	*Streptomyces* sp.	*C. thermosulfurogenes*	*Streptococcus mitis*
B. megaterium	*Thermomyces lanuginosus*	*Cladosporium resinae*	*Streptococcus pneumoniae*
B. stearothermophilus	*Trichosporon beigelii*	*Coniophora cerebella*	*Streptococcus pyogenes*
B. subtilis		*Endomyces* sp.	*Thermoactinomyces thalpophilus*
B. halmapalus		*Flavobacterium* sp.	*Thermoanaerobacter ethanolicus*
Clostridium acetobutylicum		*Fusidium* sp.	*Thermoanaerobacter finnii*
C. butricum		*Halobacter sodamense*	*Thermoanaerobacter saccharolyticum*
C. thermohydrosulfuricum		*Humicola lanuginosa*	*Thermoanaerobacter thermohydrosulfuricus*
C. thermosulfurogenes		*Lactobacillus brevis*	*Thermoanaerobacter thermosulfurogenes*
Filobasidium capsuligenum		*Magnaporthe grisea*	*Thermoanaerobium* sp.
Halobacterium halobium		*Mucor rouxianus*	*Thermococcus hydrothermalis*
H. salinarium		*Neurospora crassa*	*Thermotoga maritima*
Humicola insolens		*Penicillium italicum*	*Thermus aquaticus*
H. lanuginosa		*P. oxalicum*	*Thermus caldophilus*
H. stellata		*Piricularia oryzae*	*Thermus thermophilus*
Lactobacillus brevis		*Rhizoctania solani*	
Malbrachea pulchella var. *sulfurea*		*Rhizopus* sp.	
Micrococcus luteus		*R. delemar*	
M. varians		*R. javanicus*	
Micromonospora vulgaris		*R. niveus*	
Mucor pusillus		*R. oligospora*	
Myceliophthora thermophila		*R. oryzae*	
Myxococcus coralloides		*Sachharomyces diastaticus*	
Nocardia asteroides		*Thermomyces lanuginosus*	

TABLE 7.1
Microbial Sources of Starch Saccharifying Enzymes

α-Amylase	β-Amylase	Glucoamylase	Pullulanase
Penicillium brunneum		*Trichoderma reesei*	
Pseudomonsa stutzeri		*T. viride*	
Pyrococcus woesei			
Rhizopus sp.			
Scytalidium sp.			
Talaromyces thermophilus			
Thermus sp.			
Thermoactinomyces sp.			
T. vulgaricus			
Thermococcus profundus			
Thermoascus aurantracus			
Thermomonospora viridis			
Thermonospora curvata			
T. vulgaris			
Thermomyces lanuginosus			
Thermotoga maritime			
Torula thermophila			

7.7 PRODUCTION

The starch saccharifying enzymes can be produced by submerged fermentation (SmF) and solid-state fermentation (SSF). For ease of handling and greater control of the physicochemical factors such as temperature and pH, submerged fermentation traditionally has been the preferred method of production. However, SSF has been considered superior in several aspects to SmF (Pandey 1992, 1994, 2003). It is cost effective due to the use of simple fermentation media comprised mainly of agro-industrial residues, uses little water, which consequently releases negligible or considerably less effluent, thus reducing pollution. The SSF processes are simple, provide easy aeration, use low volume equipment (lower cost), and are yet effective by providing high product titers (concentrated products) (Pandey et al. 2004). Both natural and synthetic media are used for the production of amylolytic enzymes. Because the synthetic components are expensive, alternative cheaper sources such as agricultural by-products for the reduction of the cost of the medium are useful. In general, the production of these enzymes is affected by a variety of physicochemical factors, notably the composition and pH of the growth medium, inoculum size and age, temperature, aeration and agitation, and the carbon and nitrogen sources.

7.7.1 α- AND β-AMYLASES

To counter the increasing commercial need for the amylases, continuous efforts are underway to reduce the cost of production of α-amylase. Employing SSF to utilize agricultural polymeric wastes is one of the most significant ways of reducing the cost of amylase production. These wastes provide both support and nutrition to the microbes, and include wheat bran, spent brewing grain, maize bran, rice bran, rice husk, coconut oil cake, mustard oil cake, corn bran, amaranths grains, gram bran, palm oil cake, sunflower meal, pearl millet bran, soy meal, etc. (Pandey et al. 2000a; Ramachandran et al. 2004a, 2004b;, Bogar et al. 2002; Francis et al. 2002, 2003; Satyanarayana et al. 2004; Sivaramakrishnan et al. 2006). Among these, wheat bran is generally considered the most suitable substrate. Filamentous fungi are generally reported to produce high titers of extracellular enzymes, and several strains of *Aspergillus* sp. and *Rhizopus* sp. are commonly used. A strain of *Thermomyces lanuginosus*, a thermophilic fungus, was reported as an excellent producer of α-amylase (Jensen and Olsen 1992). The α-amylase from *Pycnoporus sanguine* by cultivation in SSF resulted in fourfold higher enzyme production than in SmF. Several yeast strains such as *Saccharomycopsis capsularia* and *Cryptococcus* sp. are also employed in α-amylase production using SSF (Satyanarayana et al. 2004; Sivaramakrishnan et al. 2006).

The industrial submerged fermentation process is generally carried out in batch or in fed-batch mode. Simple and cheap media containing corn steep liquor and soybean meal or whey-based media offer benefits of cost reduction and higher yields. The enzyme production is induced by the presence of some natural materials such as beet pulp, corn cob, rice husk, wheat bran, and wheat straw in the production medium and is generally affected by the pattern of growth of the microorganism and any morphological changes (in case of fungus). The cell growth and α-amylase production patterns are generally similar regardless of the limiting nutrient, suggesting that there exists stationary phase gene control of α-amylase production as opposed to a direct response to nutrient limitation. The dissolved oxygen tension is an imperative factor for α-amylase production and higher aeration rates improve the yields. *Bacillus* sp. is the most preferred and important source for the production for most amylolytic enzymes. Most applications of α-amylase require it to be thermotolerant, which is generally obtained from *B. licheniformis;* thermostable fungal α-amylase is obtained from *A. niger* and *A. oryzae.*

As compared to α-amylase, not much is known about the production of β-amylase using microorganisms. Some microorganisms that produce β-amylase include *B. polymyxa, B. cereus, B. megatarium, Streptomyces* sp., *Pseudomonas* sp., and *Rhizopus japonicus* (Crueger and Crueger 1989; Pandey et al. 2000a).

The technique of immobilization has also been employed for the production of the amylases (Pandey et al. 2000b). Various immobilization techniques can be utilized for α-amylase production, such as entrapment in gels using calcium alginate, kappa-carrageenan, agar and their combinations with polyethylene oxide, adsorption on cut disks of polymerized polyethylene oxide, and fixation on formaldehyde activated acrylonitrile-acrylamide membranes.

7.7.2 Glucoamylase

Most of the commercial production of glucoamylase (GA) is carried out by submerged fermentation. The production is generally characterized by the simultaneous production of small quantities of other enzymes, referred to as associated activities. These include transglucosidase, which, however, is an undesirable phenomenon and should be controlled. The production of α-amylase in very small quantities is considered desirable as it acts on the starch to catalyze the formation of saccharides of lower molecular weight, which are broken down to dextrose with relative ease by the glucoamylase. However, larger amounts of α-amylase may be detrimental to the production of dextrose as it produces saccharides, which may polymerize to unfermentable dextrose polymers by transglucosidase (Pandey 1995; Pandey et al. 2000a; Soccol et al. 2004; Sandhya and Pandey 2005).

The fungal strains of *Aspergillus* and *Rhizopus* are the main sources of the glucoamylase, although bacterial cultures such as *Lactobacillus brevis* and yeasts can also be used. The commercial production of GA is carried out by SmF as well as SSF using fungal strains and strains of *A. awamori* are frequently employed. Several agro-industrial residues, in combination and individually, imparted different patterns of GA induction in *Aspergillus*. In SSF, particle size of the substrate, moisture content, and water activity influence GA production (Pandey and Radhakrishnan 1992, 1993; Selvakumar, Ashakumary, and Pandey 1998). The type of bioreactor used, such as flasks, trays, rotary reactors, and columns (vertical and horizontal), influence the production of glucoamylase. SSF often yields higher GA titers than SmF (Pandey et al. 1995).

The technique of immobilization is also widely employed in the production of the GA. Strains of *A. niger*, *Candida* sp. and *Endymycopsis* sp. have been used for this purpose (Pandey et al. 2000b, Sandhya and Pandey 2005).

7.7.3 Pullulanase

The pullulanases are either cell bound, intracellular, or extracellular. The enzyme can be produced by different microorganisms. The selection of an appropriate carbon source is critical in pullulanase production, since the enzyme production is controlled by substrate induction and catabolite repression. Most saccharides, such as soluble starch, potato starch, amylopectin, potato dextrin, maltodextrin, and maltose, induce varying levels of pullulanase production. This could be attributed to the presence of α-(1,6) linkages, present in complex polysaccharides, that can induce production of pullulanase. Soluble starch is a good source of carbon for the production of pullulanase, since it can induce the release of the enzyme in the medium (Nair 2006; Nair, Singhal, and Kamat 2006, 2007). Several other starches, dextrins or maltosaccharides, polypeptone, yeast extract, manganese also induce the production of pullulanase.

7.8 PURIFICATION

Conventional techniques for the purification of amylases involve several steps, which include centrifugation of the culture to separate the solid media followed by selective concentration of the supernatant usually by ultrafiltration or precipitation by ammonium sulfate or ethanol cold; the crude enzyme is then subjected to different chromatographic techniques such as affinity, hydrophobic, or ion exchange chromatography and gel filtration. Table 7.2A summarizes some of the purification strategies employed in amylase purification from various microorganism (Pandey et al. 2000a; Satyanarayana et al. 2004; Patel et al. 2005). The bacterial α-amylases generally show higher temperature optima than the fungal α-amylases, which could be between 45 and 115°C in the former and around 55°C in the latter. Most of the α-amylases from the *Bacillus* sp. possess optimal activity around 60 to 70°C. Similarly, they are active in a wide range of pH, generally close to neutrality or on the alkaline side (Table 7.2A).

Several modified methods can also be used for the purification of α-amylases. The enzyme from an archaebacterium *Thermococcus profundus* was purified to homogeneity by ammonium sulfate precipitation, DEAE-Toyopearl chromatography, gel filtration on Superdex 200 HR and its thermostability was enhanced using Ca^{2+} ions. The α-amylase from *Thermus* sp. was purified by ion-exchange chromatography and also by affinity adsorption on starch granules; the enzyme from a thermophilic and photosynthetic strain of *Chloroflexux aurantiacus* was purified to homogeneity by means of ultrafiltration, ammonium sulfate fractionation, DEAE-cellulose, hydroxyapatite, and high-performance liquid chromatography (HPLC). This enzyme was stable at alkaline pH up to 12 and high temperatures up to 55°C and produced

TABLE 7.2A
Biochemical Characteristics of α-Amylase

Microorganism	Optimum Temperature (°C)	Optimum pH	Molecular Mass (KDa)	K_m (mg/ml)
Bacillus coagulans	70	6–7	66	1.53
B. stearothermophilus	55–70	4.6–5.1	53	1.0
B. subitilis	60–65	6.8	55	–
B. amyloliquefaciens	65	5–9	–	2.63
B. circulans	60	8.0	76	0.65
B. thermoamyloliquefaciens	70	5.6	78	0.21
B. brevis	80	5–9	58	0.8
Escherichia coli	50–70	6.5–7.0	48	–
Lactobacillus plantarum	65	3.0–8.0	50	2.38
Pyrococcus furiosus	115	5.6	48	–
Aspergillus oryzae	50	4.0	54	–
A. flavus	60	6.0	52.5	.013%
A. awamori	55	5	54	1.0
A. niger	50	5.0	–	–
A. fumigatus	55	6.0	–	–
A. foetidus	45	5.0	41.50	2.19
Thermomyces lanuginosus	80	4.0	61	0.68

Adapted from Satyanarayana et al. (2004).

maltotriose and maltotetraose from starch (Table 7.2A). Amylases from *B. licheniformis* were purified by enhanced two-phase separation in a PEG-Dextran system, followed by gel filtration and ion-exchange chromatography; *B. subtilis* amylase gene was cloned into a plasmid and transferred to *Escherichia coli*. The α-amylase so produced was purified to homogeneity. Its molecular weight (48,000 Da) was lower than the molecular weight calculated from the derived amino acid sequences of the *B. subtilis* complete α-amylase (57,700 Da). The α-amylase produced by the recombinant cells was purified by specific elution from the anti-peptide antibodies corresponding to the C-terminal region of target α-amylase, and then specifically eluted by the eluent containing low concentration of the antigen peptide used for immunization (Pandey et al. 2000a; Satyanarayana et al. 2004).

Some other methods adopted for the purification of α-amylase include autofocusing the enzyme produced by *B. subtilis,* followed by gel filtration and ion-exchange chromatography; sequential steps of amylopectin affinity chromatography, DEAE-ion-exchange chromatography, and Sephacryl S-200 HR gel filtration for an *A. oryzae* α-amylase; anion exchange (DEAE-cellulose) and affinity (α-cyclodextrin-Sepharose) chromatography for α-amylase from *A. fumigatus;* ammonium sulfate treatment, affinity binding on cross-linked starch, and DEAE-Biogel A chromatography for α-amylase from the yeast *Lipomyces kononenkoae,* etc. (Pandey et al. 2000a; Satyanarayana et al. 2004).

The purification of β-amylase is also done generally in the same way as that of α-amylase. A β-amylase from *Hendersonula toruloidea* was separated by ammonium sulfate fractionation, ion-exchange chromatography on DEAE-cellulose, and gel filtration on Sephadex G-75. An extracellular β-amylase from a new isolate of *B. polymyxa* was purified by adsorption on raw corn starch to 22.5-fold. A thermostable β-amylase was purified with ammonium sulfate, DEAE-cellulose column chromatography, and gel filtration using Sephadex G-200 (Reddy, Swamy, and Seenayya 1998).

The purification and characterization of the GA also follows generally the conventional methods. Table 7.2B cites a few such examples. The enzyme from different sources shows varied properties; for example, two to six isoforms can exist, with pH and temperature optima between 4.0 and 6.2 and 50 and 60°C, respectively, and molecular weight of 34 to 141 kDa. It may contain a protein moiety as well, generally a glycoprotein, and the carbohydrate content varies. For example, the GA purified to homogeneity from the anaerobic thermophilic bacterium *Clostridium thermosaccharolyticum* was made of a single subunit with a molecular weight of 75 kDa. The GA from *A. niger* contained four different forms and was purified to 32.4-fold with a final specific activity of 49.25 U/mg protein. Two forms of the GA, homogeneous in nature, were purified from *Monascus kaofiang* nov. sp. F-I, which exhibited pH optima at 4.5 and 4.7. A commercial preparation of the GA from *A. niger* showed six different forms of GA, having apparently different molecular weights and optimum pH. Another two forms of the GA from *A. niger* showed the molecular weight of GA I as 99,000 and that of GA II as 112,000. Both forms of the GA were glycoenzymes and showed identical amino acid composition but differed in carbohydrate content. The GA of *Paecifomyces varioti* AHU 9417 was purified by precipitation with ethanol, chromatography on DEAE-Sepharose CL-6B, gel filtration, and preparative disc electrophoresis. The GA from *Schizophyllum commune* was purified by ammo-

TABLE 7.2B
Biochemical Characteristics of β-Glucoamylase

Microorganism	Nature of Protein	Carbohydrate Content, %	Molecular Mass (kDa)	Optimum pH	Optimum Temperature (°C)
Neosartorya fischeri	–	–	34	4.0–4.4	55–60
T. lanuginosus	Glycoprotein	10–12	57	–	70
T. lanuginosus	Glycoprotein	3.3	75	4.4–5.6	70
R. oryzae	–	–	67	4.8	60
A. terreus	–	–	70	5.0	60
Mucor rouxianus	–	–	49, 59	4.7	55
A. saitri	Glycoprotein	18	90	4.5	50
A. niger	–	–	–	4.5	50
Acremonium sp.	–	–	22, 39	5.5	55
Aspergillus fumigatus	Glycoprotein	23	42	4.5–5.5	65
A. niger	Glycoprotein	16	112, 104	4.4	60
Clostridium sp.	–	–	4, 61	5.0	70
C. thermosaccharolyticum	–	–	75	4.5	65
Thermococcus hydrothermalis	–	–	77	5.5	95
Thermoplasma acidophilum			110	2.0	95
Neurospora crassa	Glycoprotein	5.1	141, 95	5.0–5.4	60
Saccharomycopsis fibuligera	–	–	82 62	5.0–6.2	40-50

nium sulfate precipitation, acid-clay treatment, and Sephadex G-100 gel filtration. Two kinds of GA were obtained by fractionation of *Rhizopus* sp.; one has a strong debranching activity while the other has a weak debranching activity. Three kinds of GA were reported from *Rhizopus* sp., Gluc 1, Gluc 2, and Gluc 3, which showed similar pH optima and their molecular weight was estimated to be 74,000, 58,600, and 61,400, respectively (Sandhya and Pandey 2005).

The purification and characterization of the pullulanases produced from microbial sources has traditionally followed conventional methods as for alpha-amylase and glucoamylase, that is, precipitation, ion-exchange chromatography, and gel filtration, although hydrophobic interaction and affinity-based chromatography can also be used. Table 7.2C lists the different methods of purification of pullulanase from various microbial sources. It can be observed that most of the purification processes involve the multistep sequences of the chromatographic methods, which are expensive and time consuming. In most cases, these are coupled with a lower recovery percentage. A critical overview shows that no particular method or sequence of methods should be followed to purify pullulanase. Purification using affinity chromatography usually involves maltotriose, pullulan sepharose, octyl sepharose, and schardinger dextrin. Monoclonal antibodies for specific selection of pullulanase were also described (Enevoldsen, Reimann, and Hansen 1977). These methods describe

TABLE 7.2C
Biochemical Characteristics of Pullulanases

Organism	Optimum pH	Optimum Temperature (°C)	pH Stability	Temperature Stability (°C)	No. of Amino Acids	Mol. Weight (Da)
Aerobacter aerogenes	6	50	4.5–12	50	–	–
Anaerobranca gottschalkii	8	70		70	865	98,973
Bacillus acidopullulyticus	5	65	4–8.5	55	–	–
Bacillus cereus	6	50	5–9	50	853	97,502
Bacillus circulans	7	50	5–8	50	–	–
Bacillus flavocaldarius	–	–	–	–	475	53,875
Bacteroides thetaiotaomicron	6.5	–	–	–	–	75,000
Bacillus macerans	–	55	–	50	–	–
Caldicellulosiruptor saccharolyticus	–	–	–	–	825	95,700
Clostridium thermohydrosulfuricum	5.5	60	3–5	85	–	–
Clostridium thermosaccharolyticum	5	90	4.5–10.5	90	–	–
Desulfurococcus mucosus	5	85	–	90	–	–
Fervidobacterium pennavorans	6	85	–	90	–	
Bacillus stearothermophilus	6	60	6–8.5	50	–	62,000
Klebsiella aerogenes	6	60	3–11	55	1,096	119,336
Klebsiella pneumoniae	–	–	3–11	55	1,090	118,098
Klebsiella oxytoca	–	–	–	–	–	11,600
Micrococcus sp.	8	50	–	–	–	–
Pyrococcus woesei	6	100	–	100	404	46,875
Rhodothermus marinus	–	80	–	85	–	–
Streptococcus pneumoniae	–	–	–	–	759	86,517
Streptococcus pyogenes	–	–	–	–	1,165	128,898

Thermoactinomyces thalpophilus	7	70	–	80	–	–
Thermoanaerobacter ethanolicus	5	80	5–5.5	75	1,481	166,363
Thermoanaerobacter finnii	5	90	–	–	–	–
Thermoanaerobium sp.	5.5	–	3.5–8	80	–	–
Thermobacteroides acetoethylicus	5	90	–	–	–	–
Thermoanaerobacter saccharolyticum	–	–	–	–	1,279	142,431
Thermoanaerobacter thermohydrosulfuricus	–	–	–	–	113	13,763
Thermus aquaticus	6.5	–	–	–	–	–
Thermus caldophilus	5.5	75	–	95	–	–
Thermus thermophilus	–	70	–	80	416	47,041
Thermococcus aggregans	6.5	95	–	–	–	–
Thermotoga maritime	–	–	–	–	843	96,262

Adapted from Nair (2006).

one-step purification of pullulanase and amylases from standard preparation of the enzyme. Techniques like macroaffinity ligand facilitate three-phase partitioning for the pullulanase and glucoamylase purification with alginates (Nair, Singhal, and Kamat 2006, 2007).

REFERENCES

Bertoldo, C. and G. Antranikian. 2001. Amylolytic enzymes from hyperthermophiles. *Hyperthermophilic Enzymes* 330: 269–289.

Bogar, B., G. Szakacs, R. P. Tengerdy, J. C. Linden, and A. Pandey. 2002. Production of alpha amylase with *Aspergillus oryzae* on spent brewing grains by solid-state fermentation. *Applied Biochemistry and Biotechnology* 102-103: 453–463.

Crueger, W. and A. Crueger. 1989. *Industrial Microbiology*, 2nd ed. Sunderland, MA: Sinauer Associates.

Doman-Pytka, M. and J. Bardowski. 2004. Pullulan degrading enzymes of bacterial origin. *Critical Reviews in Microbiology* 30: 107–121.

Enevoldsen, B., L. Reimann, N. L. Hansen. 1977. Biospecific affinity chromatography of pullulanase. *FEBS Letters*. 79(1): 121–124.

Francis, F., A. Sabu, K. M. Nampoothiri, G. Szakacs, and A. Pandey. 2002. Synthesis of alpha-amylase by *Aspergillus oryzae* in solid-state fermentation. *Journal of Basic Microbiology* 42(3): 322–326.

Francis, F., A. Sabu, K. M. Nampoothiri, S. Ramachandran, S. Ghosh, G. Szakacs, and A. Pandey. 2003. Use of response surface methodology for optimising process parameters for the production of alpha amylase by *Aspergillus oryzae*. *Biochemical Engineering Journal* 15: 107–115

Gupta, R., P. Gigras, H. Mohapatra, V. K. Goswami, and B. Chauhan. 2003., Microbial amylases: A biotechnological perspective. *Process Biochemistry* 38: 1599–1616.

Guzman-Maldonado, M. and O. Paredas-Lopez. 1995. Amylolytic enzymes and products derived from starch: A review. *Critical Reviews in Food Science and Nutrition* 35(5): 373–403.

Henrissat, B. 1991. A classification of glycosyl hydrolases based on amino acid sequence similarities. *Biochemical Journal* 280(2): 309–316.

Jensen, B., and J. Olsen. 1992. Physicochemical properties of a purified α-amylase from the termophilic fungus *Thermomyces lanuginosus*. *Enzyme and Microbial Technology*. 14: 112–116.

Kuriki, T. and T. Umanaka. 1999. The concept of the α-amylase family: Structural similarity and common catalytic mechanism. *Journal of Bioscience and Bioengineering* 87(5): 557–565.

Nair, S. U. 2006. Studies on microbial pullulanase. PhD thesis, University of Mumbai, Mumbai, India.

Nair, S. U., R. S. Singhal, and M. Y. Kamat. 2006. Enhanced production of pullulanase-type 1 using new isolate of *Bacillus cereus* FDTA 13 and its mutant. *Food Technology and Biotechnology* 44(2): 275–282.

Nair, S.U., R. S. Singhal, and M. Y. Kamat. 2007. Induction of pullulanase production in *Bacillus cereus* FDTA-13. *Bioresource Technology* 98(4): 856–859.

Pandey, A. 1992. Recent developments in solid-state fermentation. *Process Biochemistry* 27: 109–117.

Pandey, A. 1994. Solid-state fermentation: An overview. In *Solid-State Fermentation*, ed. A. Pandey, 3–10. New Delhi, India: Wiley Eastern.

Pandey, A. 1995. Glucoamylase research: An overview. *Starch/Starke* 47(11): 439–445.

Pandey, A. 2003. Solid-state fermentation. *Biochemical Engineering Journal* 13: 81–84.

Pandey, A. and S. Radhakrishnan. 1992. Packed bed column bioreactor for enzyme production. *Enzyme and Microbial Technology* 14: 486–488.

Pandey, A. and S. Radhakrishnan. 1993. The production of glucoamylase by *Aspergillus niger* NCIM 1245. *Process Biochemistry* 28: 305–309.

Pandey, A., L. Ashakumary, P. Selvakumar, and K. S. Vijaylakshmi. 1995. Effect of yeast extract on glucoamylase synthesis by *Aspergillus niger* in solid-state fermentation. *Indian Journal of Microbiology* 35: 335–338.

Pandey, A., P. Nigam, C. R. Soccol, V. T. Soccol, D. Singh, and R. Mohan. 2000a. Advances in microbial amylases. *Biotechnology and Applied Biochemistry* 31: 135–152.

Pandey, A., C. R. Soccol, and V. T. Soccol. 2000b. Biopotential of immobilized amylases. *Indian Journal of Microbiology* 40(1): 1–14.

Pandey, A., F. Francis, A. Sabu, and C. R. Soccol. 2004. General aspects of solid-state fermentation. In *Concise Encyclopedia of Bioresource Technology*, ed. A. Pandey, 702–708. New York: Haworth Press.

Patel, A. K., K. M. Nampoothiri, S. Ramachandran, G. Szakacs, and A. Pandey. 2005. Partial purification and characterization of alpha amylase produced by *Aspergillus oryzae* using spent-brewing grains. *Indian Journal of Biotechnology* 4(3): 336–341.

Ramachandran, S., A. K. Patel, K. M. Nampoothiri, C. Sandhya, G. Szakacs, C. R. Soccol, and A. Pandey. 2004a. Alpha amylase from a fungal culture grown on oil cakes and its properties. *Brazilian Archives of Biology and Technology* 47(2): 309–318.

Ramachandran, S., A. K. Patel, K. M. Nampoothiri, F. Francis, V. Nagy, G. Szakacs, and A. Pandey. 2004b. Coconut oil cake: A potential raw material for the production of alpha amylase. *Bioresource Technology* 93(2): 169–174

Reddy, N. S., A. Nimmagadda, and K. R. S. S. Rao. 2003. An overview of the microbial α-amylase family. *African Journal of Biotechnology* 2(12): 645–648.

Reddy, P. R. M., M. Y. Swamy, and G. Seenayya. 1998. Purification and characterization of thermostable β-amylase and pullulanase from high yielding clostridium thermosulfurogenesis. *World Journal of Microbiology and Biotechnology* 14: 89–94.

Sandhya, C. and A. Pandey. 2005. Microbial glucoamylase. In *Microbial Diversity*, ed. T. Satyanarayana and B. N. Johry, 679–694. New Delhi: IK International Publishers.

Satyanarayana, T., Uma Maheswar, J. L. Rao, and M. Ezhilvannan. 2004. ⊠-Amylases. In *Enzyme Technology*, ed. A. Pandey, C. Webb, C. R. Soccol, and C. Larroche, 221–238. New Delhi: Asiatech Publishers.

Selvakumar, P., L. Ashakumary, and A. Pandey. 1998. Biosynthesis of glucoamylase by *Aspergillus niger* in solid-state fermentation using tea waste as the basis of a solid substrate. *Bioresource Technology* 65: 83–85.

Sivaramakrishnan, S., D. Gangadharan, K. M. Napoothiri, C. R. Soccol, and A. Pandey. 2006. Alpha amylase from microbial sources: An overview on recent developments. *Food Technology and Biotechnology* 44(2): 173–184.

Soccol, C. R., P. J. Rojan, A. K. Patel, A. L. Woiciechowski, L. P. S. Vandenberghe, and A. Pandey. 2004. Glucoamylase. In *Enzyme Technology*, ed. A. Pandey, C. Webb, C. R. Soccol, and C. Larroche, 221–238. New Delhi: Asiatech Publishers.

8 Bioethanol from Starchy Biomass

Part II Hydrolysis and Fermentation

Sriappareddy Tamalampudi,
Hideki Fukuda, and Akihiko Kondo

CONTENTS

ABSTRACT

Bioethanol, which is derived from starchy and cellulosic biomass, is becoming important as an alternative fuel due to diminishing petroleum resources and environmental impacts. Acid and enzymatic methods have been developed for the hydrolysis of starchy biomass in order to release fermentable sugars. Acid hydrolysis results in the production of unnatural compounds that have adverse effects on yeast fermentation. In enzymatic hydrolysis of starch, the biomass has to be cooked at high temperatures and large amounts of amylolytic enzymes have to be added to hydrolyze the starchy biomass prior to fermentation. Recent advances in yeast cell surface engineering developed the strategies to genetically immobilize amylolytic enzymes like α-amylase and glucoamylase on the yeast cell surface. As a means of reducing the cost of ethanol production, flocculent and nonflocculent yeast strains co-displaying amylolytic enzymes have been developed and used successfully for direct ethanol production from raw starch. Hence, the cell surface engineered yeast appears to have great potential in industrial application.

8.1 INTRODUCTION

The utilization of biomass as the starting material for various chemicals and for the production of biofuels has received considerable interest in recent years. Starchy and cellulosic materials of plant origin are the most abundant utilizable biomass resources. Starchy biomass has to be hydrolyzed either by enzymatic or acid hydrolysis to release fermentable sugars. However, acid hydrolysis results in the formation of by-products such as levulinic acid and formic acid which have adverse effects on yeast growth during the fermentation process (Kerr 1944). The enzymatic hydrolysis of starchy material for ethanol production via fermentation consists of two or three steps and requires improvement if it is to realize efficient production at low cost. There are two main reasons for the present high cost: one is that starchy materials need to be cooked at a high temperature (140 to 180°C) to obtain high ethanol yield and the other is that large amounts of amylolytic enzymes, namely glucoamylase (EC 3.2.1.3) and α-amylase, need to be added. To reduce the energy cost of cooking starchy materials, previously reported noncooking and low-temperature cooking fermentation systems have succeeded in reducing energy consumption by approximately 50% (Matsumoto et al. 1982), but it is still necessary to add large amounts of amylolytic enzymes to hydrolyze the starchy materials to glucose.

Many researchers have reported attempts to resolve this problem by using recombinant glucoamylase-expressing yeasts with the ability to ferment starch to ethanol directly (Ashikari et al. 1989; Inlow, McRae, and Ben-Bassat 1988). Recombinant yeast that co-produces glucoamylase and α-amylase has been developed to further improve the efficiency of starch fermentation (Birol et al. 1998; De Moreas, Astolfi-Filho, and Oliver 1995; Eksteen et al. 2003). Recent advances in yeast cell surface engineering provided the tools for the display of amylolytic enzymes which allows the utilization of yeast whole-cell biocatalyst for direct ethanol production from starch. Moreover, integration of hydrolysis and fermentation steps by arming yeast cells can reduce the unit operations compared to that of hydrolysis by acids and isolated enzymes (Figure 8.1). This review summarizes the work on cell surface engineering systems that demonstrated direct ethanol production from soluble starch, low-temperature cooked starch, and raw starch.

8.2 YEAST CELL SURFACE ENGINEERING: A TOOL FOR DIRECT ETHANOL PRODUCTION FROM STARCH

The cell surface is a functional interface between the inside and outside of the cell. Some surface proteins extend across the plasma membrane and others are bound by noncovalent interactions to the cell surface components. Cells have systems for anchoring surface-specific proteins and for confining surface proteins to particular

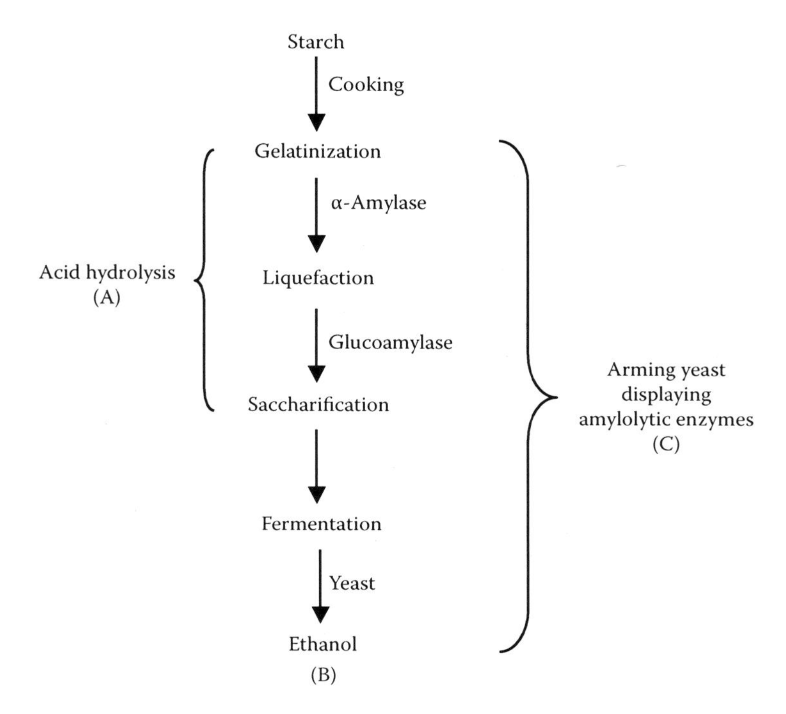

FIGURE 8.1 Schematic diagram of starch hydrolysis and ethanol fermentation using different methods. (a) Acid hydrolysis, (b) enzymatic hydrolysis, and (c) arming yeast displaying amylolytic enzymes.

domains on the cell surface. In biotechnology, the cell surface can be exploited by making use of known mechanisms for the transport of proteins to the cell surface. In particular, *Saccharomyces cerevisiae* is useful as a host for genetic engineering, because it allows the folding and glycosylation of expressed heterologous eukaryotic proteins and can be subjected to many genetic manipulations. Moreover, the yeast can be cultivated to a high density in an inexpensive medium, so that the display of enzymes on yeast cell surface has several applications in bioconversion processes.

Many glucoamylase-extractable proteins on the yeast cell surface, for example, agglutinin (Agα1 and Aga1) and Flocculin Flo1, Sed1, Cwp1, Cwp2, Tip 1, and Tir 1/Srp 1 have glycosylphosphotidylinositol (GPI) anchors which play an important role in the expression of cell surface proteins (Roy et al. 1991; Watari et al. 1994). GPI anchored proteins contain hydrophobic peptides at their C-termini. After the completion of protein synthesis, the precursor protein remains anchored in the endoplasmic reticulum (ER) membrane by the hydrophobic carboxyl-terminal sequence, with the rest of the protein in the ER lumen. Within less than a minute, the hydrophobic carboxyl-terminal sequence is cleaved at the site and concomitantly replaced with a GPI anchor, presumably by the action of a transamidase (Ueda and Tanaka 2000).

Among the GPI anchor proteins α-agglutinin and Flocculin anchors are demonstrated to be suitable for the expression of hydrolytic enzymes. The molecular-level information on α-agglutinin is utilized to target the heterologous proteins of biotechnological importance to the outermost glycoprotein layer of the cell wall. In the α-agglutinin system, the C-terminal half of the α-agglutinin containing the GPI anchor attachment signal was used to anchor the heterologous proteins on the yeast cell surface (Capellaro et al. 1991). In the case of the flocculin system two types of cell surface display methods were developed. In one system, the C-terminal region of Flo1p, contains a GPI-attachment signal; the second system, by contrast, attempts to utilize the ability of the flocculation functional domain of Flo1p to create a novel surface display apparatus (Kondo and Ueda 2004).

8.3 ETHANOL PRODUCTION FROM SOLUBLE STARCH

8.3.1 Displayed Glucoamylase

Surface expression of the amylolytic enzymes was initiated by the pioneering work of Murai et al. (1997). They reported the strategy of developing recombinant *S. cerevisiae* displaying amylolytic enzymes. The multi-copy plasmid pGA11 (Figure 8.2) was used for the expression of glucoamylase/α agglutinin fusion gene containing the secretion signal sequence of the glucoamylase under the control of the GAPDH promoter and was introduced into the *S. cerevisiae* MT8-1 as host strain. The displayed glucoamylase is from *Rhizopus oryzae*, an exo-type amylolytic enzyme, cleaving α-1,4-linked and α-1,6-linked glucose effectively from starch. The anchoring of the fusion gene on the cell wall of recombinant yeast harboring the plasmid pGA11 was demonstrated by immunofluorescence labeling of the cells with anti-glucoamylase IgG (Murai et al. 1997; Ueda et al. 1998).

Kondo et al (2002) used flocculating yeast strain YF207 for the surface expression of glucoamylase. The yeast strain YF207 is a tryptophan auxotroph with a strong flocculation ability which was obtained from *Saccharomyces diastaticus* ATCC60712 and *S. cerevisiae* W303-1B by tetrad analysis and was transformed with pGA11

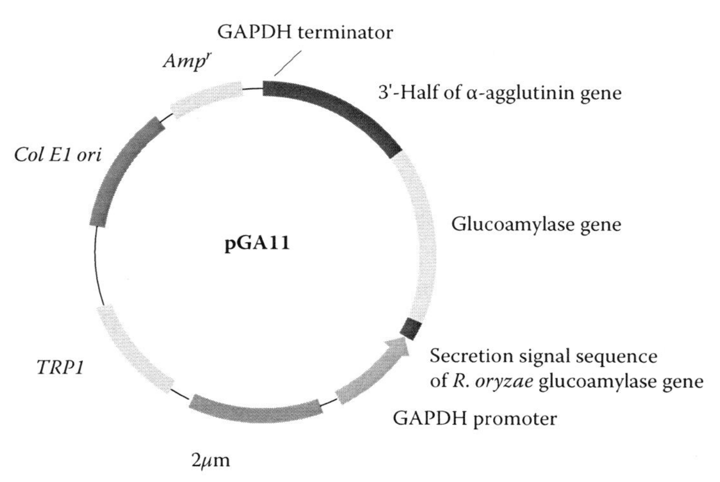

FIGURE 8.2 Schematic representation of the expression plasmid for glucoamylase/α-agglutinin fusion gene.

constructed in the previous study (Murai et al. 1997). The cell surface glucoamylase does not show any effect on flocculation ability during growth and ethanol fermentation phases. Moreover, the glucoamylase activity displayed on the surface of flocculent yeast strain was similar to that displayed on nonflocculent yeast cells. Therefore, the flocculent yeast cells displaying glucoamylase possess both strong flocculation ability and glucoamylase activity; and hence they are considered more advantageous in industrial processes for ethanol production from starchy materials.

The results shown in Figures 8.3a and 8.3b demonstrate that the cell-surface glucoamylase is effective for direct ethanol fermentation from soluble starch, because high ethanol fermentation from soluble starch was obtained. In previous studies using recombinant *S. cerevisiae* secreting glucoamylase (Nakamura et al. 1997; Briol et al. 1998) both cell growth and fermentation were performed under anaerobic or minimal aerobic conditions; and hence over 150 h was necessary to attain ethanol concentrations of 20 to 30 g/l. Ideally, a large cell mass should be obtained by high-density cell culture under aerobic conditions and cells harvested by sedimentation were used for the ethanol fermentation. However, in secretory expression of amylolytic enzymes, this approach is not suitable because inoculated cells should produce a sufficient amount of amylases before ethanol fermentation. In the study by Kondo et al. (2002), recombinant yeast strain YF207/pGA11 displaying glucoamylase gene maintained a high ethanol production rate (approximately 0.6 to 0.7 g l^{-1} h^{-1}) during repeated utilization for fermentation over 300 h. This is attributable to high plasmid stability during growth and fermentation phases, even though pGA11 is a multi-copy-type plasmid, based on pYE22m. The plasmid stability in cells cultivated in YPS medium was found to be higher than in cells cultivated with YPD medium. Since host cells could not metabolize soluble starch, the utilization of soluble starch as the carbon source would be a selection pressure for the yeast cells bearing plasmids. In the case of glucoamylase-displaying yeast cells, glucose was maintained at a very low concentration and, at the same time, a high ethanol production rate was achieved. This might be because the recombinant yeast cells metabolize the glucose as soon as glucose is released from soluble starch by the glucoamylase displayed yeast cells. However, a high ethanol production rate was obtained because local glucose concentration near the yeast cell surface was probably higher than that in the fermentation medium. This low concentration of glucose in the fermentation medium is advantageous in minimizing the risk of contamination.

8.3.2 Co-Displayed Glucoamylase and Amylase

Studies show the display of only glucoamylase leads to the accumulation of insoluble starch during fed-batch fermentation, because of the lack of liquefying enzyme α-amylase. In order to overcome this problem, Shigechi et al. (2002), developed two recombinant yeast strains co-expressing glucoamylase and α-amylase. Plasmids for the surface expression (pAA12) and secretory expression (pSAA11) of *Bacillus stearothermophilus* α-amylase were constructed and co-transformed into the flocculent yeast strain YF207 along with the plasmid pGA11 for cell surface display of *R. oryzae* glucoamylase. The ethanol productivity by these two strains was examined by fed-batch fermentations using soluble potato starch as substrate. The amylolytic

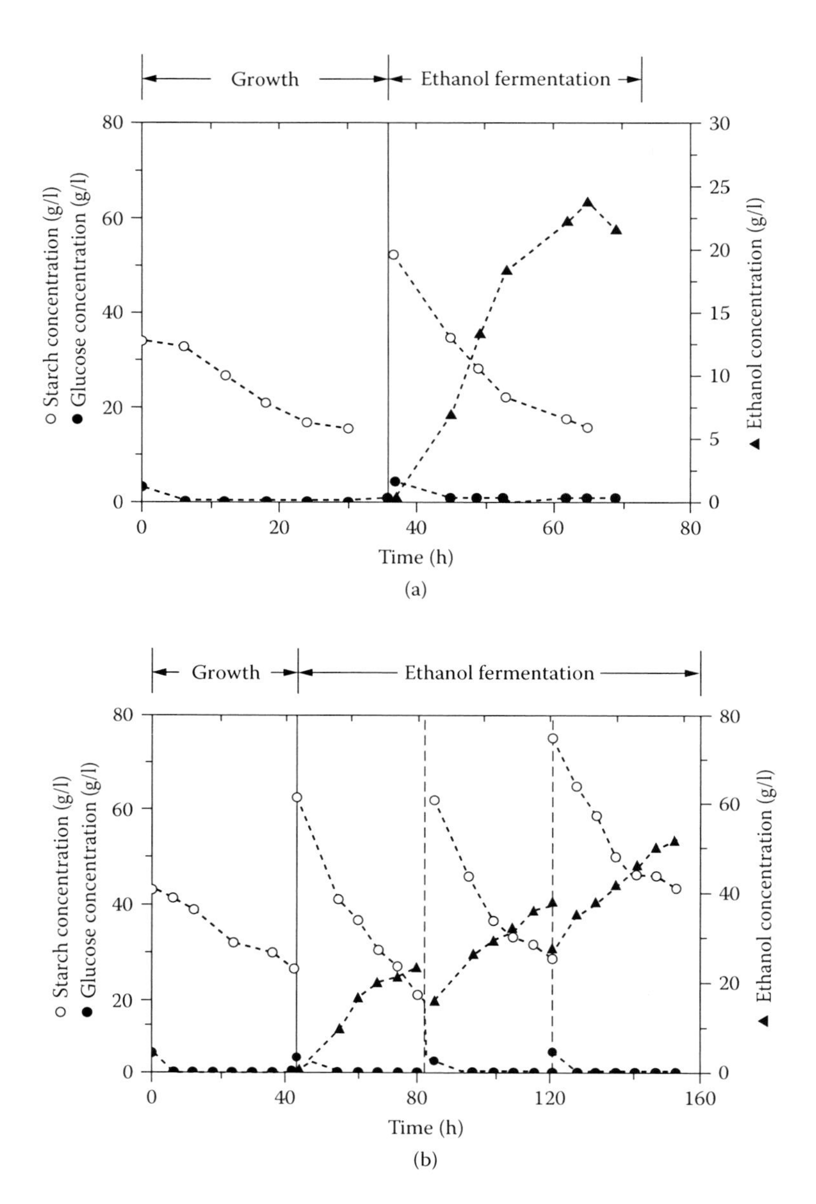

FIGURE 8.3 (a) Batch fermentation of starch to ethanol by YF207/pGA11. YF207/pGA11 cells were grown under aerobic conditions (2.0 ppm), harvested, and used for batch fermentation. The *left side* of the *solid line* in the figure is the growth phase and the *right side* is the ethanol-fermentation phase. (b) Fed-batch fermentation by YF207/pGA11. YF207/pGA11 cells were grown under aerobic conditions (2.0 ppm), harvested, and used for fed-batch fermentation under anaerobic conditions. The *left side* of the *solid line* in the figure is the growth phase and the *right side* is the ethanol-fermentation phase.

activity was detected by both the strains and flow cytometric analysis confirmed the successful co-expression of glucoamylase and α-amylase in strains YF207/ [pGA11, pAA12] and YF207/ [pGA11, pSAA11].

As shown in Figures 8.4a and 8.4b, both recombinant strains YF207/ [pGA11, pAA12] and YF207/ [pGA11, pSAA11] grew faster in the growth phase than the glucoamylase displaying yeast YF207/pGA11. The activities of glucoamylase and α-amylase displayed on the cell surface were maintained with YF207/ [pGA11, pAA12] during the ethanol fermentation phase, whereas in the case of YF207/ [pGA11, pSAA11] strain the secreted α-amylase was accumulated. The ethanol concentration produced reached 60 g l^{-1} after 100 h of fermentation by both strains. But glucose concentration is slightly higher in the culture medium of YF207/ [pGA11, pSAA11] strain. This is probably due to the secretion of α-amylase, which decomposes starch in the culture medium.

In addition, the flocculation ability of the yeast strain co-expressing glucoamylase and α-amylase did not change during the fed-batch fermentation and was almost the same as that of the yeast strains YF207 and YF207/pGA11. This finding suggested the co-display of two amylolytic enzymes on the cell surface does not influence the flocculation ability of yeast cells.

8.4 ETHANOL PRODUCTION FROM LOW-TEMPERATURE COOKED CORN STARCH

In direct ethanol production, noncooking and low-temperature cooking fermentation systems have several advantages over conventional high-temperature cooking (140 to 180°C) process (Matsumoto et al. 1982) because high-temperature cooking requires high energy and the addition of large amounts of amylolytic enzymes. Shigechi et al. (2000) performed direct ethanol production in a single step using corn starch cooked at low temperature (80°C) as the sole carbon source instead of soluble starch using yeast strains displaying amylolytic enzymes. The productivity of ethanol from corn starch cooked at low temperature was investigated by using the recombinant yeast strains that were developed in their previous study, that is, yeast strains displaying only glucoamylase on the cell surface (YF207/pGA11) or yeast strains displaying glucoamylase and either co-displaying (YF207/pGA11/pAA12) or secreting (YF207/pGA11/pSAA11) α-amylase.

The ethanol production rate increased markedly under co-expression of glucoamylase and α-amylase compared with the yeast strain displaying only glucoamylase (Figure 8.5). Specifically, by co-displaying two amylolytic enzymes on the cell surface, strain YF207/pGA11/pAA12 was able to produce ethanol more rapidly than strain YF207/pGA11/pSAA11 and without time lag. These results indicated that α-amylase, which hydrolyzes α-1,4 linkages of starch in a random fashion, plays a very important role in efficient hydrolyzation of corn starch. It is probable that the cooperative and sequential reaction of two enzymes is crucial for efficient utilization of corn starch. Because the yeast strain YF207/pGA11/pAA12 possesses enough glucoamylase and α-amylase activity in the initial stages of cultivation and fermentation, it grows fast, produces ethanol without time lag, and achieves maxi-

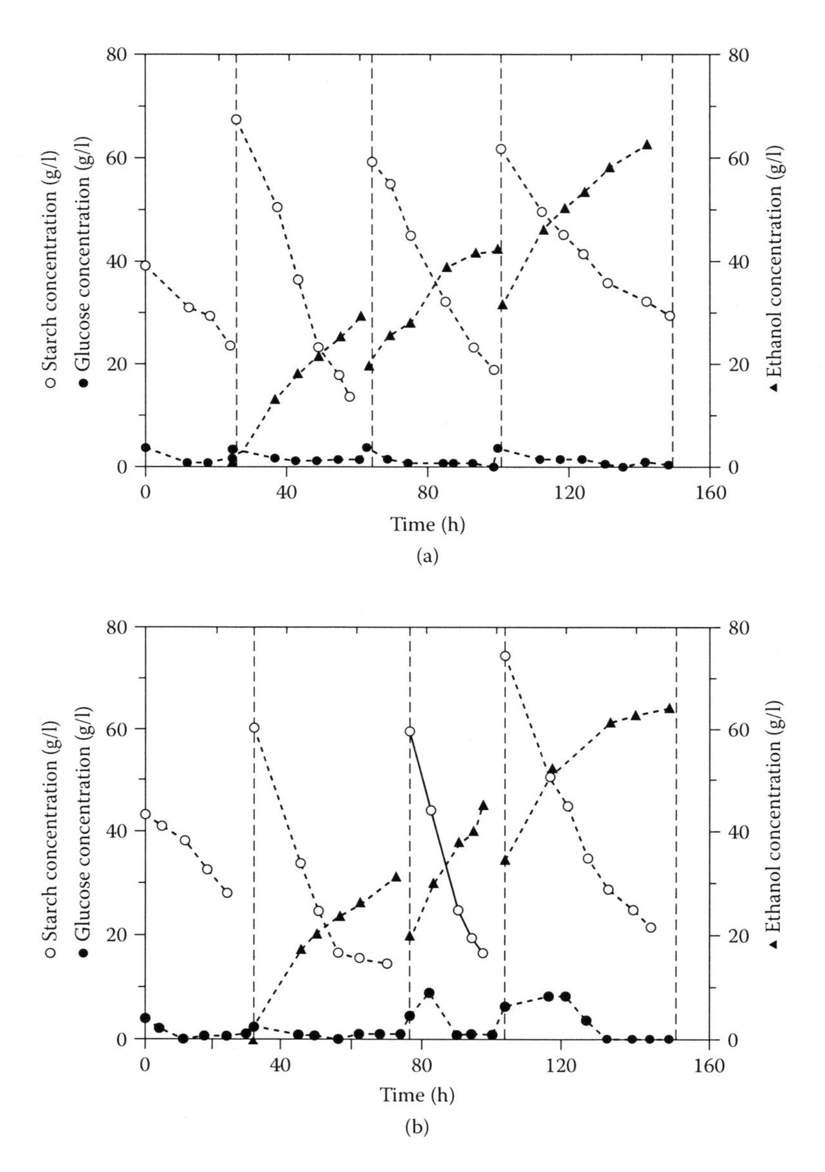

FIGURE 8.4 (a) Fed-batch fermentation of starch to ethanol by YF207/[pGA11, pAA12]. YF207/[pGA11, pAA12] cells were grown under aerobic conditions (2.0 ppm), harvested, and used for fed-batch fermentation under anaerobic conditions. To the left of the solid line in the figure is the growth phase and to the right the ethanol fermentation phase. (b) Fed-batch fermentation of starch to ethanol by YF207/[pGA11, pSAA11]. YF207/[pGA11, pSAA11] cells were grown under aerobic conditions (2.0 ppm), harvested, and used for fed-batch fermentation under anaerobic conditions.

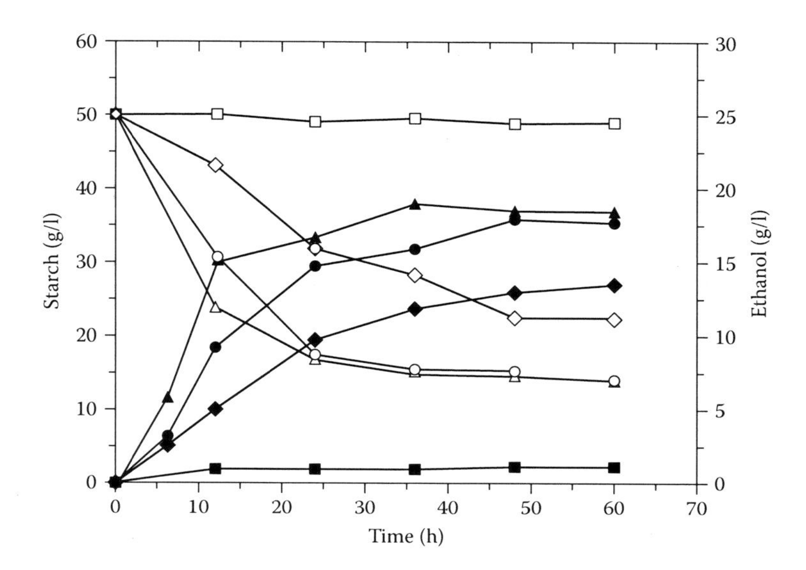

FIGURE 8.5 Time course of anaerobic ethanol fermentation from 50 g/l corn starch cooked at low temperature. Each group of cells was aerobically cultivated for 48 h on SDC medium, harvested, and used in ethanol fermentation with YPS medium cooked at 80°C for 5 min. (□,■). YF207; (◇,◆) YF207/pGA11; (△,▲). YF207/pGA11/pAA12; (○,●) YF207/pGA11/pSAA11. Open and closed symbols show the starch and ethanol concentrations, respectively.

mum ethanol concentration (18 g/l) within the short time (36 h) of the recombinant yeast strains.

A comparison of conventional high-temperature and low-temperature cooking fermentation systems using the yeast strain YF207/pGA11/pAA12 co-displaying glucoamylase and α-amylase (Figure 8.6) shows maximum ethanol concentration, ethanol-production rate, and substrate consumption rate were almost the same in the two fermentation systems. In high- and low-temperature cooking systems, the yield of ethanol produced was 0.50 g per gram of carbohydrate consumed. This corresponds to 97.2% of theoretical yield (0.51 g of ethanol per gram glucose). This indicates that the low-temperature cooking fermentation system based on YF207/pGA11/pAA12 is cost effective in direct fermentation of corn starch.

8.5 ETHANOL PRODUCTION FROM RAW CORN STARCH

The isolation of amylase enzyme from lactic acid bacteria *Streptococcus bovis* opened new horizons for the efficient hydrolysis of raw starch (Satoh et al. 1993). Shigechi et al. developed a novel noncooking fermentation system for direct ethanol production from raw corn with yeast strain YF207 that co-displayed *R. oryzae* glucoamylase and *S. bovis* 148 α-amylase by using the C-terminal half of α-agglutinin (pBAA1) and the flocculation domain of Flo1p (pUFLA) as anchor proteins (Figures 8.7a and 8.7b).

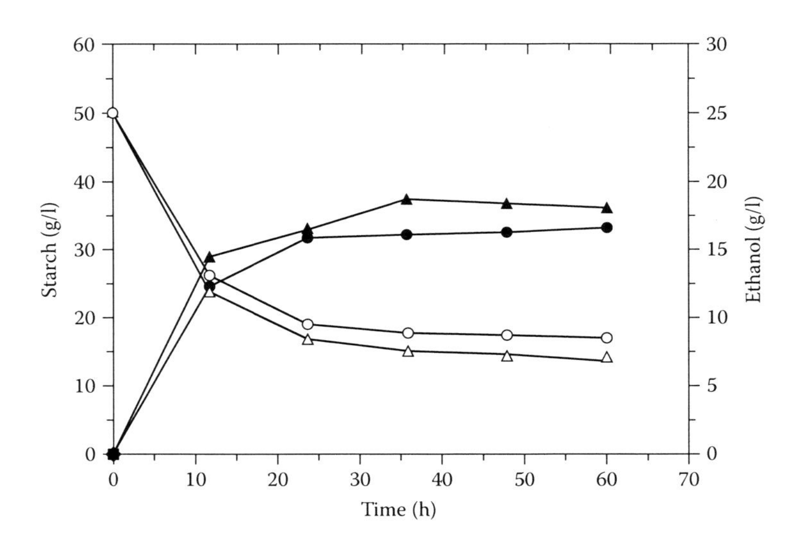

FIGURE 8.6 Comparison of ethanol production between high- and low-temperature cooking fermentation systems using yeast strain YF207/ pGA11/pAA12. YPS medium was cooked at 120°C for 20 min or at 80°C for 5 min. Ethanol fermentation started with initial starch concentration of 50 g/l. (○,●).High-temperature cooking fermentation system; (△,▲)low-temperature cooking fermentation system. Open and closed symbols show starch and ethanol concentrations, respectively.

The α-amylase and glucoamylase activities confirmed the display of both amylolytic enzymes (Table 8.1). In glucoamylase-displaying yeast strains there is not much difference in the activities; whereas in the case of α-amylase-displaying yeast strains, the activity is dependent on the anchor protein. The yeast strains YF207/pBAA1 and YF207/pGA11/pUFLA which uses Flo1 anchor showed 40 times higher α-amylase activity than the yeast strains using a-agglutin anchor. It has been reported that several α-amylases have raw starch binding abilities and that the starch digesting domain is located in the C-terminal region (Lo et al. 2002). The two recombinant yeast strains YF207/pGA11/pBAA1 and YF207/pGA11/pUFLA were used in direct ethanol production from raw corn starch. The raw corn starch, which corresponds to 200 g of total sugar per liter, was used as the sole carbon source. As shown in Figure 8.8, strain YF207/pGA11, displaying only glucoamylase, and strains YF207/pBAA1 and YF207/pUFLA, displaying only α-amylase, produced almost no ethanol, while soluble sugar accumulated in the fermentation medium of strain YF207/pUFLA due to degradation of corn starch to oligosaccharides by the surface-displayed α-amylase. Although strain YF207/pGA11/pBAA1, co-displaying glucoamylase and α-amylase via α-agglutinin, did produce ethanol from the raw corn starch, the ethanol yield was low (23.5 g l^{-1}) after 72 h of fermentation.

On the other hand, the yeast strain co-displaying glucoamylase and α-amylase using α-agglutinin and Flo1P (YF207/pGA11/pUFLA) was able to produce ethanol directly from the raw corn starch without the addition of commercial enzymes. The concentration of raw corn starch decreased drastically during the fermentation, as the ethanol concentration increased to 61.8 g l^{-1} after 72 h of fermentation. A

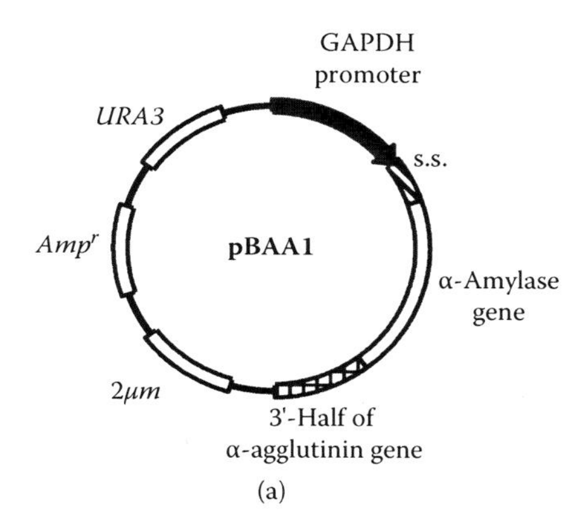

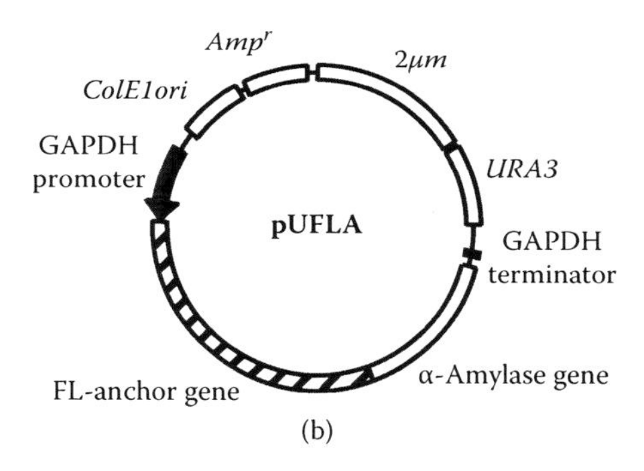

FIGURE 8.7 Expression plasmids for cell surface display of *S. bovis* α-amylase. (a) Plasmid pBAA1 for C-terminal immobilization using the α-agglutinin-based surface display system; (b) plasmid pUFLA for N-terminal immobilization using the Flo1p-based surface display system. s.s., secretion signal sequence of *R. oryzae* glucoamylase gene.

TABLE 8.1
Glucoamylase and α-Amylase Activities of Yeast Strains Carrying Different Plasmids

Strains	Glucoamylase Activity[a]	α-Amylase Activity[a]
YF207	ND[b]	ND[b]
YF207/pGA11	42.5	ND[b]
YF207/pBAA1	ND[b]	2.52
YF207/pUFLA	ND[b]	90.1
YF207/pGA11/pBAA1	45.9	2.38
YF207/pGA11/pUFLA	57.0	114

[a] Both activities shown as U/g (wet weight) of cells; values are averages of three independent experiments.

[b] ND, not detected.

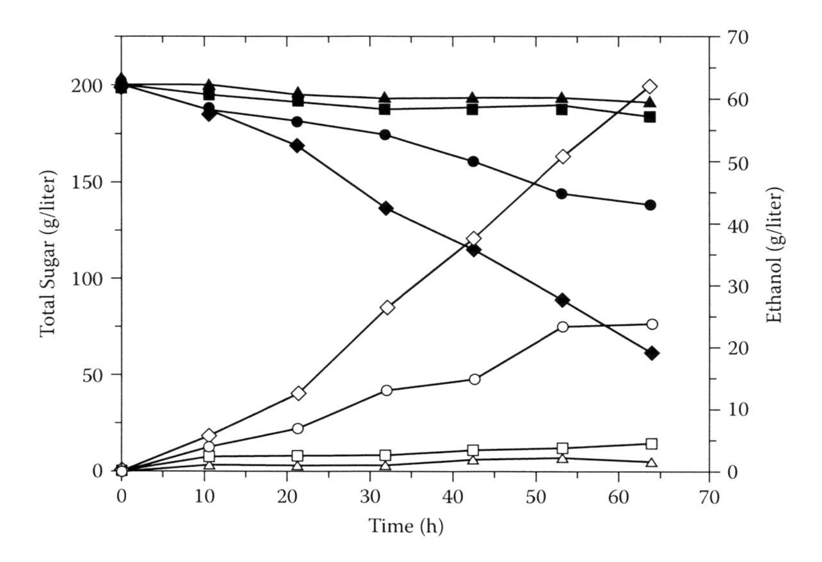

FIGURE 8.8 Time course of direct ethanol production via fermentation from raw corn starch, which corresponds to 200 g of total sugar as the sole carbon source per liter using 100 g (wet weight) of cells of yeast strains *S. cerevisiae* YF207/pGA11 (squares), YF207/pBAA1 (triangles), YF207/pUFLA (inverted triangles), YF207/pGA11/pBAA1 (circles), and YF207/pGA11/pUFLA (diamonds) per liter. Open and closed symbols show ethanol and total sugar concentrations, respectively. Data are averages from three independent experiments.

reduction in the particle size and the number of corn starch granules during fermentation was observed by microscopy (Figure 8.9). The yield in terms of grams of ethanol produced per gram of sugar consumed was 0.44 g/g, which corresponds to 86.5% of theoretical yield (0.51 g of glucose consumed per gram). No glucose was detected in the fermentation medium. The yeast strain YF207/pGA11/pUFLA maintained almost the same glucoamylase and α-amylase activities during fermentation.

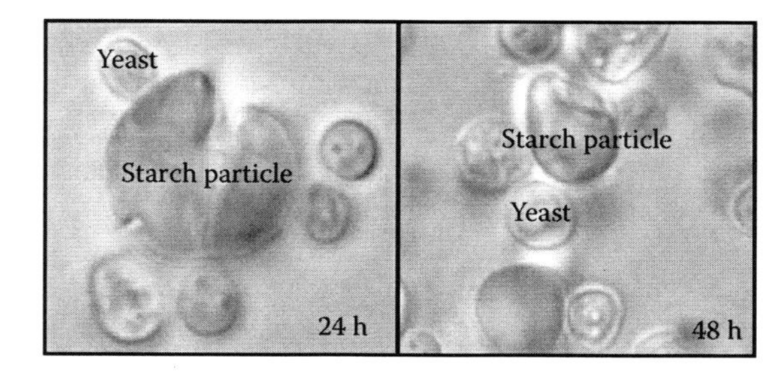

FIGURE 8.9 Time course of starch morphology: microscopic analysis of starch granules on treatment with the yeast cells displaying amylolytic enzymes at 24 and 48 h.

8.6 EVALUATION OF SURFACE ENGINEERED YEAST STRAINS

In order to evaluate the surface engineered yeast strains, Seong et al. (2005, 2006) compared ethanol production from raw corn starch using different yeast strains. They employed yeast strains displaying glucoamylase and co-displaying or secreting α-amylase for the direct conversion of raw starch because the conversion of starch to oligosaccharides by α-amylase is the rate-limiting step in the direct fermentation of raw starch to produce ethanol by arming yeast. In their study, the nonflocculent and flocculent strains that either display or secrete α-amylase were compared with respect to their performance.

The nonflocculent yeast strains secreting α-amylase (0.18 g g-dry cell^{-1} h^{-1}) showed threefold higher specific ethanol production rate than the α-amylase-displaying non-flocculent yeast strain (0.06 g g-dry cell^{-1} h^{-1}). But the specific starch consumption rate in the third batch of fermentation was decreased significantly compared with the first two batches. The decrease in the activity was due to the removal of α-amylase from the culture supernatant at the end of each batch, which leads to a reduction in the productivity of α-amylase in subsequent batches. In contrast, nonflocculent yeast strains displaying α-amylase do not show significant decrease in the specific starch consumption rate throughout the repeated batch fermentations, whereas the specific α-amylase activity decreased gradually. On the other hand, flocculent yeast strains secreting and displaying α-amylase also acted efficiently on the raw starch with a specific ethanol production rate of approximately 0.06 and 0.04 g g-dry cell^{-1} h^{-1}, respectively. The comparatively high ethanol production rate in α-amylase-secreting nonflocculent yeast is because the diameter of the displayed α-amylase is 10^{-6} m which is three orders of magnitude lower than the secreted α-amylase with a diameter of 10^{-9} m (typical diameter of globular proteins) and the rate of association of the raw starch granule is (10^{-5}) with displayed α-amylase is expected to be much lower than secreted α-amylase.

8.7 CONCLUSIONS

The yeast cells displaying amylolytic enzymes have been proved as potential biocatalysts for direct conversion of starchy materials to ethanol. The genes encoding glucoamylase and α-amylase were fused with the anchor genes and were introduced into *S. cerevisiae*. The yeast cells harboring these fused genes were successfully utilized raw and low-temperature cooked starch as the sole carbon source. Moreover, flocculent yeast renders the ethanol fermentation more economical because the recovery of cells from the fermentation medium is easy to accomplish.

It was demonstrated that the specific ethanol production rate of α-amylase-displaying or -secreting yeasts depends on the size and nature of starch granules. In soluble or low-temperature cooked starch, yeast cells displaying α-amylase showed higher ethanol yield than the α-amylase secretion systems since the displayed α-amylase can access most of the small starch molecules. Raw starch yeast cells secreting α-amylase showed better performance in batch fermentations than the α-amylase-displaying yeast cells. But the starch consumption rate of the α-amylase-secreting systems is significantly decreased in the third batch. Even though the yeast cells displaying α-amylase show low performance in batch culture, their efficiency is almost comparable to the α-amylase-secreting yeast strains during repeated batch fermentations. In conclusion, the choice of α-amylase secretion/display depends on the nature of the substrate and on the type of process operation.

REFERENCES

Ashikari, T., N. Kunisaki, T. Matsumoto, T. Amachi, and H. Yoshizumi. 1989. Direct fermentation of raw corn starch to ethanol by yeast transformants containing a modified *Rhizopus* glucoamylase gene. *Appl. Microbiol. Biotechnol.* 32:129–133.

Birol, G., Z. I. Onsan, B. Kirdar, and S. G. Oliver. 1998. Ethanol production and fermentation characteristics of recombinant *Saccharomyces cerevisiae* strains grown on starch. *Enzyme. Microb. Technol.* 22:672–677.

Cappellaro, C., K. Hauser, V. Mrsa, M. Watzele, G. Watzele, C. Gruber, and W. Tanner. 1991. *Saccharomyces cerevisiae* a- and α-agglutinin: Characterization of their molecular interaction. *EMBO J.* 10:4081–4088.

De Moreas, L., S. Astolfi-Filho, and S. G. Oliver. 1995 Development of yeast strains for the efficient utilization of starch: Evaluation of constructs that express alpha amylase and glucoamylase separately or as bifunctional fusion proteins. *Appl. Microbiol. Biotechnol.* 43:1067–1076.

Eksteen, J.M., P. Van Rensburg, R. R. C. Otero, and I. S. Pretorius 2003. Starch fermentation by recombinant *Saccharomyces cerevisiae* strains expressing the α-amylase and glucoamylase genes from *Lipomyces kononenkoae* and *Saccaromyces fibuligera. Biotechnol. Bioeng.* 84:639–646.

Inlow, D., J. McRae, and A. Ben-Bassat. 1988. Fermentation of corn starch to ethanol with genetically engineered yeast. *Biotechnol. Bioeng.* 32:227–234.

Kerr, R. W. 1944. *Chemistry and Industry of Starch: Starch, Sugar and Related Compounds.* New York: Academic Press.

Kondo, A., M. Shigechi, K. Abe, K. Uyama, T. Matsumoto, S. Takahashi, M. Ueda, A. Tanaka, M. Kishimoto, and H. Fukuda. 2002. High level ethanol production from starch by a flocculent *Saccharomyces cerevisiae* strain displaying cell surface glucoamylase. *Appl. Microbiol. Biotechnol.* 58:291–296.

Kondo, A. and M. Ueda. 2004. Yeast cell-surface display: Applications of molecular display. *Appl. Microbiol. Biotechnol.* 64:28–40.

Lo, H. F., L. L. Lin, W. Y. Chiang, M. C. Chie, W. H. Hsu, and C. T. Chang. 2002. Deletion analysis of the C-terminal region of the α-amylase of Bacillus sp. strain TS-23. *Arch. Microbiol.* 178:115–123.

Matsumoto, N., O. Fukunishi, M. Miyanaga, K. Kakihara, E. Nakajima, and H. Yoshizumi. 1982. Industrialization of a non-cooking system for alcoholic fermentation from grains. *Agric. Boil. Chem.* 46:1549–1558.

Murai, T., M. Ueda, Y. Yamamura, H. Atomi, Y. Shibasaki, N. Kamasawa, M. Osumi, T. Amachi, and A. Tanaka. 1997. Construction of a starch utilizing yeast by cell surface engineering. *Appl. Environ. Microbiol.* 63:1362–1366.

Nakamura, Y., F. Kobayashi, M. Ohnaga, and T. Sawada. 1997. Alcohol fermentation of starch by a genetic recombinant yeast having glucoamylase activity. *Biotechnol Bioeng.* 53:21–25.

Roy, A., C. F. Lu, D. L. Marykwas, P. N. Lipke, and J. Kurjan. 1991. The AGA1 product is involved in cell surface attachment of the *Saccharomyces cerevisiae* cell adhesion glycoprotein α-agglutinin. *Mol. Cell Biol.* 11:4196–4206.

Satoh, E., Y. Niimura, T. Uchimura, M. Kozaki, and K. Komagata. 1993. Molecular cloning and expression of two α-amylase genes from *Streptococcus bovis* 148 in *Escherichia coli*. *Appl. Environ. Microbiol.* 59:3669–3673.

Seong, K., Y. Katakura, J. Koh, A. Kondo, M. Ueda, and S. Shioya. 2005. Evaluation of performance of different surface engineered yeast strains for direct ethanol production from raw starch. *Appl. Microbiol. Biotechnol.* 70(5):573–579.

Seong, K., Y. Katakura, K. Ninomiya, Y. Bito, S. Katahira, A. Kondo, M. Ueda, and S. Shioya. 2006. Effect of flocculation on performance of arming yeast in direct ethanol fermentation. *Appl. Microbiol. Biotechnol.* 73:60–66.

Shigechi, H., Y. Fujita, J. Koh, M. Ueda, H. Fukuda, and A. Kondo. 2000. Energy saving direct ethanol production from low temperature cooked corn starch using a cell surface engineered yeast strain co-displaying glucoamylase and α-amylase. *Biochem. Eng. J.* 350:477–484.

Shigechi, H., J. Koh, Y. Fujita, T. Matsumoto, Y. Bito, E. Ueda Satoh, H. Fukuda, and A. Kondo. 2004. Direct ethanol production from raw corn starch via fermentation by use of novel surface engineered yeast strain co-displaying glucoamylase and α-amylase. *Appl. Environ. Microbiol.* 70(8):5037–5040.

Shigechi, H., K. Uyama, T. Fujita, T. Matsumoto, M. Ueda, A. Tanaka, H. Fukuda, and A. Kondo. 2002. Efficient ethanol production from starch through development of novel flocculent yeast strains displaying glucoamylase and co-displaying or secreting α-amylase. *J. Mol. Cat. B.* 17:179–187.

Ueda, M., T. Murai, Y. Shibasaki, N. Kamasawa, M. Osumi, and A. Tanaka. 1998. Molecular breeding of polysaccharide-utilizing yeast cells by surface engineering. *Ann. NY Acad. Sci.* 13(864):528–537.

Ueda, M. and A. Tanaka. 2000. Genetic immobilization of proteins on the yeast cell surface. *Biotechnol. Adv.* 18:121–140.

Watari, J., Y. Takata, M. Ogawa, H. Sahara, M. Koshino, M.-L. Onnela, U. Airaksinen, R. Jaatinen, M. Penttila, and S. Keranen. 1994. Molecular cloning and analysis of the yeast flocculation gene FLO1. *Yeast* 10:211–225.

9 Bioethanol from Lignocellulosic Biomass

Part I Pretreatment of the Substrates

Ryali Seeta Laxman and Anil H. Lachke

CONTENTS

ABSTRACT

In nature except in cotton bolls, cellulose fibers are embedded in a matrix of other structural biopolymers, primarily hemicelluloses and lignin. Crystallinity and presence of lignin in most of the natural celluloses are major impediments towards development of an economically viable process technology for enzymatic hydrolysis of cellulose. Most of the β-glucosidic bonds in naturally occurring lignocellulosic materials are inaccessible to cellulase enzymes by virtue of the small size of the pores in the multicomponent biomass. The molecules of individual microfibrils in crystalline cellulose are packed so tightly that not only enzymes but even small molecules like water cannot enter the complex structure. Suitable pretreatment to remove these blocks is necessary to obtain hydrolysis rates for the process to be viable. Pretreatment is a process that converts lignocellulosic biomass from its native form, in which it is recalcitrant to cellulase enzyme systems, into a form for which enzymatic hydrolysis is effective. Many different pretreatments have been attempted. Some have been demonstrated to be effective in disrupting lignin cellulose complex, while others are responsible for breaking down the highly ordered cellulose crystalline structure, which is a prerequisite for enzyme action. Sometimes a combination of two or more methods has been used in parallel or in sequence. This chapter describes the various physical, chemical, physicochemical, and biological pretreatments reported to date.

9.1 INTRODUCTION

Cellulose is typically found in the walls of plant cells, which have secondary thickening. These cell walls also contain pectin, lignin, and hemicellulose. It is now well established that lignocellulose-containing biomass is a potential renewable resource for the production of single cell protein, glucose, or ethanol. For example, one ton of dry sugarcane bagasse is theoretically reported to generate 112 gallons of ethanol (Knauf and Moniruzzaman 2004). However, the hydrolysis of cellulose by enzymes is a complex phenomenon and is affected both by the structure and reaction conditions. Unlike a homopolymer like starch, which is easily hydrolyzed, lignocellulose contains cellulose (23–53%), hemicellulose (20–35%), polyphenolic lignin (10–25%), and other extractable components (Knauf and Moniruzzaman 2004). The biodegradation of heterogeneous insoluble substrates like lignocellulosic materials is a slow process. Reducing the time to achieve satisfactory sugar yields will therefore have a large impact on the process economy. For this purpose, the lignocellulosics require specific pretreatments to overcome both the physical and chemical barriers to increase their accessibility to enzymes for hydrolysis. Pretreatment refers to the solubilization and separation of one or more of the four major components of biomass, hemicellulose, cellulose, lignin, and extractives, and make the remaining solid biomass accessible to further chemical or biological treatment. This chapter gives general aspects of various pretreatments worked out by earlier investigators.

9.2 ENZYMATIC HYDROLYSIS OF LIGNOCELLULOSIC MATERIALS: THE BARRIERS

The enzymatic hydrolysis of a solid substrate is a slow process. For example, the cellulose in the lignocellulosic materials is normally not easily degradable by the extracellular hydrolytic enzymes to any appreciable extent. This is because the cellulose molecules are not found individually but are linked together to form microfibrils. The separate molecules are linked by hydrogen bonding into a highly ordered crystalline structure. Some parts of the microfibrils have a less ordered, noncrystalline structure and are referred to as amorphous regions. The high molecular weight and ordered tertiary structure make natural cellulose insoluble in water. The crystalline regions of the cellulose are more resistant to biodegradation compared to amorphous regions. Another important factor is the degree of polymerization (DP). Cellulose of low DP will obviously be more susceptible to cellulolytic enzymes, particularly exocellulases. Cellulose does not occur alone but is associated with lignin and hemicelluloses. Lignin is heterogeneous in bond type and most of the bonds are not amenable to hydrolytic cleavage. It is insoluble and difficult to wet. Thus, the presence of lignin is always deleterious to cellulose degradation. The rate of cellulolysis is inversely related to the lignin content and is also related to the type of lignin and its association with cellulose. In general, the plant cell walls are subdivided as primary (PW) and secondary walls (SW). The distribution of the cellulose, hemicelluloses, and lignin varies considerably among these layers. The secondary wall is composed of SW1, SW2, and SW3 where SW2 is usually thicker than the others and contains the major portion of cellulose. The middle lamella, which binds the adjacent cells, is almost entirely composed of lignin (Figure 9.1).

The major structural barriers for the biodegradation of cellulose are its association with the lignin and hemicellulose, crystallinity, degree of polymerization, and surface area. When enzymes degrade the lignocellulosic substrate, there is always a residual fraction that survives the attack. This fraction absorbs a significant amount of the original enzyme and restricts the reuse of these enzymes on added, fresh substrate. All these factors serve to limit the availability of the glycoside bonds to the hydrolytic enzymes.

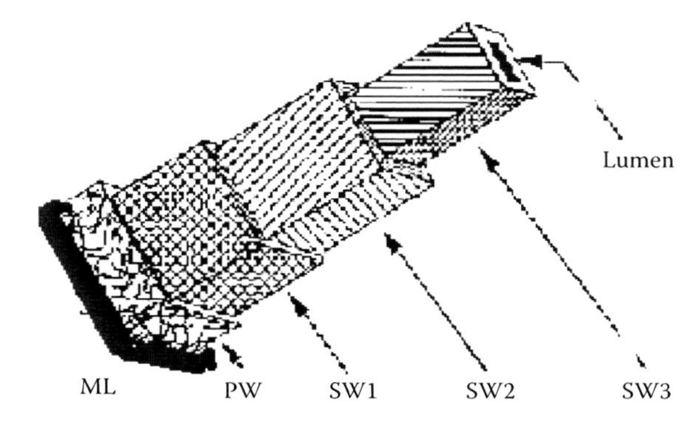

FIGURE 9.1 Diagrammatic sketch of wood cell wall showing the thin primary cell wall (PW) and the three layers of secondary cell wall (SW-1, SW-2, and SW-3) and the middle lamella (ML).

Most potential substrates for cellulose bioconversion are heavily lignified. Thus, most of the cellulose in nature is unsuitable for bioconversion unless effective and economically viable procedures are developed to remove or modify lignin. The essential feature of any successful pretreatment is to decrease the protective association between the lignin and the cellulose. The susceptibility of cellulosic substrates has to be increased in order to improve the enzymatic saccharification rate in a bioreactor. Many investigators have examined various pretreatments for improving the biodegradation of the potential substrates. The pretreatment is important from the viewpoint of utilization of natural cellulose as forage for ruminant animals or as a feedstock for the biotechnological industry. There is only a limited understanding of how these pretreatments enhance hydrolysis of the lignocellulosic substrates. The number of glucose residues that are accessible to the rather large cellulase enzymes governs the rate of hydrolysis of the cellulose. The rate of biodegradation of cellulose is not related to the concentration in terms of weight or volume but rather must be associated with the surface area. Any reduction in the time needed to obtain a satisfactory sugar yield, therefore, will have a significant impact on the process economics. For this purpose, several methods have been described in the literature for increasing the accessibility/availability and hydrolysis of cellulose (Zhang et al. 2007; Ramos 2003; Lynd et al. 2002; Cheng 2001; Gregg and Saddler 1996; Fan et al. 1982).

9.3 TYPES OF PRETREATMENT

The structure of lignocellulosics in the cell wall resembles that of a reinforced concrete pillar with the cellulose fibers being the metal rods and lignin the matrix cement. The carbohydrate polymers are tightly bound to the lignin, mainly by hydrogen bonds but also by some covalent bonds. The biodegradation of the native untreated lignocellulose is slow and the extent of the degradation is often low and does not exceed 20%. Hence, treatment of the biomass is essential in order to increase the accessibility and enzymatic hydrolysis. A large number of pretreatments have been tried by many investigators, which can be broadly classified into physical, chemical, physicochemical, and biological (Table 9.1). Sometimes a combination of two or more pretreatments is employed. These pretreatments open the structure of the potential cellulose substrate. An efficient pretreatment method is one that increases accessibility to the cellulase and enhances the complete solubilization of the polymer to monomer sugars without formation of degradation products. In addition, the process should be inexpensive, less energy intensive, and not cause any serious pollution.

9.3.1 Physical Pretreatments

The crystalline structure excludes water molecules and other large molecules such as enzymes. The smaller particles have a large surface-to-volume ratio. The surface area available for the enzyme-substrate interaction is influenced by the pore size and shielding effect by the hemicelluloses. Physical treatments such as grinding, milling, high temperature, freeze/thaw cycles, and radiation are aimed at size reduction and mechanical decrystallization. Mechanical methods such as ball milling,

TABLE 9.1
Pretreatment Methods

Physical	Ball milling, two-roll milling, hammer milling, colloid milling, vibro energy milling, pyrolysis, γ-irradiation, microwave irradiation
Chemical	Alkali—NaOH, NH_3, ammonium sulfite
	Acid—H_2SO_4, H_3PO_4, HCl
	Gases—ClO_2, NO_2, SO_2
	Oxidizing agents—H_2O_2, ozone
	Cellulose dissolving agents—cadoxen, CMCS, phosphoric acid/acetone, ionic liquids
	Solvent extraction—ethanol-water, benzene-ethanol, butanol-water, ethylene glycol
Physicochemical	Steam explosion (SE), SO_2-catalyzed SE, CO_2 explosion, SC-CO_2 explosion, ammonia freeze explosion (AFEX)
Biological	Fungi

two roll milling, colloid milling, and nonmechanical methods such as α-irradiation, high-pressure steaming, and pyrolysis have all been attempted to change one or more structural features of the cellulose and enhance the hydrolysis. Most of these methods are limited in their effectiveness and often expensive.

9.3.1.1 Milling

Milling reduces the particle size and crystallinity and increases the surface area and the bulk density. This method can be used for a variety of substrates but is highly energy intensive. Ball milling and two-roll milling have been found to increase the susceptibility of the cellulose to enzyme action. Fitz milling results in size reduction without changing the crystallinity and wet milling results in fibrillation and delamination of the cellulose with no change in the chain length and crystallinity due to the plastisizing action of water.

Ball milling: The shearing and compressive forces of ball milling cause reduction in the crystallinity, decrease in the degree of polymerization (DP), decrease in the particle size, increase in the bulk density, and increase in the external surface area. Increase in the bulk density allows use of high substrate concentrations, and reduces the reactor volume and the capital cost. Milling at elevated temperatures shows an increase in the rate of enzymatic hydrolysis compared to milling at room temperature. Ball-milled cellulose can be completely hydrolyzed to sugars. However, the effectiveness of the milling varies with the cellulosic source, and softwood shows the least response. Although this is an effective treatment, time and cost make it prohibitive for use on a large scale.

Two-roll milling: This mill consists of two cast-iron tempered surface rolls placed horizontally, with roll clearance that can be adjusted by screws. The cellulose substrates are fed into the roll and masticated for a specific period of time. The pretreated material is then scraped off. A variety of substrates, including cotton, maple chips, white pine chips, newspaper, etc., have been subjected to two-roll milling, also called differential roll milling. This method reduces the crystallinity as well as the

DP and increases the bulk density. Sometimes, the surface area of the treated material can decrease due to agglomeration of the particles and collapse of the capillary structure. Important parameters are roll clearance, roller speed, and processing time. Two-roll milling of maple wood and newspaper showed 17- and 2.5-fold increase in production of reducing sugars over the untreated samples (Fan et al. 1982). The sedimentation volume is lower for two-roll-milled newspaper than ball-milled newspaper. This facilitates reduction in the reactor volume to reduce capital cost. Advantages of this method are short pretreatment time and increase in the bulk density of the pretreated material.

Hammer milling: A hammer mill consists of a rotor with a set of attached hammers. As the rotor turns, the hammers impact the substrate against a breaker plate. The hammer milling of cellulose improves the digestibility of newsprint to a limited extent. Prolonged hammering of the substrate is not recommended as it reduces the susceptibility of the cellulose to enzymatic hydrolysis.

Colloid milling: A colloid mill consists of two disks set close to each other, revolving in opposite directions, while the substrate slurry is passed between the disks. There is only marginal improvement in the susceptibility of the cellulose to enzymatic hydrolysis and the method is uneconomical owing to the high operating cost.

Vibro energy milling: Vibro energy milling resembles ball milling, except that the mill is vibrated instead of rotated. Increase in the reducing sugar by 1.7 times was reported for 24 to 48 h Sweco milled Solka Floc over the untreated control (Fan et al. 1982). Increase in the reducing sugars was obtained when the substrate was heated to 200°C before or after the pretreatment.

Simultaneous milling and saccharification: This method combines milling with saccharification in a single step. Simultaneous ball milling and enzymatic hydrolysis could improve the rate of saccharification and/or reduce the enzyme loading required to attain total hydrolysis. The effectiveness of the method depends on the lignified matrix of the cellulose microfibrils, the grinding elements, and the oscillation frequency of the shaker. While glass beads are effective for pure cellulose, stainless steel beads are more effective for lignocellulosics. At lower substrate concentrations and with more beads during milling, Mais et al. (2002) reported up to 100% hydrolysis of lignocellulosics with enzyme loading of 10 filter paper units per gram of the cellulose. This method was more effective than separate milling and hydrolysis, or ball milling.

9.3.1.2 Effect of Temperature

Freezing cellulosic materials in water suspension at -75°C is reported to enhance chemical reactivity (as measured by dye absorption). The effect was more pronounced with repeated freezing and thawing cycles. The cryomilled cotton cellulose obtained by hammer milling in liquid nitrogen showed 36% more hydrolysis compared to untreated sample.

Pyrolysis involves heating the biomass at 200°C and is reported to increase hydrolysis. The type of gaseous atmosphere during pyrolysis affects the reaction. Pyrolysis in the presence of oxygen results in depolymerization, oxidation, and dehydration. In inert atmosphere, depolymerization is slow and by-product formation decreases.

Though negligible change in crystallinity and surface area were observed on pyrolysis of Solka Floc at 170°C in air/helium, marked increase in the hydrolysis of the treated cellulose in helium atmosphere has been reported (Fan et al. 1982). This method is usually successful only in combination with others, such as acid pretreatment. Reducing the particle size and reaction time and lowering the pressure and temperature minimized the amount of phenolics produced by pyrolysis (Williams 2006).

9.3.1.3 Effect of γ-Irradiation

High energy radiation was found to enhance in vitro digestibility as well as acid/enzymatic hydrolysis of the cellulose. The radiation treatments are effective in breaking the lignin-cellulose complex as evidenced by the increased presence of phenolics in the irradiated samples. The irradiation is reported to cause increase in the surface area, while its effect on the crystallinity of the cellulose is controversial. Irradiation in the presence of oxygen, milling, or the addition of nitrate salts, or treatment with acid or alkali prior to irradiation increased the digestibility of the treated sample. The amount of reducing sugar produced by the enzymatic hydrolysis of the samples irradiated with 100 Mrad was about three times higher than that from the untreated bagasse. At or above 50 Mrad, the crystallinity of the sugarcane bagasse decreased, and in vitro rumen digestibility increased (Han et al. 1983). The irradiation of rice straw at 100 Mrad gave 19% higher glucose yield than the unirradiated sample. The combination of the irradiation with low concentration of the alkali gave higher glucose yield (Xin and Kumakura 1993). Considerable improvement in the hydrolysis of wheat straw was obtained when gamma radiolysis was used in the presence of dilute sulfuric acid (Ramos 2003). Though *α-irradiation* has superior penetrating power and ionization action, which breaks cellulose chains, the method is slow and expensive. At higher dosages, this treatment results in oxidation, degradation of the molecules, dehydration, and destruction of anhydroglucose units to yield CO_2.

9.3.1.4 Effect of Irradiation with Microwaves

A 240 W microwave irradiation pretreatment of ground rice straw released 2% to 4% of reducing sugars (Williams 2006; Kitchaiya et al. 2003). Irradiation with microwaves singly or in combination with alkali treatment significantly accelerated the hydrolysis rate.

9.3.2 Chemical Pretreatments

There are two types of swelling of cellulose, intercrystalline and intracrystalline. Intercrystalline swelling can be affected by water and is a prerequisite for any microbial reaction to occur. Intracrystalline swelling requires a chemical agent that is capable of breaking the hydrogen bonds of the cellulose. Aqueous solutions of acid and alkali belong to this group of chemical agents.

Chemical pretreatment approaches have gained significant attention to increase the accessibility to hydrolytic attack. A wide variety of chemicals as pretreatment agents have been reported in the literature, which include cellulose solvents, sodium hydroxide, aqueous ammonia, calcium hydroxide plus calcium carbonate, phosphoric acid, alkaline hydrogen peroxide, sulfur dioxide, carbon dioxide, inorganic salts with acidic properties, ammonium salts, Lewis acids and organic acid anhydrides, acetic acid, formic acid, sulfuric acid, organic solvents, *n*-butylamine, *n*-propylamine, and alcohols such as methanol, ethanol, or butanol in the presence of an acid or alkaline catalyst (Ramos 2003). Chemical pretreatments are generally more effective in solubilizing a greater fraction of lignin while leaving behind much of the hemicellulose in an insoluble polymeric form and opening up the crystalline cellulosic substrate. The pulping of wood by the paper industry is one of the earliest methods used for delignification; however, pulping is an expensive method to use as a pretreatment for lignocellulose. A few of the most commonly used pretreatment methods are discussed below.

9.3.2.1 Cellulose Dissolving Agents

The cellulose dissolving agents fall into four groups: strong mineral acids such as H_2SO_4 and H_3PO_4, quaternary ammonium bases, transition metal complexes, and organic solvents. Strong mineral acids and transition metal complexes are commonly used as cellulose dissolving agents. Solvents such as cadoxen and CMCS are able to swell and transform solid cellulose into a soluble state. This ability to dissolve the cellulose has been exploited as a means of pretreatment. The crystalline structure of the native cellulose can be completely destroyed by dissolving in a solvent and on reprecipitation the cellulose is regenerated as a soft floc and is highly reactive. Enhancement in reactivity is observed both with acid and enzymatic hydrolysis and quantitative yields of sugar are obtained. Solvent pretreatment results in higher moisture regain values, larger pore size distribution, and lower crystallinity. The most common solvents are cadoxen, CMCS, H_2SO_4 and H_3PO_4. Only concentrated acids act as cellulose solvents.

9.3.2.1.1 Cadoxen

Cadoxen is an alkaline solution containing ethylene diamine and cadmium oxide/cadmium hydroxide. At room temperature, cadoxen can dissolve 10% cellulose by weight, which precipitates into a soft floc when excess water is added. In the soft floc form, it can be hydrolyzed with either acid or enzyme, with 90% conversion based on the amount of reprecipitated cellulose. This reagent dissolves cellulose with little or no degradation. The DP of treated cellulose does not change. Cadoxen brings about transformation of the crystalline structure in cellulose from I to II and causes decrease in fold length that is, leveling of the degree of polymerization (LODP). Cadoxen has little chance of commercial use because cadmium is highly toxic.

9.3.2.1.2 CMCS

CMCS is made up of sodium tartarate, ferric chloride, and sodium sulfite in alkaline solution and is generally recognized as safe. This solvent dissolves up to 4% cellulose at room temperature, which can be reprecipitated by the addition of water and methanol. This pretreatment resulted in increase in the surface area of Solka Floc, which was attributed to intracrystalline swelling. The reprecipitated cellulose can be completely hydrolyzed with 95% glucose yield (Fan et al. 1982).

9.3.2.1.3 Concentrated Sulfuric Acid

The strong mineral acid acts as swelling agent only in a particular concentration range. Sulfuric acid is a strong swelling as well as a hydrolyzing agent. Swelling at acid concentrations below 55% is similar to that of water but between 55% and 75%, swelling of the cellulose occurs and above 75%, dissolution and decomposition of the cellulose takes place. Dissolved cellulose is reprecipitated by the addition of methanol or ethanol. Intracrystalline swelling occurs in the concentration range of 62.5 to 70%. The DP of the treated cellulose with 75% sulfuric acid falls from 2,150 to 300. The reprecipitated cellulose is easily hydrolyzed by acid or enzyme with high conversions. As either methanol or ethanol can be distilled from the concentrated acid stream, the acid can be reused. Though this process appears attractive, large-scale testing is needed to determine the permissible recycling of sulfuric acid without building up impurities.

9.3.2.1.4 Concentrated Phosphoric Acid

Walseth (1952) employed 85% phosphoric acid as a cellulose solvent and observed a tenfold increase in the extent of the conversion by the cellulase. Increase in acid concentration increases the extent of the swelling. The phosphoric acid causes less degradation of the cellulose than other acids. Swelling of cellulose with phosphoric acid reduces the DP from 2,150 to 1,700. Although this method is effective, the large quantity of acid that must be used makes the process uneconomical.

A novel lignocellulose fractionation method using concentrated phosphoric acid/acetone was reported recently by Zhang et al. (2007). This new technology is applicable to hardwoods as well as softwoods. The main features are moderate reaction conditions (50°C, atmospheric pressure), fractionation of lignocellulose into highly reactive amorphous cellulose, hemicellulose sugars, lignin, and acetic acid and cost effective reagent recycling. Enzymatic hydrolysis of Avicel and α-cellulose was completed within 3 h while corn stover and switch grass were hydrolyzed to the extent of 94%. In the case of Douglas fir, a softwood, hydrolysis was only 73% due to inefficient removal of lignin (Zhang et al. 2007).

Dadi et al. (2006) reported a pretreatment method where cellulose was dissolved in an ionic liquid (IL) and was subsequently regenerated as an amorphous precipitate by rapidly quenching the solution with an anti-solvent such as water, ethanol, or methanol. These solvents can be recovered by distillation. Hydrolysis of the regenerated cellulose was significantly enhanced and the initial rates of the enzymatic hydrolysis were approximately an order of magnitude greater than those of untreated cellulose. The authors claimed that due to the extremely low volatility of ionic liquids, the method could be expected to have minimal environmental impact.

9.3.2.2 Organic Solvents

Delignification using organic solvents with mineral acids as catalysts has also been reported as a pretreatment method. This method breaks the internal lignin and hemicellulose bonds and separates the lignin and hemicellulose fractions that can be potentially converted to useful products. Methanol, ethanol, butanol, *n*-butylamine, acetone, ethylene glycol, etc., have been used in the organosolv process. Organic acids such as oxalic, acetylsalicylic, and salicylic acid can be used as catalysts. The hardwoods are readily delignified in acid-catalyzed systems, whereas softwoods require higher temperature. At high temperatures (above 185°C), the addition of catalyst was unnecessary for satisfactory delignification (Sun and Cheng 2002). Fifty percent aqueous butanol can extract about half of the lignin content and can change wood structures sufficiently, resulting in 80 to 90% cellulose hydrolysis by the enzymes. Phenol was more effective than aqueous butanol, with 90% delignification. Solvent recovery for phenol and butanol is 95 and 78%, respectively. Ethylene glycol was highly effective in increasing the surface area in addition to delignification, with minor reduction in crystallinity, and gave higher sugar yields on enzymatic hydrolysis. Ethylene glycol extracted most hemicellulose and *n*-butylamine selectively removed lignin from corn stover. It has high swelling action. Butylamine is advantageous in that it has a lower boiling point than water and, therefore, can be recovered for reuse by distillation of the sugar solutions. The solvents used in the process need to be drained from the reactor, evaporated, condensed, and recycled to reduce the cost. In addition to cost reduction by recycling, the removal of the solvents from the system is necessary because the solvents may be inhibitory to the growth of organisms, enzymatic hydrolysis, and fermentation.

9.3.2.3 Dilute Acids

Those pretreatments that use dilute acid result in the hydrolysis of a significant amount of the hemicellulose fraction of biomass, leading to high yields of soluble sugars from the hemicellulose fraction. The hot-wash process, a variation of the dilute acid pretreatment, involves high-temperature separation and washing of the pretreated solids, which is thought to prevent reprecipitation of the lignin and/or xylan that may have been solubilized under the pretreatment conditions. The reprecipitation of the lignin can negatively affect the subsequent enzymatic hydrolysis.

Complete removal of the hemicellulose from the lignocellulosic material during pretreatment is a necessary prerequisite for the successful enzymatic hydrolysis of the cellulosic fraction. Dilute acid pretreatment is effective in removing the hemicellulose fraction from the lignocellulose. Hemicellulose removal increases the porosity of the native lignocellulosics and, thus, enzymatic accessibility to the cellulosic fraction. The amount of lignin and cellulose dissolved during this pretreatment method is usually minor. Dilute acid is an efficient pretreatment method suitable for all kinds of lignocellulosic substrates such as corn stover, newsprint, etc., to improve the enzymatic hydrolysis of substrates.

Sulfuric acid: Acid catalyzed hydrolysis uses dilute sulfuric, hydrochloric, or nitric acids. Dilute sulfuric acid (0.5–1.5%) at temperatures above 160°C was found to be most suitable for industrial application, because of its high sugar yields from the hemicellulose hydrolysis (xylose yields of 75–90%). A dilute acid pretreatment method involving two steps was reported for hardwoods (Nguyen et al. 1998; Cheng 2001). In the first step, a temperature of 140°C was used to hydrolyze the easily degradable fraction and in the second step, the temperature was slowly increased to 170°C to hydrolyze the hemicellulose fractions that were more difficult to degrade. Treatment with 1 to 2% H_2SO_4 at less than 220°C and retention times of a few minutes reduces the DP of the cellulose, while the crystallinity does not decrease.

Enzymatic hydrolysis results in complete conversions within 24 h based on theoretical values. The acid has to be removed or neutralized before fermentation, yielding a large amount of gypsum. Although close to theoretical yields can be achieved, this process involves high capital investment, acid consumption, and acid recovery costs.

Peracetic acid: Delignification of corn stalks and sawdust with 20% peracetic acid showed significant increase in the rate of enzymatic hydrolysis. Peracetic acid resulted in 76.2% delignification with concomitant increase in the surface area of wheat straw and drastic increase in digestibility. This method causes significant reduction in crystallinity due to structural swelling and dissolution of the crystalline cellulose. The oxidizing action of peracetic acid causes relatively low decrease in hemicellulose, leaving hemicellulose-rich material.

9.3.2.4 Alkali Pretreatment

Among the various chemical pretreatments, alkali treatment with bases like sodium hydroxide is most widely used to enhance in vitro digestibility and enzymatic hydrolysis of the lignocellulose. All the lignin and part of the hemicellulose are removed and the reactivity of cellulose for hydrolysis is sufficiently increased. The success of this method depends on the amount of lignin in the biomass (Sun and Cheng 2002; McMillan 1994).

9.3.2.4.1 Sodium Hydroxide

Sodium hydroxide is used as an intracrystalline swelling agent for both crystalline and amorphous cellulose. Dilute sodium hydroxide causes separation of the structural linkages between the lignin and carbohydrate and disruption in the lignin structure; swelling leads to increase in the internal surface area and pore size, as well as decrease in the DP and crystallinity. The mechanism is believed to be saponification of intermolecular ester bonds cross-linking the hemicelluloses and lignin. The porosity of the lignocellulosic materials increases with the removal of the cross links, leading to swelling, and enhances the accessibility to the enzymes. The extent of the hydrolysis increases with increase in NaOH concentration used for pretreatment. The optimum levels range between 5 and 8 g NaOH/100 g substrate. NaOH above 20% causes extensive swelling and separation of the structural elements and transforms the cellulose from crystal structure I to II. Different cellulosic substrates respond differently to alkali treatment. It was observed that following alkali treatment, digestibility of the softwoods increases slightly as compared to hardwoods. This difference in response appears to be related to the lignin content of the wood. The digestibility of hardwoods treated with NaOH increased from 14 to 55%, while the lignin content decreased from 24–55% to 20%. However, no effect of dilute NaOH pretreatment was observed for softwoods with lignin content greater than 26%. Spruce wood treated with cold 2 N NaOH showed 80% conversion by *T. viride* enzyme.

A combination of irradiation and 2% NaOH on corn stalks doubled the glucose yield, while no significant difference was noticed for cassava bark and peanut husk (Sun and Cheng 2002). Though reactor costs are lower than those for acid technologies, some disadvantages are low bulk density of the substrate and the need for washing to recover hemicellulose and lignin. High alkali concentrations used for the treatment raise environmental concerns. Further, it may lead to prohibitive recycling, wastewater treatment, and residual handling costs.

Treatment of corn fiber, distiller's dried grains, sugarcane bagasse, and spent barley malt with 1.0% sodium percarbonate have been shown to increase sugar yields from the biomass (Williams 2006).

9.3.2.4.2 Ammonia

Treatment with liquid ammonia was first patented in 1905. Ammonia exerts a strong swelling action and brings about phase change in the cellulose crystal structure from I to III. The benefits of this method include breakage of glucuronic acid ester cross-links, solubilization of lignin, and disruption of crystalline structure, swelling and increase in the accessible surface area of cellulose. Pretreatment with liquid ammonia has been used mostly to increase in vitro digestibility of animal feed. Wheat straw treated with 50% ammonium hydroxide showed 20% delignification, with threefold increase in hydrolysis.

An improved pretreatment method involving two steps is reported by Cheng (2001). In the first step, steeping the lignocellulosic biomass in aqueous ammonia at ambient temperature removed the lignin, acetate, and extractives. This was followed by a dilute acid pretreatment that hydrolyzed the hemicellulose fraction. Finally, the cellulose fraction was collected after thorough washing. The advantage of this new method is a step-by-step separation of the lignin, hemicellulose, and cellulose from the biomass. Around 80 to 90% of lignin can be removed through the ammonia steeping step (Cheng 2001).

9.3.2.4.3 Alkaline Oxidation Pretreatment

This method is a combination of the alkaline and oxidative pretreatments. Pretreatment with H_2O_2 in an alkaline environment or combining it with a preceding alkali treatment step is an effective pretreatment for lignocelluloses. In weak alkaline media, H_2O_2 only selectively acts on phenolic compounds originated from partial scission of lignin, causing its degradation without affecting the cellulosic fraction. Only the lignin and hemicellulose are solubilized. This treatment removes approximately 50% of lignin in wheat straw and corn stover. Sugarcane bagasse treated with 2% H_2O_2 at 30°C for 8 h solubilized about 50% lignin and most of the hemicellulose. Subsequent saccharification by cellulase at 45°C gave 95% conversion to glucose in 24 h (Sun and Cheng 2002). H_2O_2 to substrate ratio of 0.25 g per gram of substrate at 25°C, pH 11.5, were the optimum treatment conditions. The lignin degradation products were not found to be toxic either for saccharification or fermentation. Reuse of the solvent six times after the pretreatment was also possible.

9.3.2.5 Gases

Pretreatment with gases has the advantage that gases can penetrate uniformly throughout the substrate. However, their recovery for reuse poses problems. Chlorine dioxide, nitrogen dioxide, sulfur dioxide, sodium hypochlorite, HCl gas, and ozone have been used as pretreatment agents to solubilize the lignin and increase in vitro digestibility. Chlorine dioxide is an active agent in chlorine pulping to solubilize the lignin. Sulfur dioxide disrupts the lignin-carbohydrate complex and depolymerizes lignin. Treatment by saturating wood particles with HCl gas under pressure in a fluidized bed reactor resulted in a twofold increase in yield after acid hydrolysis of the treated cellulose. Ozonolysis involves using ozone gas to break down the lignin and hemicellulose and increase the biodegradability of the cellulose. Ozonolysis has been shown to break down 49% of the lignin in corn stalks and 55 to 59% of the lignin in autohydrolyzed (hemicellulose free) corn stalks (Williams 2006). The gaseous ozone results in enhanced susceptibility of different woods and straws to *T. reesei* enzyme. This method is effective under mild conditions and environmentally friendly because ozone does not leave residues due to its short half-life. However, large amounts of ozone are required and the need for its onsite production makes the treatment expensive.

9.3.3 Physicochemical Pretreatments

Several pretreatment processes combine physical and chemical methods. In this regard, high pressure steaming, with or without rapid decompression (explosion), has been claimed as one of the most successful options for fractionating wood into its three major components and enhancing the susceptibility of the cellulose to enzymatic attack. Several patents have been granted to this process and many pilot plants of different capacities have been developed for either commercial or research purposes, located in various parts of the world, such as in Canada, the United States, Spain, Sweden, France, Italy, Japan, and Brazil (Ramos 2003).

9.3.3.1 Steam Treatment (Autohydrolysis)

One of the most common physicochemical pretreatment methods is steam explosion or liquid hot water (LHW) treatment. This process involves treatment with steam under high pressure and temperature, followed by quick release of the pressure, causing the biomass to undergo an explosion and shatter the structure in a popcorn-like effect. The advantages of this method are the low energy input and negligible environmental impact. However, steam explosion does not always break down all the lignin, requires small particle size, and can produce compounds that may inhibit subsequent fermentation. Despite these drawbacks, steam explosion is currently the most popular method for separating the lignin and hemicellulose from the cellulose (Sun and Cheng 2002). Wood chips are treated with saturated steam at 210 to 300°C and 500 to 1000 psi in a reactor vessel, called a gun reactor. After a few minutes, the reaction is frozen and the wood is exploded into a fine powder by sudden decompression to atmospheric pressure. Enzymatic hydrolysis of this material gives 80% of theoretical glucose but lignin remains unaffected and can be recovered in the native form. The steam explosion pretreatment of red oak wood chips removed 10% to 20% of the lignin. The steam explosion of softwood chips at 210°C and 4 minutes achieved a maximum theoretical sugar yield of 50% (Williams 2006).

Most steam treatments yield high hemicellulose solubility and low lignin solubility. Studies conducted without added catalyst reported xylose–sugar recoveries between 45% and 65%. The LHW process uses compressed, hot liquid water (at pressure above the saturation point) to hydrolyze the hemicellulose. Xylose recovery is high (88–98%), and no acid or chemical catalyst is needed in this process, which makes it economically and environmentally attractive. However, the development of the LHW process is still in the laboratory stage.

Hydrothermal treatment of lignocellulose at high temperature (80–250°C) generates acetic acid that arises from the thermally labile acetyl groups of hemicellulose and catalyzes the hydrolysis of the hemicellulose and subsequent solubilization. The treatment results in fiber fragmentation with little or no loss in crystallinity. Steaming opens up fiber, renders the hemicellulose soluble, and appears to depolymerize the lignin to some extent. Although the pretreated wood contained most of the lignin originally present in the wood, 75% of the cellulose could be enzymatically hydrolyzed. Various fractions of steam-exploded wheat straw and aspen wood chips showed substances inhibitory to enzyme activity and hydrolysis. The autohydrolysis of sunflower seed hulls and bagasse at 200°C for 4 to 5 min followed by explosive defibrillation solubilized 80 to 90% of hemicellulose and the residue is highly susceptible to hydrolysis with *T. ressei* enzymes.

9.3.3.2 Acid-Catalyzed Steam Explosion

The addition of dilute acid in the steam explosion can effectively improve enzymatic hydrolysis, decrease the production of inhibitory compounds, and lead to more complete removal of the hemicellulose. Acid-catalyzed steam explosion is one of the most cost-effective processes for hardwood and agricultural residues, but it is less effective for the softwoods. It is possible to recover around 70% potential xylose as monomer. The lignin redistribution is thought to explain why dilute acid and steam explosion is an effective pretreatment process. Although the lignin is not removed, it is thought that lignin melts during the pretreatment and coalesces upon cooling such that its properties are altered substantially (Lynd et al. 2002). Limitations include destruction of a portion of the xylan fraction, incomplete disruption of the biomass structure, and the generation of inhibitory compounds. The necessary water wash decreases the overall sugar yields.

9.3.3.3 Ammonia and Steam Explosion

The lignocellulosic materials can also be exploded using ammonia and involves liquid ammonia and steam explosion. This method is considered one of the leading biomass pretreatments. Ammonia freeze explosion (AFEX) treats the biomass with concentrated ammonia under pressure and at temperatures up to about 100°C. After a few minutes under these conditions, the pressure is rapidly released (the "explosion"). The ammonia evaporates and is recovered. AFEX disrupts the lignocellulose and reduces the cellulase requirement but removes neither hemicellulose nor lignin. The treated biomass is now much more easily converted by enzymes to sugars and then to ethanol. In a comparative economic evaluation of advanced pretreatments, AFEX performed better than all the other pretreatments studied, except for the dilute acid process. Improved understanding of the morphological changes and chemical compounds formed during AFEX may further improve the pretreatment performance. Ammonia explosion does not produce products that may inhibit fermentation but it requires that the ammonia be recycled for economic and environmental reasons (Sun and Cheng 2002).

9.3.3.4 CO_2-Catalyzed Steam Explosion

CO_2 explosion is similar to steam and ammonia explosion. The glucose yields in the later enzymatic hydrolysis are lower compared to steam and ammonia explosion. The steam explosion of wheat straw, bagasse, and eucalyptus wood chips in the presence of CO_2 at 200°C increased the digestibility above 75%. However, CO_2 explosion is more cost effective than ammonia explosion and does not cause the formation of inhibitors as in steam explosion.

9.3.3.5 SO_2-Catalyzed Steam Explosion

Martin et al. (2002) investigated SO_2 and H_2SO_4 impregnation during steam explosion (205°C, 10 min) of sugarcane bagasse and their influence on the enzymatic hydrolysis. The SO_2-impregnated bagasse gave the highest yields of xylose, arabinose, and total sugar on hydrolysis. The hydrolysates of SO_2-impregnated and non-impregnated bagasse showed similar fermentability with *S. cerevisiae*, whereas the fermentation of the hydrolysate of H_2SO_4-impregnated bagasse was considerably poor. Corn fiber that was steam exploded in a batch reactor at 190°C for 5 min with 6% SO_2 resulted in 81% conversion of all the polysaccharides in the corn fiber to monomeric sugars on enzymatic hydrolysis, which was subsequently converted to ethanol very efficiently by *S. cerevisiae*, yielding 90 to 96% of the theoretical conversion (Bura et al. 2002).

Soderstrom et al. (2002) reported a two-step steam pretreatment of softwood. In the first step, the softwood was impregnated with the SO_2 and steam pretreated at low severity to hydrolyze the hemicellulose and release the sugars into the solution. In the second step, the washed solid material from the first step was impregnated once more with SO_2 and steam pretreated in a temperature range of 180 to 220°C and residence times between 2 and 10 min to enhance enzymatic digestibility. The two-step steam pretreatment resulted in a higher yield of sugars and slightly higher yield of ethanol compared with the one-step steam pretreatment. Enzymatic hydrolysis gave an overall sugar yield of 80%, which gave 69% ethanol on subsequent simultaneous saccharification and fermentation (SSF) (Soderstrom et al. 2002).

9.3.3.6 Supercritical Carbon Dioxide (SC-CO_2)

Supercritical carbon dioxide is carbon dioxide above its critical point of 31°C and 73 atm. This pretreatment has many advantages, such as nontoxicity, low solvent cost, (CO_2), low pretreatment temperatures, easy recovery of CO_2, and high solids concentrations in the pretreated materials. However, the low effectiveness for softwood and the high capital cost for the high-pressure equipment may be obstacles for its commercialization. Supercritical carbon dioxide had no significant effect on the yield of reducing sugars or enzymatic digestibility of aspen lignocellulosic biomass (Williams 2006). SC-CO_2 explosion of Avicel enhanced the accessible surface area and increased glucose yield by 50% (Zheng et al. 1995).

9.3.3.7 Advantages and Disadvantages of the Steam Explosion

9.3.3.7.1 Advantages

1. Ability to separate the three components of wood: modifies lignocellulose to allow fractionation of hemicellulose in autohydrolysis steam, lignin in aqueous alcohol or alkali, and cellulose as insoluble biomass.
2. Cellulose is highly susceptible to acid or enzymatic hydrolysis.
3. Lignin in suitable form for conversion to chemicals.
4. Hemicellulose is easily converted to liquid fuels.
5. Inhibitors are easily extractable.

9.3.3.7.2 Disadvantages

1. Produces substrates with low bulk density.
2. Does not always break down the lignin completely.
3. Requires small particle size.
4. Some of the compounds produced can be inhibitory to the subsequent ethanol fermentation.

A limitation of all the pretreatment processes is their capital intensive nature. For example, the requirement of costly reactor materials and additional process steps for waste treatment and the recovery of the pretreatment catalysts present additional costs. Some pretreatments, such as AFEX, offer potential advantages in operating costs such as low waste generation. Thus, criteria for successful pretreatment can be narrowed to high cellulose digestibility, high hemicellulose sugar recovery, low capital and energy cost, low lignin degradation, and recoverable process chemicals.

9.3.4 Biological Pretreatments

In these pretreatments, the natural wood attacking microorganisms that can degrade lignin are allowed to grow on the biomass, resulting in lignin degradation. The main biological pretreatments include fungi and their enzymes. There is significant loss of the xylan and mannan components of the hemicellulose during the lignin hydrolysis. Reductions up to 65% in the lignin content of cotton straw have been reported using white-rot fungi. This is the most promising organism for biological pretreatment of lignocellulose. The various means to use these organisms are: use of naturally occurring white-rot fungi; use of cellulose-less mutants as efficient lignin degraders and/or to repress the enzymes that degrade wood carbohydrates.

A white-rot fungus was used to remove 42% lignin, 2% glucan (including cellulose), and 30% hemicellulose of birch wood (Fan et al. 1982). The degradation of wood lignin by white-rot is oxidative and needs an accompanying carbohydrate such as cellulose or hemicellulose. *Phanerochaete chrysosporium*, a white-rot fungus, is the most commonly used organism for delignification. It degraded 48.58% of lignin, 5.3% of cellulose, and 19.72% of hemicellulose in grape cluster stems over the course of 10 to 12 days. *Phanerochaete* did not have any effect on the enzyme digestibility of raw corn stover. However, another fungus, *Cyathus* sp., increased the digestibility by 3 to 6.9 times the control values over 29 days (Williams 2006). A 17% delignification was achieved by exposing birch wood to cellulase-less mutants of *Polyporus adustus* for 6 weeks. Similarly 72% conversion of cellulose to glucose by enzymatic hydrolysis after biological delignification of wheat straw using *Pleurotus ostreatus* has also been reported. Other organisms used for biological treatment are *Ceriporiopsis subvermispora* and *Trametes versicolor.* The rate of lignin and hemicellulose breakdown is very slow and still needs optimization in most cases to make it an effective pretreatment method (Sun and Cheng 2002). The advantages of these biological pretreatments are that they require little energy input and are environmentally friendly. The economic feasibility of a nonoptimized biological pretreatment process is still poor due to long cultivation times of 10 to 14 days. This method can be considered cost effective only if applied in conjunction with other physical and/or chemical methods such as thermomechanical pulping and steam explosion. In both cases, the removal of resins and other extractable materials can also play an important role in improving the accessibility of the lignocellulosics to bioconversion (Ramos 2003). Sometimes biological treatments are used in combination with chemical treatments (Hamelinck et al. 2005).

9.4 CONCLUSIONS

Pretreatment enables more efficient enzymatic hydrolysis of lignocellulosic substrates by removal of the surrounding hemicellulose and/or lignin along with modification of the cellulose microfibril structure. An effective pretreatment should increase the number of available sites for cellulase action and promote the extensive hydrolysis of the substrate. The choice of pretreatment influences the cost and performance in subsequent hydrolysis and fermentation. It is essential that hemicellulose and lignin are utilized for the hydrolysis to be economical. The ideal pretreatment process would produce reactive cellulose, yield pentoses in nondegraded form, exhibit no significant inhibition of fermentation, require little or no feedstock size reduction, and be simple. In addition, the effectiveness of a pretreatment method should not be judged by the initial rates of hydrolysis, as the accessibility and composition of cellulose varies, sometimes substantially, as the hydrolysis progresses.

Current research is directed toward identifying, evaluating, developing, and demonstrating different pretreatment methods that result in complete enzymatic hydrolysis to theoretical values. A single pretreatment process cannot be identified due to the diverse nature of biomass. Thus, several physical, chemical, physicochemical, and biological treatments or their combinations are under evaluation. Though the resulting composition of the treated material is dependent on the source of the biomass and the type of treatment, in general it is much more amenable to enzymatic hydrolysis than native biomass.

Each type of pretreatment has its own advantages and disadvantages. Some have been demonstrated to be effective in disrupting lignin-cellulose complex, while others are responsible for breaking down the highly ordered cellulose crystalline structure, which is a prerequisite for enzyme action. Among the promising pretreatment options, dilute acid is as yet the most developed. It also produces less fermentation inhibitors and significantly increases cellulose hydrolysis. However, acid consumption is an expensive part of the method; it requires expensive corrosion-resistant materials and disposal of solid waste generated is a problem. Steam explosion with or without added catalyst is an upcoming technology. It is environmentally friendly, less problematic but less effective than acid pretreatment. Additional research is required to make it more effective, leading to higher sugar yields. The liquid hot water (LHW) process with yields projected to be higher than for dilute acid or steam explosion is still at the laboratory stage. Ammonia freeze explosion (AFEX) offers potential advantages in operating costs such as low waste generation. AFEX disrupts lignocellulose but removes neither hemicellulose nor lignin. In a comparative economic evaluation of pretreatments by the U.S. National Renewable Energy Laboratory, AFEX performed better than all other pretreatments studied except for the dilute acid process. A limitation of most pretreatment processes is their capital intensive nature. Thus, criteria for successful pretreatment can be narrowed to high cellulose digestibility, high hemicellulose sugar recovery, low capital and energy cost, low lignin degradation, and recoverable process chemicals.

REFERENCES

Bura, R., S. D.Mansfield, J. N. Saddler, and R. J. Bothast. 2002. SO_2-catalyzed steam explosion of corn fiber for ethanol production. *Appl. Biochem. Biotechnol.* 98-100: 59–72.

Cheng, W. 2001. Pretreatment and enzymatic hydrolysis of lignocellulosic materials. M.Sc Thesis, West Virginia University, Morgantown, WV.

Dadi, A. P., S. Varanasi, and C. A. Schall. 2006. A novel ionic liquid pretreatment strategy to achieve enhanced cellulose saccharification kinetics. Paper presented at the 2006 AIChE Annual Meeting, San Francisco, CA, November 13–17, Abstract 672c.

Fan, L.-T., Y.-H. Lee, and M. M. Gharpuray. 1982. Nature of lignocellulosics and their pretreatments for enzymatic hydrolysis. *Adv Biochem Eng.* 23: 157–187.

Gregg, D. and J. N. Saddler. 1996. Factors affecting cellulose hydrolysis and potential of enzyme recycle to enhance the efficiency of an integrated wood to ethanol process. *Biotechnol. Bioeng.* 51: 375–383.

Hamelinck, C. N., G. van Hooijdonk, and A. P. C. Faaij. 2005. Ethanol from lignocellulosic biomass: Techno-economic performance in short, middle and long-term. *Biomass and Bioenergy* 28: 384–410.

Han, Y. W., E. A. Catalano, and A. Ciegler. 1983. Chemical and physical properties of sugarcane bagasse irradiated with γ–rays. *J. Agric. Food Chem.* 31: 34–38.

Kitchaiya, P., P. Intanakul, and M. Krairish. 2003. Enhancement of enzymatic hydrolysis of lignocellulosic wastes by microwave pretreatment under atmospheric pressure. *J. Wood Chem. Technol.* 23: 217–225.

Knauf, M. and M. Moniruzzaman. 2004. Lignocellulosic biomass processing: A perspective. *Int. Sugar J.* 106(1263): 147–150.

Lynd, L. R., P. J. Weimer., W. H. van Zyl, and I. S. Pretorius. 2002. Microbial cellulose utilization: Fundamentals and biotechnology, *Microbiol. Mol. Biol. Rev.* 66: 506–577.

Mais, U., A. R. Esteghlalian, J. N. Saddler, and S. D. Mansfield. 2002. Enhancing the enzymatic hydrolysis of cellulosic materials using simultaneous ball milling. *Appl. Biochem. Biotechnol.* 98-100: 815–832.

Martin, C., M. Galbe, N.O. Nilvebrant, and L. J. Jonsson. 2002. Comparison of the fermentability of enzymatic hydrolyzates of sugarcane bagasse pretreated by steam explosion using different impregnating agents. *Appl. Biochem. Biotechnol.* 98-100: 699–716.

McMillan, J. D. 1994. Pretreatment of lignocellulosic biomass. In *Enzymatic Conversion of Biomass for Fuels Production*, ed. M. E. Himmel, J. O. Baker, and R. P. Overend, 292–324. ACS Symposium Series, vol. 566. Washington, DC: American Chemical Society.

Nguyen, Q. A., M. P. Tucker, B. L. Boynton, F. A. Keller, and D. J. Schell. 1998. Dilute acid pretreatment of softwoods. *Appl. Biochem. Biotechnol.* 70–72: 77–87.

Ramos, L. P. 2003. The chemistry involved in the steam treatment of lignocellulosic materials. *Quim Nova*, 26: 863–871.

Soderstrom, J., L. Pilcher, M. Galbe, and G. Zacchi. 2002. Two-step steam pretreatment of softwood with SO_2 impregnation for ethanol production. *Appl. Biochem. Biotechnol.* 98-100: 5–21.

Sun, Y. and J. Cheng. 2002. Hydrolysis of lignocellulosic materials for ethanol production: A review. *Bioresource Technol.* 83: 1–11.

Walseth, C. S. 1952. Influence of fine structure of cellulose on the action of cellulases. *TAPPI* 35: 233–238.

Williams, K. C. 2006. Subcritical water and chemical pretreatments of cotton stalk for the production of ethanol. M.Sc. thesis, North Carolina State University.

Xin, L. Z. and M. Kumakura. 1993. Effect of radiation pretreatment on enzymatic hydrolysis of rice straw with low concentrations of alkali solution. *Bioresource Technol.* 43:13–17.

Zhang, Y.-H. P., S.-Y. Ding, J. R. Mielenz, J.-B. Cui, R. T. Elander, M. Laser, M. E. Himmel, J. R. McMillan, and L. R. Lynd. 2007. Fractionating recalcitrant lignocellulose at modest reaction conditions. *Biotech. Bioeng.* 97: 214–223.

Zheng, Y., H.-M. Lin, J. Wen, C. Ningjun X. Yu, and G. T. Tsao. 1995. Supercritical carbon dioxide explosion as a pretreatment for cellulose hydrolysis. Biotech. Lett. 17: 845.

10 Bioethanol from Lignocellulosic Biomass

Part II Production of Cellulases and Hemicellulases

Rajeev K Sukumaran

CONTENTS

ABSTRACT

Creating ethanol from biomass is considered to be one of the most valuable solutions to the increasing liquid fuel demand. The technology for generating fermentable sugars from lignocellulosic biomass is still not mature and is largely dependent on developments in cellulase enzyme technology since the most promising scheme for biomass hydrolysis involves the use of cellulose- and hemicellulose-degrading enzymes. The technology is receiving a renewed interest in the current scenario with increasing efforts to improve its efficiency and cost effectiveness. Currently the major limiting factor in the commercialization of biomass to ethanol technology is the cost of cellulase enzymes, which is the major contributor to the production cost of bioethanol. Innumerable research efforts are directed towards understanding the fundamentals of microbial enzymes involved in biomass hydrolysis, and their production and applications. Proper exploitation of microbial sources for biomass-degrading enzymes requires in-depth understanding of their physiology, molecular biology, and strategies for fermentation. This chapter summarizes some of the current knowledge of microbial cellulase production and explores the avenues of its exploitation.

10.1 INTRODUCTION

The microbial degradation of lignocellulosic biomass is accomplished by the concerted action of several enzymes, of which cellulases form a major category. Cellulose is a linear homopolymer of β-1,4-linked glucose units, while hemicellulose is a heteropolysaccharide made of different carbohydrate monomers. The kinds of linkages are different and often there are substitutions on the monomers, making the hemicellulose structure more complex. These differences in the structure of the polymers have contributed to the existence of a wide range of enzymes capable of degrading them. Although cellulases themselves are a large group of enzymes, the complexity of hemicellulose has resulted in an even larger number of enzymes that act on it, with different specificities and modes of action. In general, both cellulases and hemicellulases can be grouped into endo-acting enzymes, which cleave the polysaccharide internally, and exo-acting enzymes, which cleave the polymer progressively from either the reducing or nonreducing end. Besides these major groups, cellulases are comprised of a third group of exo-enzymes categorized as "β-glucosidases," which cleave cello-oligosaccharides produced by the exo-acting enzymes. Correspondingly, there is an analogous group included under hemicellulases, which cleaves the oligosaccharides generated by hemicellulose hydrolysis (e.g. β-xylosidases). However, the major difference is in the existence of a different category called the "accessory enzymes" under the hemicellulases, the members of which are required for the hydrolysis of native plant biomass. This category includes a variety of acetyl esterases and esterases that hydrolyze the lignin glycoside bonds.

10.1.1 CELLULASES

Cellulases are produced by several microorganisms and include different classes of the enzymes. The β-1,4-D-glucan linkages in cellulose polymer are degraded by these enzymes and the hydrolysis of native cellulose yields glucose as the main product and also cellobiose and cello-oligosaccharides. There are three major types of cellulase enzymes: (1) exoglucanases, which include cellodextrinases (1,4-β-D-glucan-4-glucanohydrolase, EC 3.2.1.74) and cellobiohydrolases (CBH or 1,4-β-D-glucan cellobiohydrolase, EC 3.2.1.91); (2) endo-β-1,4-glucanase (EG or endo-1,4-β-D-glucan 4-glucanohydrolase, EC 3.2.14); and (3) β-glucosidases (BG-EC 3.2.1.21). The enzymes within these classifications can be separated into individual components. For example, the microbial cellulase compositions may consist of one or more CBH components, one or more EG components, and possibly β-glucosidases. The endoglucanases produce nicks in the cellulose polymer exposing reducing and nonreducing ends and the exoglucanases act upon these reducing and nonreducing ends to liberate cello-oligosaccharides, cellobiose and glucose, while the β-glucosidase cleaves the cellobiose to liberate the glucose, thereby completing the hydrolysis (Figure 10.1). The complete cellulase system comprising CBH, EG, and BG components thus acts synergistically to convert crystalline cellulose to glucose.

The majority of the cellulases have a characteristic two domain structure with a catalytic domain (CD) and a cellulose binding domain (CBD). The CDs and CBDs are connected through a linker peptide. The core domain or the catalytic domain contains the catalytic site, whereas the CBDs help in binding the enzyme to cellulose. The degradation of the native cellulase requires different levels of cooperation between the cellulases. Such synergisms exist between the endo- and exoglucanases (exo/endo synergism) and among the exoglucanases. In the first type, the endoglucanase action creates free ends on which the exoglucanases act, and in the second one, the exoglucanases cooperate by acting on the reducing and nonreducing ends to bring about effective cellulose degradation. Though the cellulases are generally identified based on their functional classification, a refined classification system based on sequence and structural similarities exists for the cellulases. These are one of the largest groups of enzymes in the structural classification of the glycosyl hydrolases. Cellulases and hemicellulases make up 15 of the 70 identified glycosyl hydrolase families and some of the families are divided to subfamilies. This classification is based on the variability of their catalytic domains and does not consider variability in the cellulose binding domains. A detailed discussion on these classifications is out

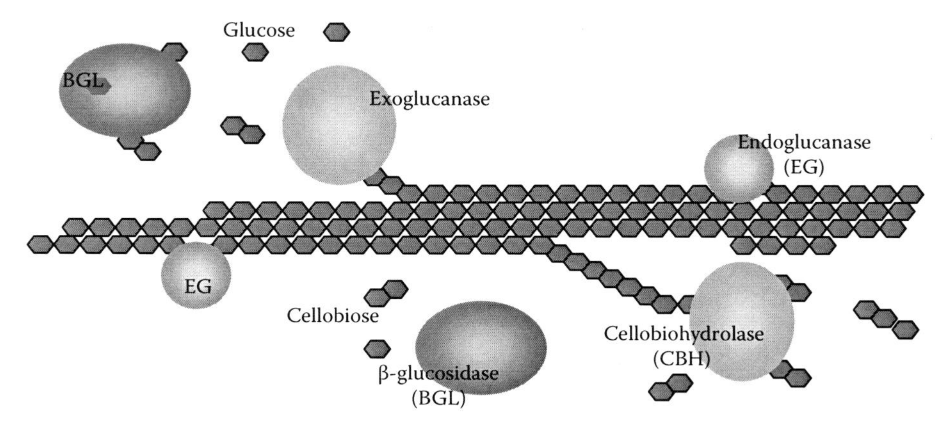

FIGURE 10.1 Schematic diagram showing the mode of action of cellulases.

of scope for this chapter; details may be found in the relevant literature (e.g., Henrissat et al. 1989; Henrissat 1992; Rabinovich, Melnik, and Bolobova 2002).

10.1.2 Hemicellulases

Unlike cellulose, hemicellulose is a heteropolysaccharide composed of various carbohydrate monomers with different linkages and substitutions on the primary branch. Though the types of chemical bonds are limited, they can be presented in different structural surroundings, leading to a greater variability. The most common hemicellulose is xylan, which has a backbone of β-1,4-linked xylopyranose units, while other hemicelluloses contain β-1,4-linked mannopyranose in combination with glucopyranose (glucomannans) as backbone. Galacto glucomannans contain β-1,6-linked galactopyranose in addition to the mannose and glucose units. The backbone xylan in the hemicelluloses is generally modified with various side chains, including 4-*O*-methyl-D glucuronic acid, β-1-2 linked to xylose and acetic acid esterified at the *O*-2 or *O*-3 positions. In addition to uronic acids, L-arabinofuranose residues may be attached by β-1,2 or β-1,3 linkages to the backbone. With the possibility of different backbone and side-chain compositions, the hemicellulose structure is rather complex and the degradation of hemicellulose necessitates the concerted action of a variety of enzymes with different specificities. The hemicellulases can be placed into three general categories.

1. Endo-acting enzymes, which cleave the polysaccharide chains internally with very little activity on short oligomers
2. Exo-acting enzymes, which cleave progressively from either the reducing or nonreducing termini
3. Side-chain-cleaving enzymes and "accessory enzymes," which include acetyl esterases and esterases that hydrolyze lignin glycosidic bonds

The major hemicellulose-degrading enzymes include enzymes that break down the xylan backbone (endo- and exo-xylanases and β-xylosidases) and the side chains (arabinofuranosidases, glucuronidases, acetyl xylan esterases, ferulic acid esterases, and β-galactosidases). Since the hemicellulases are mainly xylan-degrading enzymes, an extensive coverage of all the hemicellulases is not undertaken in this chapter and the discussion is limited to xylan-degrading enzymes. A total degradation of xylan requires the synergistic action of mainly endo-xylanases, which cleave the β-1,4-xylose linkages of the xylan backbone; exo-xylanases, which hydrolyze the β-1,4-linkages of xylan from the reducing or nonreducing ends, releasing xylobiose and xylooligosaccharides; and β-xylosidases, which cleave the xylobiose and xylooligosaccharides to release xylose. In addition, the enzymes β-arabinofuranosidase and β-arabinofuranose remove arabinose and 4-*O*-methyl glucuronic acid substituents from the xylose backbone, and the esterases acetylxylan esterase, ferulic acid esterase, and β-coumaric acid esterase hydrolyze the ester-bonded substituents acetic acid, ferulic acid, and β-coumaric acid from the xylan. Hemicellulase classifications based on structure and sequence similarities give more insights into their structure function relationships similar to those for cellulases. More detailed information may be found in Henrissat (1992), Rabinovich, Melnik, and Bolobova (2002), and Shallom and Shoham (2003).

10.2 MICROBIAL LIGNOCELLULOLYTIC MACHINERY: COMPLEXED AND NONCOMPLEXED SYSTEMS

The cellulase-hemicellulase systems of the microbes can be generally regarded as complexed or noncomplexed (reviewed in Lynd et al. 2002). Utilization of the insoluble cellulose requires the production of extracellular cellulases by the organism. The cellulase systems consist of either secreted or cell associated enzymes belonging to the class cellobiohydrolase, endoglucanase, and β-glucosidase. In the case of filamentous fungi, actinomycetes, and aerobic bacteria, the cellulase enzymes are free and mostly secreted. In such organisms, by the very nature of the growth of the organisms, they are able to reach and penetrate the cellulosic substrate and, hence, the "free" secreted cellulases are capable of efficiently hydrolyzing the substrate. The enzymes in these cases are not organized into high-molecular-weight complexes and are called noncomplexed. The polysaccharide hydrolases of the aerobic fungi are largely described based on the examples from *Trichoderma, Penicillum, Fusarium, Humicola, Phanerochaete,* etc., where a large number of the cellulases are encountered. In addition to the true cellulases, the fungal cellulase-hemicellulase systems also contain a number of xylanases, which includes endo- and exo-xylanases, β-xylosidases, and side-chain-cleaving enzymes (Rabinovich, Melnik, and Bolobova 2002). In contrast, in most of the anaerobic cellulose-degrading bacteria, the cellulase-hemicellulase systems are organized to form structures called cellulosomes and their lignocellulolytic systems are said to be complexed.

The cellulosomes are found as protuberances on the cell wall and are stable enzyme complexes capable of binding the cellulose and bringing about its degradation. Much of what is known about the cellulosomes has come though studies on the anaerobic bacterium, *Clostridium thermocellum* (Schwarz 2001). The cellulase-hemicellulase complex of *C. thermocellum* contains up to 26 polypeptides. Among them, at least 12 endo- and exo-cellulases, three xylanases, lichenase, and a noncatalytic cellulosome integrating protein (CipA) or scaffoldin have been identified. The enzymes bind through the dockerin moieties onto complementary receptors on scaffoldin, called cohesins (Bayer et al. 1998). The type of activities and the number of catalytic domains may be different in other anaerobic bacteria with complexed cellulolytic systems, but the basic architecture of the cellulosome is almost always conserved.

Noncomplexed cellulase-hemicellulase systems, however, are more common and are presently the most exploited for industrial applications. Though several filamentous fungi, actinomycetes, and aerobic bacteria are capable of producing free cellulases and xylanases that are secreted outside their cell walls, the cellulase systems of certain fungi are the most extensively studied ones. Of these, the fungus *Trichoderma reesei* has been in research focus for several decades. The noncomplexed cellulase system of *T. reesei* consists of two exo-glucanases, CBHI and CBHII, about eight endoglucanases, EGI to EGVIII, and seven β-glucosidases, BGI to BGVII (Aro, Pakula, and Penttila 2005). The cellulase system of another major cellulase producer, *Humicola insolens*, is homologous to *T. reesei* and contains at least seven cellulases. Most of the cellulases have hemicellulase activity and quite often the functional demarcation of several enzymes is difficult, except for fine differences in their ability to degrade the polymers. Many microorganisms such as *Penicillium capsulatum* and *Talaromyces emersonii* possess complete xylan-degrading enzyme systems. Though xylan has a more complex structure compared to cellulose and consequently requires several different enzymes for a complete hydrolysis, it does not form tightly packed crystalline structures like cellulose and thus is more accessible to enzymatic hydrolysis. The hemicellulases assume importance in biofuel applications mainly by facilitating cellulose hydrolysis by exposing the cellulose fibers making them more accessible to the cellulases (Shallom and Shoham 2003). The following discussions are mainly focused on the noncomplexed cellulase-hemicellulase systems, since they are the most exploited class of cellulases for industrial applications, including biofuel production.

10.3 MICROORGANISMS PRODUCING CELLULASES AND HEMICELLULASES

10.3.1 Cellulases

A large number of microorganisms, including fungi, actinomycetes, and bacteria, are capable of producing extracellular cellulases, which find applications in various industries. The ability to secrete large amounts of extracellular protein is the characteristic of certain fungi and such strains are most suited for the production of higher levels of extracellular cellulases. One of the most extensively studied fungi is *Trichoderma reesei*, which converts native as well as derived cellulose to glucose. Some other commonly studied cellulolytic organisms include the fungal species *Trichoderma, Humicola, Penicillium,* and *Aspergillus*; bacteria, *Bacilli, Pseudomonads,* and *Cellulomonas*; and actinomycetes, *Actinomucor and Streptomyces.*

Although several fungi can metabolize cellulose as an energy source, only a few strains are capable of secreting a complex of the cellulase enzymes that could have practical application in the enzymatic hydrolysis of cellulose. Besides *T. reesei,* other fungi, such as *Humicola, Penicillium,* and *Aspergillus,* and aerobic bacteria, such as *Bacillus, Cellulomonas, Cytophaga, Erwinia, Pseudomonas, Steptomyces*, etc., are capable of giving high levels of extracellular cellulases. However, the microbes commercially exploited for cellulase production are mostly limited to *T. reesei, H. insolens, A. niger, Thermomonospora fusca, Bacillus* sp., and a few other organisms.

T. reesei has a long history in industrial production of different hydrolyzing enzymes, especially cellulases and hemicellulases. The organism also has the best-characterized cellulase system and the best strains are capable of secreting up to 40 g of protein per liter of the culture (Durand, Clanet, and Tiraby 1988), most of which is cellobiohydrolase-I. However, a major limitation of *T. reesei* cellulase is the relatively lower amount of β-glucosidase activity compared to the other classes of enzymes. In the process of converting biomass to glucose, the final step in cellulose-mediated hydrolysis catalyzed by β-glucosidase is of much relevance because the substrate of this enzyme, cellobiose, which is generated by the action of cellobiohydrolases, is a very potent inhibitor of the CBH and EG enzymes if it is accumulated beyond certain limits. The cellobiose can decrease the rate of the cellulose hydrolysis by CBH and EG as much as 50% at a concentration of 3 g/l (White and Hindle 2000). This decrease in hydrolysis rate necessitates the addition of higher levels of cellulase enzymes during the biomass saccharification process, which adversely impacts the overall process economics. The goal of several research activities on cellulases has been to make the cellulose to glucose conversion process more economical by either supplying external β-glucosidase into the reaction mixture, or by enhancing the β-glucosidase production by *T. reesei*. The latter can be achieved only by understanding the cellulolytic machinery of the producers at the molecular level and targeted manipulations to obtain higher yields. Several studies have, therefore, addressed the regulation of cellulase genes.

10.3.2 Hemicellulase

A diverse array of enzymes are categorized as a specific type of hemicellulase which include glucanases, xylanases, mannanases, etc., based on their ability to hydrolyze the heteropolysaccharides composed of glucan, xylan, or mannan, respectively. It is known that the enzymes that hydrolyze hemicellulose are produced by a large number of fungi and bacteria and numerous plants. Industrial uses of the hemicellulases traditionally have been in the applications where hemicelluloses must be removed selectively to enhance the value of complex substrates such as foods, feeds, paper pulp, etc. The commercial development of the hemicellulases for the hydrolysis of lignocellulose is not as advanced as the cellulases since the current biomass to ethanol technologies have been largely developed for biomass pretreated with dilute acid where the hemicellulose is removed in the wash stream leaving behind mainly cellulose. However, with the improved outlook on pentose sugar utilization in bioethanol production and the development of nonacid pretreatment methods where the hemicellulose fraction of the biomass is recovered for alcohol fermentation, enzymes capable of hemicellulose degradation are rapidly gaining importance.

Because xylan is the second most abundant polysaccharide in any biomass (next only to cellulose) and forms a major part of the hemicelluloses, the enzymes degrading xylan assume greater importance in the context of bioethanol production from lignocellulosic biomass. Similar to cellulases, xylanase production has been reported from bacteria, fungi, and actinomycetes. Most of the cellulase producers are also capable of hemicellulase production and reports indicate the production of both the enzyme classes from several species of *Trichoderma, Aspergillus, Penicillium, Fusarium,* and *Thermomyces*. The bacterial sources are mainly species of *Bacillus*. Xylanases are also elaborated by actinomycetes like *Streptomyces and Thermoactinomyces*.

A majority of the studies on xylanase have concentrated on the production of cellulase-free xylanases for application in the paper and pulp industry where cellulases are not desired. A detailed review on the microorganisms producing xylanases and the applications of the enzymes in various industries is available in Haltrich et al. (1996) and Beg et al. (2001). Though a large number of fungi and bacteria are capable of xylanase production, the commercial sources of hemicellulases and xylanases in particular have remained species of *Trichoderma, Aspergillus, Thermomyces,* and certain *Bacilli*. Commercial sources of xylanases include the fungal strains

T. reesei and *T. viride*, while more generic industrial hemicellulase preparations are made from *A. niger*. The latter is also a source for commercial preparations of arabinase, galactosidase, and mannanase. Commercial preparations tailored for use in bioethanol production are not available at present, and unlike the cellulases, research on hemicellulases for biofuel application is only now catching up.

10.4 REGULATION OF CELLULASE AND HEMICELLULASE GENE EXPRESSION

Over several years of research, though the exact control mechanisms governing cellulase and hemicellulase expression in microbes is not fully understood, considerable information is still available on this topic, especially in the case of the cellulase genes of *Trichoderma reesei*. The *T. reesei* cellulases are inducible enzymes and the regulation of cellulase production is finely controlled by activation and repression mechanisms. The regulation of the cellulase genes has been studied to a great extent in this fungus and it is now known that the genes are coordinately regulated. The production of cellulolytic enzymes is induced only in the presence of the substrate, and is repressed when easily utilizable sugars are available. Natural inducers of cellulases have been proposed long back and the disaccharide sophorose is considered to be the most probable inducer of at least the *Trichoderma* cellulase system. It has been proposed that the inducer is generated by the trans-glycosylation activity of a basally expressed β-glucosidase. Cellobiose, δ-cellobiose-1-5-lactone, and other oxidized products of cellulose hydrolysis can also act as inducers of cellulose (reviewed in Lynd et al. 2002). Lactose is another known inducer of the cellulases and is utilized in the commercial production of the enzyme owing to economic considerations. Though the mechanism of lactose induction is not fully understood, it is believed that the intracellular galactose-1-phosphate levels might control the signaling. The glucose repression of the cellulase system overrides its induction, and de-repression is believed to occur by an induction mechanism mediated by the trans-glycosylation of glucose. Cellulase production in *T. reesei* is regulated through transcription factors (Ilmen et al. 1997). Detailed analyses performed on two cellulase promoters (*cbh1* and *cbh2*) have demonstrated the involvement of at least three transcriptional factors ACEI, ACEII, and HAP 2/3/5 and one repressor, CRE1 (reviewed in Aro, Pakula, and Penttila 2005). However, the mechanism of how the expression of these genes is turned on by the presence of cellulose is still unclear. The transcriptional activator *ACEII* binds to the promoter of *cbh1* and is believed to control the expression of *cbh1, cbh2, egl1,* and *egl2*. The *Ace1* gene also produces a transcription factor similar to *ACEII* and has binding sites in the *cbh1* promoter, but it acts as a repressor of cellulase gene expression. The *cbh1* promoter also contains the CCAAT sequence which binds the HAP 2/3/5 complex, which is another putative activator. Glucose repression of cellulase is supposed to be mediated through the carbon catabolite repressor protein *CRE1* and the promoter regions of the *cbh1, cbh2, eg1,* and *eg2* genes have *CRE1* binding sites, indicating the fine control of these genes by carbon catabolite repression. A detailed review on the induction and catabolite repression of cellulases is given by Suto and Tomita (2001).

In analogy to the cellulase systems of *T. reesei*, though many studies have been performed on the biochemistry of xylan degradation by this fungus, not much is known about the regulation of the xylanase genes. It is, however, known that most of the biomass-degrading enzymes, including cellulases and hemicellulases, in the fungus are co-regulated. Hemicellulases are also inducible enzymes and the induction is thought to be effected through low levels of certain oligosaccharides made by the enzymes that are constitutively expressed. The end products of these enzymes, especially xylobiose, is thought to be an effective inducer of xylanases. A model for the regulation of endoxylanase *xyn2* expression in *Hypocrea jecorina* (anamorph *Trichoderma reesei*) has been proposed by Wurleitner et al. (2003). The fungus elaborates two endo-xylanases, *XYN1* and *XYN2*. The expression of *xyn1* is induced by D-xylose and is repressed by glucose in a *CRE1*-dependent manner, whereas the expression of *xyn2* is partially constitutive and further induced by the xylobiose, xylan, cellulose, and sophorose. According to the model, nucleotide sequences within a 55 bp region in the promoter are responsible for the regulation of the *xyn2* gene. This region includes two adjacent *cis*-acting motifs on the noncoding strand (5′-AGAA-3′ and 5′-GGGTAAATTGG-3′, respectively), which are speculated to bind regulatory proteins. The latter sequence is believed to be the binding site of the *HAP 2/3/5* complex and *ACEII*, whereas the former is supposed to bind a repressor. It is speculated that *HAP 2/3/5* binding partially mediates repression and induction may be effected through covalent changes brought about in the complex mediated through *ACEII* phosphoryation. The regulation of the xylanolytic system is effected by a transcriptional activator called *XLNR* in *Aspergillus niger* (van Peij, Visser, and de Graaff 1998) and it is believed to control the expression of more than ten genes. Apart from the results of isolated studies on xylanase gene expression, nothing much is known about the regulation of a majority of the hemicellulases.

10.5 MOLECULAR APPROACHES IN IMPROVING PRODUCTION AND PROPERTIES OF CELLULASES AND HEMICELLULASES

Several approaches have been tried in *T. reesei* for the enhancement of cellulase production. Systematic improvements of the production strains through random mutagenesis and screening actually yielded strains with considerably enhanced levels of production reaching over 40 g/l of protein, with *CBHI* being the major component (Durand, Clanet, and Tiraby 1988). Genetic engineering techniques have been employed successfully to construct *T. reesei* strains with novel cellulase profiles. The *cbh1* promoter from *T. reesei* has been used extensively for the expression of various homologous and heterologous proteins (reviewed in Mantyla, Paloheimo, and Suominen 1998 and Pentilla 1998) in the fungus. The *cbh1* promoter is one of the best known promoters in the fungal world, which can yield an unusually high rate of expression. When the *cbh1* promoter is used for the expression of proteins in *T. reesei*, strong induction is achieved using cellulose, complex plant material, and the known inducers like sophorose. However, strong repression is also a possibility, mediated by the carbon catabolite repressor protein *CRE1*. This problem has been addressed by the finding that de-repression can be brought about by mutating a single hexanucleotide sequence at position -720 of the *cbh1* promoter which is a putative binding site for the *CRE1* repressor protein (Ilmen et al. 1996). The removal of sequences upstream of position -500 in relation to the initiator ATG also abolishes the glucose repression, and this does not affect the sophorose induction. Another major strategy employed for improving cellulase production in the presence of glucose is to use promoters that are insensitive to glucose repression. Nakari-Setala and Pentilla (1995) used the promoters of transcription elongation factors *1α* and *tef1*, and that of an unidentified cDNA (*cDNA1*) for driving the expression of endoglucanase and cellobiohydrolase in *T. reesei* with the result of de-repression of these enzymes. This implies that proper engineering of sequences to obtain expression of proteins from the *cbh1* promoter along with manipulations of the promoter to abolish repression can dramatically improve the production of the cloned protein.

A major limitation of the cellulolytic system of *T. reesei* is the relatively lower amount of β-glucosidase and its feedback inhibition by glucose. Unlike *CBH1*, which is the most abundantly expressed protein in *T. reesei* under conditions of cellulase induction, β-glucosidase is expressed to a lesser extent by the fungus. *T. reesei* has been reported to produce extracellular, cell-wall-bound, and intracellular β-glucosidases. The gene *bgl1* encodes an extracellular product that forms the major β-glucosidase in the fungus. The β-glucosidase enzyme has a transglycosylation activity that supposedly produces the inducer of the cellulase genes. Deletion of *bgl1* does not result in a complete removal of β-glucosidase activity but it results in a delayed induction of the cellulase genes by cellulose. Nevertheless, induction by sophorose is not affected, indicating that the *bgl1* gene product is involved in the formation of the soluble inducer of the cellulase enzymes. Data on the protein product of *bgl2* suggests that this second β-glucosidase is an intracellular enzyme. In an enzyme cocktail for biomass hydrolysis, the extracellular β-glucosidase plays a larger role by driving the hydrolysis to completion as well as eliminating cellobiose, which is a major inhibitor of *CBH* and *EG* enzymes. However, the commercially used cellulase producer *T. reesei* makes very little β-glucosidase and the enzyme is very sensitive to glucose inhibition. There are also reports that the enzyme is also inhibited by its own substrate, cellobiose. Considering these, a β-glucosidase that is insensitive or at least tolerant to glucose and cellobiose is highly desired for the conversion of cellulosic biomass to glucose. Research on this line has yielded potential β-glucosidases from different microorganisms such as *Candida peltata*, *Aspergillus oryzae*, and *A. niger*. However, reports on the use of these enzymes for biomass hydrolysis are rather limited. One of the major approaches taken towards improving the enzyme cocktail for biomass hydrolysis is to increase the copy number of *bgl1* and, thus, the amount of the *BGLI* enzyme in the cellulase mixture produced by *T. reesei* (Fowler, Barnett, and Shoemaker 1992). This approach, though it could enhance the production of *BGL*, is not sufficient to alleviate the shortage of β-glucosidase for cellulose hydrolysis. The amount of β-glucosidase made by natural *Trichoderma* strains must be increased several-fold to meet the requirements of cellulose hydrolysis. The *CBH1* promoter of *T. reesei* and a xylanase secretion signal was used by White and Hindle (2000) to drive the expression of the *BGL* gene and the secretion of the protein product, respectively, with some dramatic increase in the enzyme yield. This strategy can probably help to reduce the amount of cellulases needed for saccharification, but further improvements are needed in increasing the glucose tolerance of β-glucosidases.

In an effort to find novel cellulases and enhance the production and/or efficiency of the existing ones, several works have focused on the molecular cloning of the cellulases from different sources into heterologus host systems. Modification of the cellulase properties to enhance the efficiency or to impart the desired features is another major area of research. Studies on the protein engineering approaches adopted in cellulase modification are reviewed in Schulein (2000). These studies apparently give basic information about the cellulase molecular biology, which is crucial for the designing of any strategy for genetic improvement of the fungus for enhanced production of the enzyme.

Molecular approaches in improvement of the production and properties of the xylanases have been largely oriented toward developing the enzymes for the paper and pulp industry, which is currently the largest consumer of commercial xylanase preparations. The xylanases desired here are enzymes that are active at alkaline pH and/or thermotolerant. Its development for biomass conversion is rare or nonexistent. However, most of the approaches followed will be similar whether the target application is biomass conversion or other industries. There are several reports on the cloning of bacterial xylanases. The use of well-studied industrial microorganisms such as *T. reesei* or *A. niger* as hosts for the expression of desirable heterologous xylanases has the potential advantage of cost-effective industrial-scale production and bioprocess development. This potential was exploited in the expression of thermostable xylanases from *Dictyoglomus thermophilum* and *Humicola grisea* in *T. reesei* where dramatic improvements in expression were obtained (Teo et al. 2000). A review on the expression of thermostable xylanases in fungal hosts is given by Bergquist et al. (2002). The major problem associated with the expression of the heterologous proteins, especially from bacteria, is the change in codon preferences. It becomes necessary to alter the codon usage to match that of *T. reesei* while expressing the protein in this host. Cloning of *D. thermophilum* xylanases in *T. reesei* was achieved by codon optimization (Teo et al. 2000). Another impressive attempt in enzyme expression which might be suitable for biomass processing was the design of a "Xylanase–Cellulase" fusion protein. The xylanase gene from *Clostridium thermocellum* and the cellulase gene from *Pectobacterium chrysanthemi* PY35 were fused and expressed in *E. coli* to derive a bifunctional "xylanase-cellulase" (An et al. 2005). Thermostable xylanases from the fungi have also been cloned and expressed successfully in a *Pichia pastoris* expression system. Protein engineering approaches to impart desirable features to xylanases is another major area under active investigation. There have been reports on the improvement of thermotolerance by engineering of the xylanase protein in *T. reesei* and of shifting the pH optimum to alkaline pH in addition to imparting thermotolerance. The introduction of disulfide bonds has been employed successfully to impart thermotolerance in *T. reesei* and in *Bacillus circulans* xylanases. More information on the engineering of thermotolerance and pH optima of xylanases can be found in Turunen et al. (2004). With the renewed interest in hemicellulases for bioethanol production, the research on overexpression of these enzymes and their engineering to impart desirable features is expected to yield better enzymes for biomass conversion.

10.6 BIOPROCESSES FOR CELLULASE AND HEMICELLULASE PRODUCTION

Apart from organism development for cellulase and hemicellulase production, the key to a successful technology for "biomass-ethanol" production is the process for producing the enzymes itself. Numerous reports are found in the literature on aspects of cellulase or xylanase production and a majority of them aim to attain maximal specific activities at modest cost and time. Within the limits of an organism's potential for enzyme production, dramatic improvements can be made in the yield of the enzyme through the use of bioprocess optimization strategies.

10.6.1 CELLULASE PRODUCTION

Cellulase production has been the subject of active research for several decades. Probably the production of no other class of enzyme has so many choices in terms of the substrates used, ranging from pure cellulose to dairy manure. Both solid-state fermentation (SSF) and submerged fermentation (SmF) technologies have been tried successfully in cellulase production, as well as different reactor configurations (reviewed in Sukumaran, Singhania, and Pandey 2005). The majority of the studies on the microbial production of cellulases utilizes the submerged fermentation technology (SmF), and the most widely studied organism used in cellulase production is *T. reesei*. However, in nature, the growth and cellulose utilization of the aerobic microorganisms elaborating cellulases probably resembles solid-state fermentation rather than a liquid culture. That apart, the advantages of better monitoring and handling are still associated with the submerged cultures, with a range of reactor configurations to choose from.

Cellulase production in cultures is growth associated and is influenced by various factors, which alone or in interaction can affect cellulase productivity. These include the substrate used for the enzyme production, pH of the medium, fermentation temperature, aeration, inducers, etc. Agro-residues have been the major choice as substrates because they are cheap and easily available. These include lignocellulosic material such as sugarcane bagasse, rice and wheat straw, spent hulls of cereals and pulses, rice or wheat bran, paper industry waste, and various other lignocellulosic residues. Complex plant materials in the agro-residues are capable of inducing the cellulase system in the microbes just like the known inducers, or sometimes even better. Among the known inducers of cellulase genes, lactose is the only economically feasible additive in industrial fermentation media. In *T. reesei*, a basal medium after Mandels and Weber (1969) have been most frequently used with or without modifications. In the majority of reported fermentations, the pH of the medium was in the acidic range, from 4 to 6.5, and the incubation temperature ranged from 25 to 30°C. Though most of the processes are operated in batch, there have been attempts to produce cellulase in fed batch, or continuous mode, which supposedly helps to override the repression caused by the accumulation of the reducing sugar. The major technical limitation in the fermentative production of the cellulases remains the increased fermentation times with a low productivity.

SSF for the production of cellulases is rapidly gaining interest as a cost-effective technology because the enzyme preparations from SSF are more concentrated and, thus, are suitable directly for biomass saccharification (Chahal 1985), the final application for which they are needed. SSF is believed to reduce the cost of cellulase production almost tenfold (Tengerdy 1996) compared to SmF. Several researchers have proved that SSF technology results in a higher enzyme yield. Cen and Xia (1999) have reviewed the application of SSF for cellulase production, along with the microorganisms used, raw materials, pretreatment of raw materials, sterilization, and inoculation. The paper also describes bioreactors for cellulase production under SSF. Though there are a considerable number of reports on SSF production of cellulases, the process has yet to be realized at commercial levels for producing cellulase that can be used for bioethanol applications and the large-scale commercial processes still use the proven technology of SmF.

10.6.2 Xylanase Production

Commercial-scale xylanase production for the biomass to bioethanol process is very rare, and most of the existing processes have targeted production of the cellulase free xylanases, especially those with alkaline pH optima suited for applications in the paper and pulp industry. Filamentous fungi are important in production of xylanases, since they generally produce higher amounts of the enzyme compared to bacteria and yeasts. Moreover, the proteins are secreted into the medium, making the recovery rather simple. Of the microorganisms used for xylanase production, *A. niger* and *T. reesei* have been mostly used in commercial production (Haltrich et al. 1996). Other major sources of the commercial xylanases include the fungi *Humicola insolens, Thermomyces lanuginosus,* and species of *Bacilli.* A review of the different strains of *Thermomyces* and their xylanase production under various bioprocess configurations is given by Singh, Madala, and Prior (2003).

Both SmF and SSF have been used successfully for the production of xylanases from bacteria and fungi. The choice of substrate for the fermentation has been wide and varied. In general, purified xylans are good substrates for the enzyme production; in some cases cellulose can also act as a good inducer of the xylanase. However, a major problem with the use of pure xylans is the cost of the substrate. It has been noted that several cheap lignocellulosic substrates support even better production of the enzyme compared to purified xylan or cellulose. Even in cases where this is not true, the supplementation of inducers in the production medium might help to enhance the production of xylanase. The synthetic xylobiose analogue β-methyl-D-xyloside (BMX) has been used successfully as an inducer for increasing the xylanase yield from *Aspergilli.* The agro-industrial residue-based feedstock used in xylanase production is as diverse as the ones used for cellulase production and includes wheat bran, rice and wheat straw, corn cobs, xylan from different sources, sugar cane bagasse, cellulose powder, xylose, lactose, etc.

The advantages of SSF are apparent also in xylanase production and several research attempts have been oriented toward developing SSF-based processes for xylanase production. Nevertheless, as is the case with cellulases, the commercial-scale production of xylanases is mostly performed with SmF. Interested readers can find a comprehensive review on fungal xylanases and their production strategies and bioprocesses in Haltrich et al. (1996).

10.7 ASSAY OF CELLULASES AND XYLANASES

A large number of different protocols exist for the assay of cellulase and xylanase activities and different laboratories might use their own modifications. However, the International Union of Pure and Applied Chemistry (IUPAC) commission on biotechnology has recommended standard assay protocols for cellulases (Ghose 1987) and hemicellulases (Ghose and Bisaria 1987), which are now universally followed. Such standardization allows the comparison of results from different laboratories.

10.8 CELLULASES AND HEMICELLULASES FOR BIOMASS ETHANOL: CHALLENGES FOR THE FUTURE

Plant biomass is the only foreseeable renewable resource on the planet and with the depleting petroleum resources and increasing demand on energy, lignocellulose-derived ethanol seems to be the future of transportation fuels. Also, it is apparent that the integrated biorefineries, which generate chemicals from the biomass, are going to replace the current petroleum refineries, moving the world toward a carbohydrate-based economy. This being said, the major hurdle in this transition is the lack of efficient technologies for the saccharification of the biomass. Acid hydrolysis is a feasible technology but with much less efficiency and many associated problems, including pollution. Enzymatic saccharification of biomass using cellulases and hemicellulases is projected to be highly efficient with ample scope for improvement. In a process for ethanol production from lignocellulosic biomass, the enzymatic hydrolysis of the pretreated biomass is the key step and the yield of sugars from a pretreated feedstock is largely dependent on the type of enzymes and their activities. These features will largely determine the enzyme loading and the duration of the hydrolysis, which in turn determines the overall process economics. The evaluations done on the economics of bioethanol production from lignocellulosic biomass shows that the cost of the cellulase enzyme is a major contributor to the production costs and sensitivity analyses performed on the costing data indicate that at least a tenfold reduction in cellulase production costs is needed for the process to become economically attractive. Current commercial preparations of the enzymes are slow acting and are subject to problems of feedback inhibition. Major breakthroughs are needed to reduce the cost of producing the cellulases, and to bring about improvements in their activity and physical properties such as thermotolerance. Noteworthy results in this direction have been made by the U.S. National Renewable Energy Laboratory with its industrial allies Genecor and Novozymes. The NREL project has been successful in achieving more than the targeted tenfold reduction in the economics of enzymatic saccharification of biomass. Nevertheless, further improvements are needed still to make biomass ethanol competitive against gasoline as a transportation fuel.

The major goals for future cellulase research would be reduction in the cost of cellulase production and improving the performance of cellulases to make them more effective, so that less enzyme is needed. The former task may include such measures as optimizing growth conditions or processes, whereas the latter requires directed efforts in protein engineering and microbial genetics to improve the properties of the enzymes.

The key issues related to bioprocess development for cellulase production would be the use of cheaper fermentation techniques, for example, SSF, the search for cheaper inducers, the development of glucose-tolerant BGL enzymes, improving the stability, thermotolerance and resistance to shear forces of the cellulases, which are the challenges needing attention. Yet another important issue is the need for tailoring the cellulases to make them suitable for an efficient lignocellulose to bioethanol process. Important areas being explored worldwide include protein engineering to improve specific activities and overexpression of cellulase genes, as well as developing optimal cellulase mixtures and conditions for hydrolysis.

Compared to the research and development activities ongoing and initiated on cellulases for biofuel applications, the initiatives in this direction with respect to hemicellulases have been far fewer, though it needs equal attention. The wealth of knowledge gathered on xylanases developed for other applications may be effectively used for developing the enzymes for biomass conversion to ethanol.

10.9 CONCLUSIONS

The ability to utilize plant biomass, the single most abundant renewable resource for fuel, energy, and chemicals, is going to determine the future economics and probably even survival of the human population, which underlines the importance of having efficient technologies for biomass saccharification. Lignocellulose saccharification is brought about by the concerted action of a battery of enzymes, which include cellulases and hemicellulases, thereby making these enzymes the crux of the research on fuel production from biomass. After several decades of research, no gigantic leaps have been made in improving either cellulase or hemicellulase production, or the properties of these enzymes to make them more efficient and faster acting. The importance of these enzyme classes was probably underestimated, with the result that we are still far from having economically viable technologies for bioethanol production from biomass using the more energy-efficient enzymatic route. However, it seems now that this "oversight" has been realized, and active research has been reinitiated to provide the cellulases and hemicellulases the status they deserve as the crucial protein classes that is going to provide mankind with fuel, energy, and chemicals in the future. Basic knowledge of cellulase and hemicellulase molecular biology has to improve as has the application of this knowledge to improve the enzymes. The problems that warrant attention are not limited to the enzyme production and properties. A concerted effort to understand the basic physiology of lignocellulolytic microbes and the utilization of this knowledge coupled with the engineering principles is imperative to achieve a better processing and utilization of the most abundant natural resource, the plant biomass.

REFERENCES

An, J. M., Y. K. Kim, W. J. Lim, S. Y. Hong, C. L. An, E. C. Shin, K. M. Cho, B. R. Choi, J. M. Kang, S. M. Lee, H. Kim, and H. D. Yun. 2005. Evaluation of a novel bifunctional xylanase-cellulase constructed by gene fusion. *Enzyme Microb. Technol.* 36: 989–995.

Aro, N., T. Pakula, and M. Penttila. 2005. Transcriptional regulation of plant cell wall degradation by filamentous fungi. *FEMS Microbiol. Rev.* 29: 719–739.

Bayer, E. A., H. Chanzy, R. Lamed, and Y. Shoham. 1998. Cellulose, cellulases and cellulosomes. *Curr. Opin. Struct. Biol.* 8: 548–557.

Beg, Q. K., M. Kapoor, L. Mahajan, and G. S. Hoondal. 2001. Microbial xylanases and their industrial applications: A review. *Appl. Microbiol. Biotechnol.* 56: 326–338.

Bergquist, P., V. Teo, M. Gibbs, A. Cziferszky, F. P. de Faria, M. Azevedo, and H. Nevalainen. 2002. Expression of xylanase enzymes from thermophilic microorganisms in fungal hosts. *Extremophiles* 6: 177–184.

Cen, O. and L. Xia. 1999. Production of cellulase by solid-state fermentation. *Adv. Biochem. Eng. Biotechnol.* 65: 69–92.

Chahal, D. S. 1985. Solid-state fermentation with *Trichoderma reesei* for cellulase production. *Appl. Environ. Microbiol.* 49: 205–210.

Durand, H., M. Clanet, and G. Tiraby. 1988. Genetic improvement of *Trichoderma reesei* for large scale cellulase production. *Enzyme Microb. Technol.* 10: 341–345.

Fowler, T., C. C. Barnett, and S. Shoemaker. 1992. Improved saccharification of cellulose by cloning and amplification of the beta-glucosidase gene of *Trichoderma reesei*. Patent WO/1992/010581 A1 (to Genencor Int. Inc.), June 25.

Ghose, T. K. 1987. Measurement of cellulase activities. *Pure Appl. Chem.* 59: 257–268.

Ghose, T. K. and V. S. Bisaria. 1987. Measurement of hemicellulase activities. I. Xylanases. *Pure Appl. Chem.* 59: 1739–1752.

Haltrich, D., B. Nidetzky, K. D. Kulbe, W. Steiner, and S. Zupancic. 1996. Production of fungal xylanases. *Bioresour. Technol.* 58: 137–161.

Henrissat, B. 1992. Analysis of hemicellulase sequences. Relationships to other glycanases. In *Xylans and Xylanases*, ed. J. Visser, G. Beldman, M. A. Kustres-van Someren, and A. G. J. Voragen, 97–110. New York: Elsevier.

Henrissat, B., M. Claeyssens, P. Tomme, L. Lemesle, and J. P. Mornon. 1989. Cellulase families revealed by hydrophobic cluster analysis. *Gene* 81: 83–95.

Ilmen, M., M. L. Onnela, S. Klemsdal, S. Keränen, and M. Penttilä. 1996. Functional analysis of the cellobiohydrolase I promoter of the filamentous fungus *Trichoderma reesei*. *Molecular Gen. Genet.* 253: 303–314.

Ilmen, M., A. Saloheimo, M. L. Onnela, and M. E. Penttila. 1997. Regulation of cellulase gene expression in the filamentous fungus *Trichoderma reesei*. *Appl. Environ. Microbiol.* 63: 1298–1306.

Lynd, L. R., P. J. Weimer, W. H. van Zyl, and I. S. Pretorious. 2002. Microbial cellulase utilization: Fundamentals and biotechnology. *Microbiol. Mol. Biol. Rev.* 66: 506–577.

Mandels, M. and J. Weber. 1969. The production of cellulases. *Adv. Chem. Ser.* 95: 391–413.

Mantyla, A., M. Paloheimo, and P. Suominen. 1998. Industrial mutants and recombinant strains of *Trichoderma reesei*. In *Trichoderma and Gliocladium*, vol. 2, ed. G. E. Harman and C. P. Kubicek, 291–309. London: Taylor & Francis.

Nakari-Setala, T. and M. Pentilla. 1995. Production of *Trichoderma reesei* cellulases on glucose containing media. *Appl. Environ. Microbiol.* 61: 3650–3655.

Penttila, M. 1998. Heterologous protein production in *Trichoderma*. In *Trichoderma and Gliocladium*, vol. 2, ed. G. E. Harman and C. P. Kubicek, 365–382. London: Taylor & Francis.

Rabinovich, M. L., M. S. Melnik, and A. V. Bolobova. 2002. Microbial cellulases (Review). *Applied Biochem. Microbiol.* 38: 305–321.

Schulein, M. 2000. Protein engineering of cellulases. *Biochim. Biophys. Act.* 1543: 239–252.

Schwarz, W. H. 2001. The cellulosome and cellulose degradation by anaerobic bacteria. *Appl. Microbiol. Biotechnol.* 56: 634–649.

Shallom, D. and Y. Shoham. 2003. Microbial hemicellulases. *Curr. Opin. Microbiol.* 6: 219–228.

Singh, S., A. M. Madala, and B. A. Prior. 2003. *Thermomyces lanuginosus*: Properties of strains and their hemicellulases. *FEMS Microbiol. Rev.* 27: 3–16.

Sukumaran, R. K., R. R. Singhania, and A. Pandey. 2005. Microbial cellulases: Production, applications and challenges. *J. Sci. Ind. Res.* 64: 832–844.

Suto, M. and F. Tomita. 2001. Induction and catabolite repression mechanisms of cellulase in fungi. *J. Biosci. Bioeng.* 92: 305–311.

Tengerdy, R. P. 1996. Cellulase production by solid substrate fermentation. *J. Sci. Ind. Res.* 55: 313–316.

Teo, V. S. J., A. E. Cziferszky, P. L. Bergquist, and K. M. H. Nevalainen. 2000. Codon optimization of xylanase gene xynB from the thermophilic bacterium *Dictyoglomus thermophilum* for expression in the filamentous fungus *Trichoderma reesei. FEMS Microbiol. Lett.* 190: 13–19.

Turunen, O., J. Jänis, F. Fenel, and M. Leisola. 2004. Engineering the thermotolerance and pH optimum of family 11 xylanases by site-directed mutagenesis. *Meth. Enzymol.* 388: 156–167.

van Peij, N. N., J. Visser, and L. H. de Graaff. 1998. Isolation and analysis of xlnR, encoding a transcriptional activator co-ordinating xylanolytic expression in *Aspergillus niger. Mol. Microbiol.* 27: 131–142.

White, T. and C. Hindle. 2000. Genetic constructs and genetically modified microbes for enhanced production of beta-glucosidase. U.S. Patent 6015703 (to Iogen Corporation, Ottawa, CA), January 18.

Wurleitner, E., L. Pera, C. Wacenovsky, A. Cziferszky, S. Zeilinger, C. P. Kubicek, and R. L. Mach. 2003. Transcriptional regulation of *xyn2* in *Hypocrea jecorina. Eukaryot. Cell* 2: 150–158.

11 Bioethanol from Lignocellulosic Biomass

Part III Hydrolysis and Fermentation

Ramakrishnan Anish and Mala Rao

CONTENTS

ABSTRACT

Lignocellulose is the most abundant natural renewable resource and is one of the preferred choices for the production of bioethanol. As a substrate for bioethanol production it has a barrier in its complex structure, which resists hydrolysis. For lignocellulose to be amenable to fermentation, treatments are necessary that release monomeric sugars, which can be converted to ethanol by microbial fermentation. The current state of the art on acid and enzymatic hydrolysis of lignocellulose and subsequent microbial fermentation to ethanol are described in this chapter. Approaches for detoxification of the lignocellulose hydrolysate for effective fermentation to ethanol are also discussed.

11.1 INTRODUCTION

The rapid depletion of fossil fuels coupled with the increasing demands for transportation fuels has necessitated research focus on alternative renewable energy sources. Lignocellulose is the most abundant renewable resource, abundantly available for conversion to fuels. On a worldwide basis, terrestrial plants produce 1.3×10^{10} metric tons of wood per year (equivalent to 7×10^9 metric tons of coal) or about two-thirds of the world's energy requirement (Demain, Newcomb, and Wu 2005). Agriculture and other sources provide about 180 million tons of cellulosic feedstock per year. Furthermore, tremendous amounts of cellulose are available as municipal and industrial wastes causing pollution problems. Lignocellulosic biomass includes materials such as agricultural and forestry residues, municipal solid waste, and industrial wastes. Herbaceous and woody crops can also be used as a source of biomass. Lignocellulosic biomass can be used as an inexpensive feedstock for production of renewable fuels and chemicals.

Lignocellulosic biomass is made up of cellulose, hemicellulose, and a cementing material, lignin. Cellulose is a linear polymer of glucose, whereas hemicellulose is a branched heteropolymer of D-xylose, L-arabinose, D-mannose, D-glucose, D-galactose and D-glucuronic acid. Lignin is a complex, hydrophobic, cross-linked aromatic polymer that interferes with the hydrolysis process. Current processes for the conversion of biomass to ethanol involve chemical and/or enzymatic hydrolysis of cellulose and hemicellulose to the respective sugars and subsequent fermentation to ethanol. Enzymatic processes are highly specific and are carried out under mild conditions of temperature and pH and do not create a corrosion problem. The process requires the use of expensive biocatalysts. Dilute acid hydrolysis is fast and easy to perform but is hampered by nonselectivity and by-product formation.

11.2 HYDROLYSIS OF LIGNOCELLULOSIC BIOMASS

The most commonly considered hydrolysis processes are the concentrated hydrochloric acid process, the two-step dilute acid hydrolysis, and enzymatic hydrolysis. During the hydrolysis of lignocellulosic materials a wide range of compounds are released which are inhibitory to microbial fermentation. The composition of the inhibitors differs depending on the type of lignocellulosic hydrolysates.

11.2.1 ACID HYDROLYSIS

11.2.1.1 Dilute Acid Hydrolysis

Dilute acid hydrolysis of biomass is, by far, the oldest technology for converting biomass to ethanol. The first attempt at commercializing a process for producing ethanol from the wood was carried out in Germany in 1898. It involved the use of dilute acid to hydrolyze the cellulose to glucose, and was able to produce 7.6 liters of ethanol per 100 kg of wood waste (18 gal per ton).

The hydrolysis occurs in two stages to accommodate the differences between the hemicellulose and the cellulose (Harris et al. 1985) and to maximize the sugar yields from the hemicellulose and cellulose fractions of the biomass. The first stage is operated under milder conditions to hydrolyze the hemicellulose, while the second stage is optimized to hydrolyze the more resistant cellulose fraction. The liquid hydrolysates are recovered from each stage, neutralized, and fermented to ethanol.

The National Renewable Energy Laboratory (NREL), a facility of the U.S. Department of Energy (DOE) operated by Midwest Reseach Institute, Bettelle, outlined a process whereby the hydrolysis is carried out in two stages to accommodate the differences between hemicellulose and cellulose. The first stage can be operated under milder conditions, which maximize yield from the more readily hydrolyzed hemicellulose. The second stage is optimized for hydrolysis of the more resistant cellulose fraction. NREL has reported the results for a dilute acid hydrolysis of softwoods in which the conditions of the reactors were as follows: Stage 1, 0.7% sulfuric acid, 190°C, and a 3-minute residence time; Stage 2, 0.4% sulfuric acid, 215°C, and a 3-minute residence time. The liquid hydrolysates are recovered from each stage and fermented to alcohol. Residual cellulose and lignin left over in the solids from the hydrolysis reactors serve as boiler fuel for electricity or steam production. These bench-scale tests confirmed the potential to achieve yields of 89% for mannose, 82% for galactose, and 50% for glucose. Fermentation with *Saccharomyces cerevisiae* achieved ethanol conversion of 90% of the theoretical yield (Nguyen 1998).

The degradation of the lignocellulosic structure often requires two steps, first, the prehydrolysis in which the hemicellulose structure is broken down, and second, the hydrolysis of the cellulose fraction in which lignin will remain as a solid by-product. The two hydrolyzed streams are fermented to ethanol either together or separately, after which they are mixed together and distilled (Figure 11.1). During the degradation of the lignocellulosic structure, not only fermentable sugars are released, but a

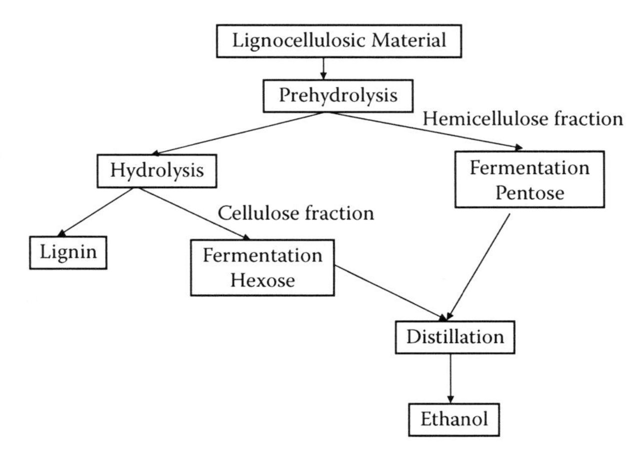

FIGURE 11.1 Flow chart for ethanol production from lignocellulosic biomass.

broad range of compounds, some of which might inhibit the fermenting microorganism. The prehydrolysis process can be performed by physical, chemical, or biological methods such as steam pretreatment, milling, freeze explosion, acid treatment (hydrochloric acid, phosphoric acid, sulfuric acid, sulfur dioxide), alkaline treatment (sodium hydroxide, ammonia), or treatment with organic solvents (ethanol, ethylene glycol) or white rot fungi (Vallander and Eriksson 1990; Saddler, Ramos, and Breuil 1993). In the prehydrolysis step, the hemicellulose is liquefied, resulting in a mixture of mono- and oligosaccharides. The hydrolysis of the cellulose is usually performed by weak acids or by enzymes (Olsson and Hahn-Hägerdal 1996).

11.2.1.2 Concentrated Acid Hydrolysis

This process is based on concentrated acid decrystallization of the cellulose followed by dilute acid hydrolysis to sugars at near theoretical yields. The separation of acid from the sugars, acid recovery, and acid reconcentration are critical operations. The fermentation converts sugars to ethanol. A process was developed in Japan in which the concentrated sulfuric acid was used for the hydrolysis. The process was commercialized in 1948. The remarkable feature of their process was the use of membranes to separate the sugar and acid in the product stream. The membrane separation, a technology that was way ahead of its time, achieved 80% recovery of acid (Wenzl 1970).

The concentrated sulfuric acid process was also commercialized in the former Soviet Union. However, these processes were only successful during times of national crisis, when economic competitiveness of ethanol production could be ignored. Concentrated hydrochloric acid has also been utilized and in this case, the prehydrolysis and hydrolysis are carried out in one step. Generally, acid hydrolysis procedures give rise to a broad range of compounds in the resulting hydrolysate, some of which might negatively influence the subsequent steps in the process. A weak acid hydrolysis process is often combined with a weak acid prehydrolysis.

In 1937, the Germans built and operated commercial concentrated acid hydrolysis plants based on the use and recovery of hydrochloric acid. Several such facilities were successfully operated. During World War II, researchers at the U.S. Department of Agriculture's Northern Regional Research Laboratory in Peoria, Illinois, further refined the concentrated sulfuric acid process for corn cobs. They conducted process development studies on a continuous process that produced a 15 to 20% xylose sugar stream and a 10 to 12% glucose sugar stream, with the lignin residue remaining as a by-product. The glucose was readily fermented to ethanol at 85 to 90% of theoretical yield. Research and development based on the concentrated sulfuric acid process studied by the USDA (and which came to be known as the "Peoria Process") picked up again in the United States in the 1980s, particularly at Purdue University and at the Tennessee Valley Authority (TVA) (Broder, Barrier, and Lightsey 1992). Among the improvements added by these researchers were recycling of dilute acid from the hydrolysis step for pretreatment, and improved recycling of sulfuric acid. Minimizing the use of sulfuric acid and recycling the acid cost effectively are critical factors in the economic feasibility of the process. (see http://www1.eere.energy.gov/biomass/printable_versions/concentrated_acid.html). The conventional wisdom in the literature suggests that the Peoria and TVA processes cannot be economical because of the high volumes of acid required (Wright and d'Agincourt 1984). The improvements in the acid sugar separation and recovery have opened the door for commercial application. Two companies, Arkenol and Masada, in the United States are currently working with DOE and NREL to commercialize this technology by taking advantage of niche opportunities involving the use of biomass as a means of mitigating waste disposal or other environmental problems (http://www1.eere.energy.gov/biomass/concentrated_acid.html). Minimizing the use of the sulfuric acid and recycling the acid cost effectively are the critical factors in the economic feasibility of the process. U.S. Patent 5,366,558 (Brink 1994) describes the use of two "stages" to hydrolyze the hemicellulose sugars and the cellulosic sugars in a countercurrent process using a batch reactor, which results in poor yields of glucose and xylose using a mineral acid. Further, the process scheme is complicated and the economic potential on a large scale to produce inexpensive sugars for fermentation is low. U.S. Patent 5,188,673 employs concentrated acid hydrolysis which has the benefit of high conversion of biomass, but suffers from low product yields due to degradation and the requirement of acid recovery and recycling. Sulfuric acid concentrations used are 30 to 70 weight percent at temperatures less than 100°C. Although 90% hydrolysis of the cellulose and hemicellulose is achieved by this process, the concentrated acids are toxic, corrosive, and hazardous and require reactors that are resistant to corrosion. In addition, the concentrated acid must be recovered after the hydrolysis to make the process economically feasible (Von Sivers and Zacchi 1995). A multifunction process for hydrolysis and fractionation of lignocellulosic biomass to separate hemicellulosic sugars using mineral acids like sulfuric acid, phosphoric acid, or nitric acid has been described (Torget et al., U.S. Patent 6,022,419). A process for treatment of hemicellulose and cellulose in two different configurations has also been described (Scott and Piskorz, U.S. Patent 4,880,473). Hemicellulose is treated with dilute acid in a conventional process. The cellulose is separated out from the prehydrolysate and then subjected to pyrolysis at high temperatures. Further, the process step between the hemicellulose and cellulose reactions requires a drying step with a subsequent high-temperature pyrolysis step at 400 to 600°C for conversion of the cellulose to fermentable products. A 70% yield of glucose was obtained from the hydrolysis of lignocellulose under extremely low acid and high temperature conditions by autohydrolysis (Ojumu and Ogunkunle 2005).

11.3 ENZYMATIC HYDROLYSIS OF LIGNOCELLULOSIC BIOMASS

The enzymatic hydrolysis or saccharification of lignocellulosic biomass is preceded by a pretreatment process in which the lignin component is separated from the cellulose and hemicellulose to make it amenable to the enzymatic hydrolysis. The lignin interferes with hydrolysis by blocking the access of the cellulases to the cellulose and by irreversibly binding the hydrolytic enzymes. Therefore, the removal of the lignin can dramatically increase the hydrolysis rate (McMillan 1994). For the efficient enzymatic hydrolysis of lignocellulosic biomass a pretreatment step is necessary. Various pretreatment processes and the enzymes involved in hydrolysis have been described in different chapters.

11.3.1 Factors Governing Enzymatic Hydrolysis

There are different factors that affect the enzymatic hydrolysis of cellulose, namely, substrates, cellulase activity, and reaction conditions (temperature, pH, as well as other parameters). To improve the yield and rate of enzymatic hydrolysis, research has been focused on optimizing the hydrolysis process and enhancing the cellulase activity. The yield and initial rate of enzymatic hydrolysis of cellulose is affected mainly by the substrate concentration. At low substrate levels, an increase of substrate concentrations yields an increase in the reaction rate of the hydrolysis and the products (Cheung and Anderson 1997). However, substrate inhibition is caused at high substrate concentration, which considerably lowers the rate of hydrolysis. The ratio of the enzyme to substrate in the hydrolysis reaction is crucial to establish the level of substrate inhibition (Huang and Penner 1991). The hydrolysis of cellulosic substrates by the enzymes depend to a large extent on the structural features of the substrate, such as cellulose crystallinity, degree of cellulose polymerization, surface area, and content of lignin (Table 11.1).

The yield and rate of hydrolysis of the cellulosic substrate can be increased to a certain extent by increasing the dosage of the cellulases in the process, but that would significantly increase the cost of the process. Cellulase dosage of 10 FPU/g cellulose is often used in laboratory studies because it provides a hydrolysis profile with high levels of glucose yield in a reasonable time (48–72 h) at a reasonable enzyme cost (Gregg and Saddler 1996). Depending on the type and concentration of the substrates, cellulase enzyme could be used in the hydrolysis (7–33 FPU/g substrate). The adsorption of the cellulase enzymes onto the surface of the cellulose, the biodegradation of cellulose to fermentable sugars, and desorption of the cellulase are three steps involved in enzymatic hydrolysis of the cellulose. The cellulase activity decreases during hydrolysis because of the irreversible adsorption of the cellulase on the cellulose (Converse et al. 1988). The cellulose surface property can be modified and the irreversible binding of the cellulase can be minimized by the addition of surfactants during the hydrolysis. The ionic surfactants Q-86W (cationic) at high concentration and Neopelex F-25 (anionic) have been shown to have an inhibitory effect (Ooshima,

TABLE 11.1
Structural Properties Potentially Limiting Enzymatic Hydrolysis of Cellulosic Fibers at Different Structural Levels

Structural Level	Substrate Factor
Microfibril	Degree of polymerization
	Crystallinity
	Cellulose lattice structure
Fibril	Structural composition (lignin content and distribution)
	Particle size (fibril dimension)
Fiber	Available surface area
	Degree of fiber swelling
	Pore structure and distribution

From Mansfield et al. 1999. Biotechnol. Progr. 15: 804–816. With permission.

Sakata, and Harano 1986), hence, the nonionic surfactants such as Tween 20, 80 (Wu and Ju 1998), polyoxyethylene glycol (Park et al., 1992), Tween 81, Emulgen 147, amphoteric Anhitole 20BS, cationic Q-86W (Ooshima, Sakata, and Harano 1986), sophorolipid, rhamnolipid, and bacitracin (Helle, Duff, and Cooper 1993) have been used to enhance the cellulose hydrolysis. Cellulose conversion with 2% (w/v) F68 and 2 g/l cellulase reached 52%, compared to 48% conversion with 10 g/l cellulase in a surfactant-free system (Wu and Ju 1998). However, Tween 20 was highly inhibitory to *D. clausenii* even at a low concentration of 0.1%. Use of a cellulase mixture from different microorganisms, or a mixture of cellulases and other enzymes, in the hydrolysis of cellulosic materials was studied by Excoffier, Toussaint, and Vignon (1991). The addition of β-glucosidases into the *Trichoderma reesei* cellulases system achieved better saccharification than the system without β-glucosidases. The β-glucosidase hydrolyzes the cellobiose, which is an inhibitor of the cellulase activity. The saccharification of the cellulose is reported to be faster when supplemented with additional β-glucosidase. There are few organisms that secrete complete cellulase, for example, *Penicillium funiculosum* with high β-glucosidases activity (Rao, Seeta, and Deshpande 1983). A mixture of hemicellulases or pectinases with cellulases exhibited a significant increase in cellulose conversion (Beldman et al. 1984). A 90% enzymatic saccharification of 8% alkali-treated sugarcane bagasse has been reported when a mixture of the cellulases (dose, 1.0 FPU/g substrate) from *Aspergillus ustus* and *Trichoderma viride* was used (Mononmani and Sreekantiah 1987). The use of the cellulase mixture of the commercial Cellucast and Novozyme preparations has achieved a nearly complete saccharification of steam-explosion pretreated *Eucalyptus viminalis* chips (substrate concentration of 6% and enzyme loading of 10 FPU/g cellulose) (Ramos, Brueil, and Saddler 1993). Baker, Adney, and Nieves (1994) reported a new thermostable endoglucanase from *Acidothermus cellulolyticus* E1 and an endoglucanase from *T. fusca* E5 that exhibited striking synergism with *T. reesei* CBH1 in the saccharification of the microcrystalline cellulose. The cellulases can be recovered from the liquid supernatant or the solid residues and recycled. Enzyme recycling can effectively increase the rate and yield of the hydrolysis and lower the enzyme cost (Mes-Hartree, Hogan, and Saddler 1987). The efficiency of cellulose hydrolysis gradually decreases with each recycling step (Ramos, Brueil, and Saddler 1993).

Recently, the enzymatic hydrolysis of lignocellulosic biomass has been optimized using enzymes from different sources and mixing in an appropriate proportion using a statistical approach of factorial design. A twofold reduction in the total protein required to reach glucan to glucose and xylan to xylose hydrolysis targets (99% and 88% conversion, respectively), thereby validating this approach toward enzyme improvement and process cost reduction for lignocellulose hydrolysis (Kim, Kang, and Lee 1997, Berlin et al. 2005).

Many studies have been presented over the years aiming to understand the inhibiting factors in enzymatic hydrolysis of lignocellulose substrates. Reasons for low yield of fermentable sugars in enzymatic conversion include reduced accessible surface area of cellulose in the lignocellulose complex, leading to restricted access for enzymes; restricted pore volume of the substrate (Eklund et al. 1990; Mooney et al. 1998); slow enzyme kinetics for crystalline cellulose (Fan et al. 1980); and obstacles in the structure of cellulose leading to unproductive enzyme binding (Eriksson, Karlsson, and Tjerneld 2002; Väljamäe et al., 1998). Lignin has also been identified to have a high binding affinity for cellulase proteins (Lu et al. 2002; Berlin et al. 2005). Both addition of lignin (Sewalt et al. 1997) and the composition of lignin have been shown to be responsible for inhibitory factors for the degradation of cellulose. It was recently found that cellulases lacking cellulose binding module (CBM) also have a high affinity for lignin, indicating the presence of lignin-binding sites on the catalytic module (Berlin et al. 2005).

An enhancement in enzymatic hydrolysis of softwood lignocellulosic by non-ionic surfactants and polymers was observed. It was suggested that ethylene oxide containing surfactants and polymers such as polyethylene glycol bind to lignin by hydrophobic interaction and hydrogen bonding and helps to reduce the unproductive binding of enzymes, thus yielding more fermentable sugars (Börjesson, Peterson, and Tjerneld 2007).

11.3.2 Detoxification

Biological, physical, and chemical methods have been employed for detoxification (the specific removal of inhibitors prior to fermentation) of lignocellulosic hydrolysates (Olsson and Hahn-Hägerdal, 1996). The methods of detoxification change depending on the source of the lignocellulosic hydrolysate and the microorganism being used. The lignocellulosic hydrolysates vary in their degree of inhibition and different microorganisms have different inhibitor tolerances. Several reports on adaptation of yeasts to inhibiting compounds in lignocellulosic hydrolysates are found in the literature (e.g., Amartey and Jeffries 1996; Buchert, Puls, and Poutanen 1988; Nishikawa, Sutcliffe, and Saddler 1988).

11.3.2.1 Biological Detoxification Methods

Biological methods of treatment make use of specific enzymes or microorganisms that act on the toxic compounds present in hydrolysates and change their composition. Treatment with the enzymes peroxidase and laccase, obtained from the ligninolytic fungus *Trametes versicolor*, has been shown to increase maximum ethanol productivity in a hemicellulose hydrolysate of willow two to three times due to their action on acid and phenolic compounds (Jönsson et al. 1998). The filamentous soft-rot fungus *Trichoderma reesei* has been reported to degrade inhibitors in a hemicellulose hydrolysate obtained after steam pretreatment of willow, resulting in around three times increased maximum ethanol productivity and four times increased ethanol yield (Palmqvist et al. 1997). Acetic acid, furfural, and benzoic acid derivatives were removed from the hydrolysate by treatment with *T. reesei*. The use of microorganism has also been proposed to selectively remove inhibitors from lignocellulose hydrolysates. Adaptation of a microorganism to the hydrolysate is another interesting biological method for improving the fermentation of hemicellulosic hydrolysate media.

11.3.2.2 Physical Detoxification Methods

Hydrolysate concentration by vacuum evaporation is a physical detoxification method for reducing the concentration of volatile compounds such as acetic acid, furfural, and vanillin present in the hydrolysate. However, physical detoxification increases moderately the concentration of nonvolatile toxic compounds and consequently the degree of fermentation inhibition.

11.3.2.3 Chemical Detoxification Methods

Chemical detoxification includes precipitation of toxic compounds and ionization of some inhibitors under certain pH values, the latter being able to change the degree of toxicity of the compounds (Mussatto 2002). Toxic compounds may also be adsorbed on activated charcoal (Dominguez, Gong, and Tsao 1996; Mussatto and Roberto 2001), on diatomaceous earth (Ribeiro et al. 2001) and on ion exchange resins (Larsson et al. 1999; Nilvebrant et al. 2001).

11.4 FERMENTATION OF LIGNOCELLULOSIC BIOMASS TO ETHANOL

The hydrolysis of lignocellulosic biomass yields reducing sugars. Once the sugars are available, its fermentation to ethanol is not a difficult task as many technologies have been developed. Essentially, there are three different types of processes by which this can be achieved, namely,

1. Separate hydrolysis and fermentation (SHF)
2. Direct microbial conversion (DMC)
3. Simultaneous saccharification and fermentation (SSF)

SSF has been shown to be the most promising approach to biochemically convert cellulose to ethanol in an effective way (Wright, Wyman, and Grohmann 1988).

11.4.1 Separate Hydrolysis and Fermentation (SHF)

This is a conventional two-step process where the lignocellulose is hydrolyzed using enzymes to form reducing sugars in the first step and the sugars thus formed are fermented to ethanol in the second step using *Saccharomyces* or *Zymomonas* (Bisaria and Ghose 1981; Philippidis 1996). The advantage of this process is that each step can be carried out at its optimum conditions.

11.4.2 Direct Microbial Conversion (DMC)

This process involves three major steps, namely, enzyme production, hydrolysis of the lignocellulosic biomass, and the fermentation of the sugars, all occurring in one step (Hogsett et al. 1992). The relatively lower tolerance of the ethanol is the main disadvantage of this process. A lower tolerance limit of about 3.5% has been reported as compared to 10% of ethanologenic yeasts. Acetic acid and lactic acid are also formed as by-products in this process in which a significant amount of carbon is utilized (Klapatch et al. 1994). *Neurospora crassa* is known to produce ethanol directly from cellulose/hemicellulose, because it produces both cellulase and xylanase and also has the capacity to ferment the sugars to ethanol anaerobically (Deshpande et al. 1986).

11.4.3 Simultaneous Saccharification and Fermentation (SSF)

The saccharification of lignocellulosic biomass by enzymes and the subsequent fermentation of the sugars to ethanol by yeast such as *Saccharomyces* or *Zymomonas* take place in the same vessel in this process (Glazer and Nikaido 1995). The compatibility of both saccharification and fermentation processes with respect to various conditions, such as pH, temperature, substrate concentration, etc., is one of the most important factors governing the success of the SSF process. The main advantages of using SSF for ethanol bioconversion are:

- Enhanced rate of lignocellulosic biomass (cellulose and hemicellulose) due to removal of the sugars that inhibit cellulase activity
- Lower enzyme loading
- Higher product yield
- Reduced inhibition of the yeast fermentation in case of continuous recovery of the ethanol
- Reduced requirement for aseptic conditions, resulting in increasing economics of the process (Deshpande, Siva Raman, and Rao 1984; Schell et al. 1988; Wright, Wyman, and Grohmann 1988; Philippidis and Smith 1995).

Because several inhibitory compounds are formed during hydrolysis of the raw material, the hydrolytic process has to be optimized so that inhibitor formation can be minimized. When low concentrations of inhibitory compounds are present in the hydrolysate, detoxification is easier and fermentation is cheaper. The choice of detoxification method has to be based on the degree of microbial inhibition caused by the compounds. As each detoxification method is specific to certain types of compounds, better results can be obtained by combining two or more different methods. Another factor of great importance in the fermentative processes is the cultivation conditions, which, if inadequate, can stimulate the inhibitory action of the toxic compounds.

SSF seems to offer a better option for commercial production of ethanol from lignocellulosic biomass. *Penicillium funiculosum* cellulase and *Saccharomyces uvarum* cells have been reported to be used for SSF (Deshpande et al. 1981).

11.5 RECOMBINANT DNA APPROACHES

Recombinant DNA methods are being used currently for lignocellulosic hydrolysis and fermentation to ethanol. Genetic manipulations of *Saccharomyces cerevisiae* and *Z. mobilis* have been explored for improving their ability to utilize lignocellulosic biomass. *S. cerevisiae* has been engineered with arabinose metabolizing genes from yeasts such as *Candida aurigiensis* (Jeffries and Shi, 1999) and xylose transporting gene from *P. stipitis*. *Z. mobilis*, an ethanologenic microorganism has been engineered to utilize glucose through the Entner-Doudoroff pathway and possess elevated levels of glycolytic and ethanologenic enzymes (pyruvate decarboxylase, PDC, and alcohol dehydrogenase, ADH), resulting in high ethanol yields, around 97% of theoretical value (Zhang et al. 1995). A recombinant strain of *Z. mobilis* has been constructed wherein the xylose and arabinose utilization genes have been inserted.

11.6 CONCLUSIONS AND FUTURE PROSPECTS

The effective hydrolysis of cellulosic biomass requires the synergistic action of cellulases such as exocellulase, endocellulase, and β-glucosidase. Even though soluble substrates have been developed for measuring endoglucanase and β-glucosidase activities there are very few substrates available for the estimation of exoglucanase activity. The hydrolysis data from soluble substrates cannot yield useful information on the hydrolysis of insoluble substrates. The heterogeneity of the cellulosic biomass, the dynamic interactions between insoluble substrates, and the complexity of cellulase components result in formidable problems in extrapolating the activity measured on one solid substrate to other solid substrates, especially those with significance for biorefinery processes. This point is critical to the eventual improvement of cellulases for the conversion of pretreated plant cell walls in energy crops and agricultural residues. Realistic methods must be based on physically and chemically relevant industrial substrates.

For an economically viable bioconversion process it is necessary to utilize both the cellulosic and hemicellulosic fractions of biomass. Extensive work has been conducted on xylose fermenting yeasts, such as *Pachysolen tannophilus*, *Candida shehatae*, and *Pichia stiptis*. However, low ethanol tolerance and catabolite repression in xylose conversion due to glucose need to be addressed. Another potential ethanologen is recombinant *Zymomonas mobilis* in view of its ability to ferment both xylose and glucose. An *Escherichia coli* strain developed by Ingram's group deserves special recognition, as it not only ferments all five sugars (glucose, xylose, arabinose, mannose, and galactose) present in synthetic sugar mixtures to ethanol but also performs competently in real hydrolysates like that of *Pinus*. *Saccharomyces* LNH-ST is another promising recombinant capable of fermenting dilute acid-treated corn fiber hydrolysates.

Although bioethanol production has been greatly improved by new technologies there are still challenges that need further investigation. These challenges include maintaining a stable performance of the genetically engineered yeasts in commercial-scale fermentation operations and integrating the optimal components into economic ethanol production systems.

Metabolic engineering and other classical techniques such as random mutagenesis address the further enhancement of microorganism capabilities by adding or modifying traits such as tolerance to ethanol and inhibitors, efficient hydrolysis of cellulose/hemicellulose, thermotolerance, reduced need for nutrient supplementation, and improvement of sugar transport. The improvement achieved in the fermentation step with the help of metabolic engineering is just one of the aspects of an integrated process. Keeping a realistic perspective one can conclude that several pieces still remain to be properly assembled and optimized before an efficient industrial configuration is acquired.

REFERENCES

Amartey, S. and T. Jeffries. 1996. An improvement in *Pichia stipitis* fermentation of acid-hydrolysed hemicellulose achieved by over-liming (calcium hydroxide treatment) and strain adaptation. *Wo. J. Microbiol. Biotechnol.* 12: 281–283.

Baker, J. O., W. S. Adney, and R. A. Nieves. 1994. A new thermostable endoglucanase, *Acidothermus cellulolyticus* E1: Synergism with *Trichoderma reesei* CBH1 and comparison to *Thermomonospora fusca* E5. *Appl. Biochem. Biotechnol.* 45/46: 245–256.

Berlin, A., N. Gilkes, A. Kurabi, R. Bura, M. B. Tu, D. Kilburn, and J. N. Saddler. 2005. Weak lignin binding enzymes: A novel approach to improve activity of cellulases for hydrolysis of lignocellulosics. *Appl. Biochem. Biotechnol.* 121: 163–170.

Bisaria, V. S. and T. K. Ghose. 1981. Biodegradation of cellulosic materials: Substrates, microorganisms, enzymes and products. *Enz. Microb. Technol.* 3: 90–104.

Börjesson, J., R. Peterson, and F. Tjerneld. 2007. Enhanced enzymatic conversion of softwood lignocellulose by polyethylene glycol addition. *Enz. Microb. Technol.* 40: 754–762.

Brink, D. L. 1994. Methods of treating biomass material. U.S. Patent 5366558, Nov. 22.

Broder, J. D., J. W. Barrier, and G. R. Lightsey. 1992. Conversion of cotton trash and other residues to liquid fuel. In *Liquid Fuels from Renewable Resources: Proceedings of an Alternative Energy Conference,* ed. J. S. Cundiff, 189–200. St. Joseph, MI: American Society of Agricultural Engineers.

Buchert, J., J. Puls, and K. Poutanen. 1988. Comparison of *Pseudomonas fragi* and *Gluconobacter oxydans* for production of xylonic acid from hemicellulose hydrolysates. *Appl. Microbiol. Biotechnol.* 28: 367–372.

Cheung, S. W. and B. C. Anderson. 1997. Laboratory investigation of ethanol production from municipal primary wastewater. *Bioresour. Technol.* 59: 81–96.

Converse, A. O., R. Matsuno, M. Tanaka, and M. Taniguchi. 1988. A model for enzyme adsorption and hydrolysis of microcrystalline cellulose with slow deactivation of the adsorbed enzyme. *Biotechnol. Bioeng.* 32: 38–45.

Demain, A. L., M. Newcomb, and J. H. D. Wu. 2005. Cellulase, clostridia, and ethanol. *Microbiol. Mol. Biol. Rev.* 69: 124–154.

Deshpande, V. V., S. Keskar, C. Mishra, and M. Rao. 1986. Direct conversion of cellulose/hemicellulose to ethanol by *Neurospora crassa*. *Enz. Microb. Technol.* 8: 149.

Deshpande, V. V., H. Sivaraman, and M. Rao. 1981. Simultaneous saccharification and fermentation of cellulose to ethanol using *P. funiculosum* cellulase and free or immobilized *Saccharomyces uvarum* cells. *Biotech. Bioeng.* 25: 1679–1684.

Dominguez, J. M., C. S. Gong, and G. T. Tsao. 1996. Pretreatment of sugar cane bagasse hemicellulose hydrolyzate for xylitol production by yeast. *Appl. Biochem. Biotechnol.* 57–58: 49–56.

Eklund, R., M. Galbe, and G. Zacchi. Optimization of temperature and enzyme concentration in the enzymatic saccharification of steam-pretreated willow. *Enzyme and Microbial Technology.* 12 (3): 225-228.

Eriksson, T., J. Karlsson, and F. Tjerneld. 2002. A model explaining declining rate in hydrolysis of lignocellulose substrates with cellobiohydrolase I (Cel7A) and endoglucanase I (Cel 7B) of *Trichoderma reesie*. *Appl. Biochem. Biotechnol.* 101: 41–60.

Excoffier, G., B. Toussaint, and M. R. Vignon. 1991. Saccharification of steam-exploded poplar wood. *Biotechnol. Bioeng.* 38: 1308–1317.

Fan, L. T., Y. H. Lee, and D. H. Beardmore. 1980. Mechanism of the enzymatic hydrolysis of cellulose: Effects of major structural features of cellulose on enzymatic hydrolysis. *Biotechnol. Bioeng.* 22: 177-199.

Fan, L. T., Y. H. Lee, and M. M. Gharpuray. 1982. The nature of lignocellulosics and their pretreatments for enzymatic hydrolysis. *Adv. Biochem. Eng.* 23: 158–187.

Glazer, A. N. and H. Nikaido. 1995. From biomass to fuel. In *Microbial Biotechnology: Fundamentals of Applied Microbiology,* 325–391. New York: W.H. Freeman.

Gregg, D. J. and J. N. Saddler. 1996. Factors affecting cellulose hydrolysis and the potential of enzyme recycle to enhance the efficiency of an integrated wood to ethanol process. *Biotechnol. Bioeng.* 51: 375–383.

Harris, J. F., A. J. Baker, A. H. Conner, T. W. Jeffries, J. L. Minor, R. C. Patterson, R. W. Scott, E. L. Springer, and J. Zorba. 1985. *Two-Stage Dilute Sulfuric Acid Hydrolysis of Wood: An Investigation of Fundamentals.* General Technical Report FPL-45, U.S. Forest Products Laboratory, Madison, Wisconsin.

Helle, S. S., S. J. B. Duff, and D. G. Cooper. 1993. Effect of surfactants on cellulose hydrolysis. *Biotechnol. Bioeng.* 42: 611–617.

Hogsett, D. A., H. J. Ahn, T. D. Bernardez, C. R. South, and L. R. Lynd. 1992. Direct microbial conversion: Prospects, progress and obstacles. *Appl. Biochem. Biotechnol.* 34/35: 527–541.

Huang, X. L. and M. H. Penner. 1991. Apparent substrate inhibition of the *Trichoderma reesei* cellulase system. *J. Agric. Food Chem.* 39: 2096–2100.

Jeffries, T. W. and N. Q. Shi. 1999. Genetic engineering for improved xylose fermentation by yeasts. *Adv. Biochem. Eng. Biotechnol.* 65: 117–161.

Jönsson, L. J., E. Palmqvist, N. O. Nilvebrant, and B. Hahn-Hägerdal. 1998. Detoxification of wood hydrolysates with laccase and peroxidase from the white-rot fungus *Trametes versicolor. Appl. Microbiol. Biotechnol.* 49: 691–697.

Kim, S. W., S. W. Kang, and J. S. Lee. 1997. Cellulase and xylanase production by *Aspergillus niger* KKS in various bioreactors. *Bioresource Technol.* 59: 63–67.

Klapatch, T. R., D. A. L. Hogsett, S. Baskaran, S. Pal, and L. R. Lynd. 1994. Organism development and characterization for ethanol production using thermophilic bacteria. *Appl. Biochem. Biotechnol.* 45/46: 209–213.

Larsson, S., A. Reimann, N. Nilvebrant, and L. J. Jönsson. 1999. Comparison of different methods for the detoxification of lignocellulose hydrolysates of spruce. *Appl. Biochem. Biotechnol.* 77–79: 91–103.

Lu, Y., B. Yang, D. Gregg, J. N. Saddler, and S. D. Mansfield. 2002. Cellulase adsorption and an evaluation of enzyme recycle during hydrolysis of steam exploded softwood residues. *Appl. Biochem. Biotechnol.* 98: 641–654.

Mansfield, S. D., Mooney, C., and Saddler, J. N. 1999. Substrate and enzyme characteristics that limit cellulose hydrolysis. *Biotechnol. Progr.* 15: 804–816.

McMillan, J. D. 1994. Pretreatment of lignocellulosic biomass. In *Enzymatic Conversion of Biomass for Fuels Production*, ed. M. E. Himmel, J. O. Baker, and R. P. Overend, 292–324.Washington, DC: American Chemical Society.

Mes-Hartree, M., C. M. Hogan, and J. N. Saddler. 1987. Recycle of enzymes and substrate following enzymatic hydrolysis of steam pretreated aspenwood. *Biotechnol. Bioeng.* 30: 558–564.

Mononmani, H. K. and K. R. Sreekantiah. 1987. Saccharification of sugar-cane bagasse with enzymes from *Aspergillus ustus* and *Trichoderma viride. Enzyme Microb. Technol.* 9: 484–488.

Mooney, C. A., S. H. Mansfield, M. G. Touhy, and J. N. Saddler. 1998. The effect of initial pore size and lignin content on the enzymatic hydrolysis of softwood. *Biores. Technol.* 64: 113–119.

Mussatto, S. I. 2002. Influencia do Tratamento do Hidrolisado Hemicelulosico de Palha de Arroz na Producao de Xilitol por Candida guilliermondii. M.Sc. thesis, Faculdade de Engenharia Quımica de Lorena, Brasil.

Mussatto, S. I. and I. C. Roberto. 2001. Hydrolysate detoxification with activated charcoal for xylitol production by *Candida guilliermondii. Biotechnol. Lett.* 23: 1681–1684.

Nguyen, Q. 1998. *Milestone Completion Report: Evaluation of a Two-Stage Dilute Sulfuric Acid Hydrolysis Process.* Internal Report, National Renewable Energy Laboratory, Golden, Colorado.

Nilvebrant, N. O., A. Reimann, S. Larsson, and L. J. Jönsson. 2001. Detoxification of lignocellulose hydrolysates with ion exchange resins. *Appl. Biochem. Biotechnol.* 91–93: 35–49.

Nishikawa, N. K., R. Sutcliffe, and J. N. Saddler. 1988. The influence of lignin degradation products on xylose fermentation by *Klebsiella pneumoniae. Appl. Microbiol. Biotechnol.* 27: 549–552.

Ojumu, T. V. and O. A. Ogunkunle. 2005. Production of glucose from lignocellulosic under extremely low acid and high temperature in batch process, auto-hydrolysis approach. *J. Appl. Sci.* 5: 15–17.

Olsson, L. and B. Hahn-Hägerdal. 1996. Fermentation of lignocellulosic hydrolysates for ethanol production. *Enz. Microb. Technol.* 18: 312–331.

Ooshima, H., M. Sakata, and Y. Harano. 1986. Enhancement of enzymatic hydrolysis of cellulose by surfactant. *Biotechnol. Bioeng.* 28: 1727–1734.

Palmqvist, E., B. Hahn-Hägerdal, Z. Szengyel, G. Zacchi, and K. Reczey. 1997. Simultaneous detoxification and enzyme production of hemicellulose hydrolysates obtained after steam pretreatment. *Enz. Microb. Technol.* 20: 286–293.

Park, J. W., Y. Takahata, T. Kajiuchi, and T. Akehata. 1992. Effects of nonionic surfactant on enzymatic hydrolysis of used newspaper. *Biotechnol. Bioeng.* 39: 117-120.

Philippidis, G. P. 1996. Cellulose bioconversion technology. In *Handbook on Bioethanol: Production and Utilization*, Ed. C. E. Wyman, 253–285. Washington, DC: Taylor & Francis.

Philippidis, G. P. and T. K. Smith. 1995. Limiting factors in the simultaneous saccharification and fermentation process for conversion of cellulosic biomass to fuel ethanol. *Appl. Biochem. Biotechnol.* 51/52: 117–124.

Ramos, J. P., C. Breuil, and J. N. Saddler. 1993. The use of enzyme recycling and the influence of sugar accumulation on cellulose hydrolysis by *Trichoderma* cellulases. *Enzyme Microb. Technol.* 15: 19–25.

Rao, M., R. Seeta, and V. Deshpande. 1983. Effect of pretreatment on the hydrolysis of cellulose by *Penicillium funiculosum* cellulase and recovery of enzyme. *Biotech. Bioeng.* 25: 1863–1871.

Ribeiro, M. H. L., P. A. S. Lourenc, J. P. Monteiro, and S. Ferreira-Dias. 2001. Kinetics of selective adsorption of impurities from a crude vegetable oil in hexane to activated earths and carbons. *Eur. Food Res. Technol.* 213: 132–138.

Saddler, J. N., L. P. Ramos, and C. Breuil. 1993.Steam pretreatment of lignocellulosic residues. In *Bioconversion of Forest and Agricultural Plant Residues,* ed. J. N. Saddler, 73–91. Wallingford, UK: CAB International.

Schell, D. J., N. D. M. Hinman, C. E. Wyman, and P. J. Werdene. 1988. Whole broth cellulase production for use in simultaneous saccharification and fermentation of cellulose to ethanol. *Appl. Biochem. Biotechol.* 17: 279–291.

Scott, D. S. and J. Piskorz. 1989. Process for the production of fermentable sugars from biomass. U.S. Patent 4880473, Nov. 14.

Sewalt, V. J. H., W. G. Glasser, and K. A. Beauchemin. 1997. Lignin impact on fiber degradation. 3. Reversal of inhibition of enzymatic hydrolysis by chemical modification of lignin and by additives. *J. Agric. Food Chem.* 45 (5): 1823-1828.

Sherrard, E. C. and F. W. Kressman. 1945. Review of processes in the United States prior to World War II. *Industr. Eng. Chem.* 37: 5–8.

Torget, R. W., N. Padukone, C. Hatzis, and C. E. Wyman. 2000. Hydrolysis and fractionation of lignocellulosic biomass. U.S. Patent 6022419, Feb. 8.

Väljamäe, P., V. Sild, G. Pettersson, and G. Johansson. 1998. The initial kinetics of hydrolysis by cellobiohydrolases I and II is consistent with a surface-erosion model. *Eur. J. Biochem.* 253: 469–475.

Vallander, L. and K. E. L. Eriksson. 1990. Production of ethanol from lignocellulosic materials: State of the art. *Adv. Biochem. Eng. Biotechnol.* 42: 63–95.

Van Zyl, C., B. A. Prior, and J. C. du Preez. 1991. Acetic acid inhibition of D-xylose fermentation by Pichia stipitis. *Enzyme Microb. Technol.* 13: 82–86.

Von Sivers, M. and G. Zacchi. 1995. A techno-economical comparison of three processes for the production of ethanol from pine. *Bioresour. Technol.* 51: 43–52.

Wenzl, H. F. J. 1970. The acid hydrolysis of wood. In *The Chemical Technology of Wood*, 157–252. New York: Academic Press.

Wright, J. D. and C. G. d'Agincourt. 1984. Evaluation of sulfuric acid hydrolysis processes for alcohol fuel production. In *Biotechnology and Bioengineering Symposium*, No. 14, 105–123. New York: John Wiley & Sons.

Wright, J. D., C. E. Wyman, and K. Grohmann. 1988. Simultaneous saccharification and fermentation of lignocellulose. *Appl. Biochem. Biotechnol.* 18: 75–90.

Wu, J., and L. K. Ju. 1998. Enhancing enzymatic saccharification of waste newsprint by surfactant addition. *Biotechnol. Prog.* 14: 649–652.

Wyman, C. E. 2001. Twenty years of trials, tribulations, and research progress in bioethnol technology: Selected key events along the way. *Appl. Biochem. Biotechnol.* 91–93: 5–21.

Zhang, M., C. Eddy, K. Deanda, M. Finkelstein, and S. Picataggio. 1995. Metabolic engineering of a pentose metabolism pathway in ethanologenic *Zymomonas mobilis*. *Science* 267: 240–243.

Section III

Production of Biodiesel

12 Biodiesel
Current and Future Perspectives

Milford A. Hanna and Loren Isom

CONTENTS

ABSTRACT

Biodiesel, a fuel comprised of mono-alkyl esters of long chain fatty acids derived from vegetable oils, animal fats, or mixtures thereof, is produced by transesterification, with glycerol being produced as a co-product. Worldwide, 1 billion tons of diesel fuel are consumed annually. The total feedstocks available for food, feed, and industrial applications are 115 million tons. This represents less than 12% of diesel fuel use. The opportunities for the future for biodiesel include improvements in the conversion technology, which appears promising, and expanding the amount of available feedstock through various plans to increase oil yields or oilseed production.

12.1 INTRODUCTION

Biodiesel, a fuel comprised of mono-alkyl esters of long chain fatty acids derived from vegetable oils, animal fats, and mixtures thereof, is produced by transesterification, with glycerol being produced as a co-product. Overall, the energy ratio for biodiesel production is in the range of 3.2 to 3.6 (Sheehan et al. 1998). Blending biodiesel with diesel fuel offers improved lubricity and reduced emissions. Worldwide, one billion tons of diesel fuel are consumed annually (Hanna, Isom, and Campbell 2005). The total feedstocks produced that can be used, competitively, for food, feed, and industrial applications are approximately 115 million tons. Assuming a 1:1

conversion on a weight basis, this represents a maximum of 12% of diesel fuel use (Hanna, Isom, and Campbell 2005). The opportunities for the future for biodiesel include improvements in the conversion technology, expanding the amount of available feedstock, and adding value to the glycerol by-product.

12.2 DIESEL FUEL MARKETS

Diesel fuel use worldwide is estimated to be 1.14 billion tons (330 billion gallons) per year. The United States uses an estimated 18% of that, or 205 million tons (60 billion gallons) (USEIA 2004). Because diesel (compression ignition) engines are more efficient than gasoline (spark ignition) engines (45% versus 30%), there is the possibility that the use of diesel engines in vehicles will increase, thereby increasing the demand for diesel fuel (DOE 2003).

12.3 BIODIESEL STANDARDS

If biodiesel is to be accepted as a fuel for diesel engines, it will have to be produced and handled in such a way that the variations in its properties and its performance characteristics will be less than, or equal to, what the consumer is used to with (petroleum-based) diesel fuel. A single, widely accepted standard for biodiesel is not available. Instead, the standard ASTM D6751 was developed for the United States, the CEN EN14214 Standard was developed for the European Union (EU), and other regions also have established alternative standards. One effort being made to ensure biodiesel fuel quality is the National Biodiesel Boards BQ 9000 certification program. Numerous producers and marketers have been certified.

12.4 BIODIESEL FEEDSTOCKS

The annual production of biodiesel is increasing rapidly worldwide, from 10 thousand tons (4 million gallons) in 2000 to 3.5 million tons (1 billion gallons) in 2005 (Gubler 2006; EBB 2006). In the United States, most new production assumes that soybean oil will be the oil of choice, with animal fats coming in as a distant second choice.

Soybean production in the United States was 81.8 million tons (3 billion bushels) in 2005 (Ash and Dohlman 2006). Even though soybean oil is the single most available oil worldwide, predicting a significant increase in soybean production in the United States seems to be in conflict with the projected increase in ethanol production and the concomitant increase in corn production. Typical yields, on a per hectare basis, would be 568 l (150 gallons) of biodiesel based on 18% oil content from 2.9 tons/ha of soybeans (43 bu/ac) at a 1:1 conversion rate of oil to biodiesel versus 3,760 l (1,000 gallons) of ethanol based on a yield of 9.4 tons/ha of corn (150 bu/ac) at a conversion rate of 400 l (106 gallons) per ton of corn (2.7 gallons/bu). The 2.4 tons/ha of soybean meal remaining after oil extraction will have to compete for market share with the 2.9 tons/ha of distillers grain remaining after fermentation of corn starch to ethanol.

Although soybean oil currently is the feedstock of choice in the United States and Brazil, it is not an obvious feedstock for biodiesel production from a global perspective. Other materials (oil palm in Southeast Asia, jatropha in India, rapeseed in Eastern Europe, and crops such as sunflowers, camelina, and hazelnuts in the United States) have greater potential on an oil yield basis (Kurki, Hill, and Morris 2006). Of course, various oilseed varieties have preferred growing conditions and in most cases are not widely adaptable.

Animal fats are a viable feedstock for biodiesel production but are available, in large quantities, on a very limited basis. Total fat production is estimated at 15 million tons per year worldwide (Rossell 2001). Again, assuming a 1:1 conversion on a weight basis, all of the animal fat worldwide could replace less than 2% of the diesel fuel used on an annual basis.

The demand for vegetable oils and animal fats for biodiesel production will quickly deplete any surpluses. The demands represented by competing uses, primarily food, will result in increased feedstock prices, which are known to be the single largest cost in biodiesel production.

The forecasts for worldwide biodiesel demand and the production capacity vary greatly, reflecting the rapid development associated with this industry. SRI Consulting's Marketing Research Report on biodiesel (Gubler 2006) forecasts the worldwide demand will surpass 40 million tons and the production capacity will surpass 80 million tons by 2010. Another study, Biofuel Market Worldwide (2006), estimated that worldwide biodiesel production would continue to grow rapidly and would reach 11 million tons (3.17 billion gallons) by 2010. Considering the variance in these forecasts, it is difficult to estimate a specific rate of growth. However, it is clear that extreme pressure will be exerted on the available feedstock. The feedstock supplies will need to expand beyond the estimated 115 million tons currently produced, or biodiesel production costs will rise, thus reducing its competitiveness with petroleum diesel and increasing its dependence on government incentives.

12.5 OPPORTUNITIES

The feedstock accounts for 70% to 80% of the cost of biodiesel production, and clearly is the key factor to evaluate when considering the competitiveness of biodiesel with petroleum-based diesel fuel. However, within the biodiesel industry, the processing technologies will play a key role in determining industry leaders and maximizing profitability. By-product utilization will be another key factor as the rapid expansion of the biodiesel industry has greatly outpaced glycerin markets at both the crude- and pharmaceutical-grade levels.

Raw material availability for biodiesel production, worldwide, is significant. Additional sources include expanded oilseed production, higher oil content varieties, and higher oil content crops. In the United States, it is estimated that expanding oilseed production by releasing 4 million hectares of productive land from government set-aside programs and switching another 8 million hectares from export grains could produce additional vegetable oil feedstocks of 2.1 million and 4.2 million tons, respectively. If the average soybean oil yield could be increased to 20%, versus the current 18%, which has been proven possible with several improved varieties, an additional 800 thousand tons of vegetable oil would be available. Or if future improvements could increase oil yields to an average of 22% oil, an additional 1.6 million tons would be available. If the oil demand outpaces the protein demand, the soybean production could be replaced with higher yielding oil crops such as sunflower and canola, which produce as much as 500 l/ha more oil than soybeans. Assuming soybean production at 29 million hectare and a 20% conversion to higher yielding oil crops, an additional 2.6 million tons would be available. Overall, the conversion of all existing and potential feedstocks in the United States will not be able to generate more than 12% of the domestic diesel demand. Therefore, it seems reasonable to project that biodiesel will be consumed primarily in niche markets: 20% biodiesel blends for emission benefits and 5% or lower blends as a lubricity additive in ultra-low-sulfur diesel fuel (Hanna, Isom, and Campbell 2005).

New processing technologies are anticipated to improve operating efficiencies. The most significant improvement is expected to come from the use of heterogeneous catalysts systems. Such catalysts have the potential to eliminate the need for water washing to remove the excess catalysts, thus, reducing both capital and operating costs. It also is perceived that the free fatty acids present in feedstocks can be converted concurrently to alcohol esters rather than being separated out and used for the lower value purposes. Yet another potential advantage is that a higher quality glycerol will be obtained.

Glycerol, a co-product of biodiesel production, has many commercial uses. The current oversupply of glycerin may make it a burden to the industry in the near term until new markets are developed for crude glycerin, or alternative markets are developed for refined glycerin. If cost-effective markets are not available, the disposal costs could be significant. In the long run, the most profitable facilities will identify the markets to capture the significant value from the glycerin by-product. Currently, the opportunity most often cited is that of using the glycerol to produce 1,4-propanediol, a commodity chemical and a precursor for many products.

Adding value to other co-products of biodiesel production, or at least maintaining their current values, will enhance the economic viability of the biodiesel industry. The oilseed meals/press cakes are used predominately as animal feeds. Their values as feeds, or feed supplements, are functions of their protein content and quality and the need for additional processing to inactivate or remove the anti-growth factors such as trypsin inhibitors and glucosinolates. Opportunities for adding value to the meals include industrial uses for the proteins, such as adhesives, and the extraction of higher value materials, such as policosanols.

12.6 CONCLUSIONS

The opportunities for expanded biodiesel production are great considering the high demand for petroleum-based products in both the industrialized and the developing regions of the world. The increased awareness of the negative environmental factors associated with petroleum fuels and a desire to move to renewable fuels has caused significant growth in the biodiesel industry. However, it represents only a small fraction of the current diesel fuel demands. Even as biodiesel production and available feedstocks expand, that growth will be challenged to keep pace with the growing demand for fuel. Biodiesel feedstock availability, the price, and the resulting by-products, combined with government incentives for economic or environmental issues, ultimately will determine the competitiveness of biodiesel as a direct substitute for petroleum diesel. It is anticipated that biodiesel will drive the industrial applications for vegetable oils and animal fats but will not displace the food applications which will continue to determine the vegetable oil price for the most desirable oils, while the lower grade oils will become the industrial feedstocks. The conversion technologies will continue to improve and will allow government incentives to be phased out as the biodiesel feedstock supplies and the prices balance out compared to petroleum diesel prices and supplies. The trend toward the development of larger production facilities and consolidation of ownership is expected to continue as the competition for the feedstock supplies and fuel markets increase.

REFERENCES

Ash, M. and E. Dohlman. 2006. *Oil Crops Outlook*. Washington, DC: United States Department of Agriculture, OCS-06j, http://usda.mannlib.cornell.edu/ers/OCS/2000s/2006/OCS-11-13-2006.pdf (accessed May 2008)

Biofuel Market Worldwide. 2006. RNCOS, Research & Markets, Ireland, http://www.researchandmarkets.com/reports/340050 (accessed May 2008)

DOE (Department of Energy). 2003. *Just the Basics: Diesel Engine*. Washington, DC: U.S. Department of Energy. http://www1.eere.energy.gov/vehiclesandfuels/pdfs/basics/jtb_diesel_engine.pdf (accessed May 2008)

EBB (European Biodiesel Board). 2006. http://www.ebb-eu.org/prev_stat_production.php (accessed May 2008)

Gubler, R. 2006. Marketing Research Report – Biodiesel. SRI Consulting.

Hanna, M., L. Isom, and J. Campbell. 2005. Biodiesel: Current perspectives and future. *Journal of Scientific & Industrial Research* 64: 854–857.

Kurki, A., A. Hill, and M. Morris. 2006. *Biodiesel: The Sustainability Dimensions*. National Sustainable Agriculture Information Service. http://www.attra.ncat.org/attra-pub/PDF/biodiesel_sustainable.pdf (accessed May 2008)

Rossell, B. 2001. *Oils and Fats. Vol. 2. Animal Carcass Fats*. Leatherhead Food International. http://www.leatherheadfood.com/lfi/pdf/animalcar.pdf. (accessed May 2008)

Sheehan, J., V. Camobreco, J. Duffield, M. Graboski, and H. Shapouri. 1998. *An Overview of Biodiesel and Petroleum Diesel Life Cycles*. NREL/YP-580-24772. Golden, CO: National Renewable Energy Laboratory.

USEIA (United States Energy Information Administration). 2004. *World Apparent Consumption of Refined Petroleum Products – 2003, International Energy Annual 2004*. Washington, D.C: USEIA. http://www.eia.doe.gov/pub/international/iea2004/table35.xls (accessed May 2008)

13 Biodiesel Production Technologies and Substrates

Arumugam Sakunthalai Ramadhas

CONTENTS

ABSTRACT

Biodiesel is an emerging alternative to diesel fuel, which has received much attention with respect to environmental concerns and fuel security of the world. Vegetable oils and animal fats are the major feedstock for biodiesel production. The quality of the feedstock is the vital criterion in selection of a suitable biodiesel production technology. The purification of the end products and production plant economics play an important role in the commercial evaluation of biodiesel. The various biodiesel production technologies, that is, alkaline, acid, two-step, ultrasonic, lipase, and supercritical alcohol are discussed in this chapter. Process parameters such as molar ratio of the alcohol to oil, the catalyst amount, reaction temperature, and water content with respect to the yield are also analysed. The comparison of various biodiesel production technologies, properties of biodiesel and their testing methods, the influence of chemical composition of biodiesel on storage, and its use in engines are discussed.

13.1 INTRODUCTION

The fossil fuels, such as petroleum products and coal, are a major source of energy in the world but these are nonrenewable in nature and have a great impact on the environment. Renewable energy sources, such as biomass, are more advantageous in terms of their reproduction, cyclic, and carbon neutral properties. Significant research work on the production and application of biomass energy for fuel purposes is being carried out all around the world. Alcohols, vegetable oils, and their derivatives are promising biomass sources for use in engines. The concept of using vegetable oil as fuel dates back to 1895 when Dr. Rudolf Diesel developed the first diesel engine to run on vegetable oil. Dr. Diesel demonstrated his engine at the World Exhibition in Paris in 1900 using peanut oil. The advent of petroleum and its appropriate fractions, low cost petroleum products, caused the replacement of vegetable oils for use in engines. However, during the energy crisis periods (1970s), vegetable oils and alcohol were widely used as engine fuel. Due to the ever-rising crude oil prices and environmental concerns, there has been a renewed focus on vegetable oils and their derivatives for use as engine fuel (Shaheed and Swain 1998).

Biodiesel is defined as the mono-alkyl esters of fatty acids derived from vegetable oils and animal fats. It can be made by chemically reacting the vegetable oils or fat with an alcohol, with or without the presence of a catalyst. Catalysts are used to increase the transesterification reaction rate and move the reaction in a forward direction. Biodiesel contains no petroleum, but can be blended with petroleum diesel to make a biodiesel-diesel blend. In general, Bxx represents xx% of biodiesel in a biodiesel-diesel blend; for example, B100 and B20 are neat biodiesel and a blend of 20% biodiesel and 80% petroleum diesel, respectively.

Biodiesel is derived from renewable and domestic resources and, hence, is capable of relieving reliance on petroleum fuel. Moreover, it is biodegradable, nontoxic, and environmentally friendly. The physiochemical properties of biodiesel are very close to that of diesel. Hence, biodiesel or its blends can be used in diesel engines with a few or no modifications. Biodiesel has a higher cetane number than petroleum diesel, no aromatics, and contains about 10 to 11% oxygen by weight. These characteristics of biodiesel reduce emissions of carbon monoxide (CO), hydrocarbon (HC), and particulate matter (PM) in the exhaust gas compared with diesel. The carbon dioxide produced by the combustion of biodiesel is recycled during photosynthesis, thereby minimizing the impact of biodiesel combustion on the greenhouse effect (Ramadhas, Jayaraj, and Muraleedharan 2005b; Barnwal and Sharma 2004).

13.2 VEGETABLE OIL CHARACTERIZATION

The fatty acid composition of vegetable oils depends on the soil conditions, moisture content in the seeds, and oil expelling method. The fatty acid composition determines its fuel properties, such as oxidation stability, cetane number, and specific gravity, and its distillation characteristics. Oils higher in unsaturated bonds are more prone to oxidation and the formation of sludge on storage for longer periods. The important physiochemical properties and the fatty acid composition of different vegetable oils are given in Table 13.1. Their physiochemical properties are almost similar to each

TABLE 13.1

Physiochemical Properties and Fatty Acid Composition of Vegetable Oils

Vegetable Oils	KV(mm²/s)	CN	HCV(MJ/kg)	Ash (wt %)	IV(mg of I / g oil)	C16:0(%)	C18:0(%)	C18:1(%)	C18:2(%)	C18:3(%)
Cottonseed	33.7	33.7	39.4	0.02	113.2	28.33	0.89	13.27	57.5	0.0
Rapeseed	37.3	37.5	39.7	0.006	108.05	3.49	0.85	64.4	22.3	8.23
Sunflower	34.4	36.7	39.6	0.01	132.32	6.08	3.26	16.93	73.76	0.0
Linseed	28.0	27.6	39.3	0.01	156.74	5.1	2.5	18.9	18.1	55.1
Castor	29.7	42.3	37.4	0.01	88.72	1.1	3.1	4.9	1.3	89.3
Soybean	33.1	38.1	39.6	0.006	69.82	11.75	3.15	23.26	55.53	6.31
Peanut	40.0	34.6	39.5	0.02	119.55	11.4	2.4	48.3	32.0	0.9

Reprinted from Demirbas, A. (2003), Biodiesel fuels from vegetable oils via catlytic and non-catalytic supercritical alcohol transesterification and other methods: a survey, Energy Conversion and Management, 44: 2039–2109, Elsevier Publications, with permission.

other but the fatty acid composition varies (Demirbas 2003). Vegetable oils have higher viscosity (about 10 to 15 times higher than that of diesel fuel), higher flash point (about 3 to 5 times), and lower calorific value (about 10% less).

Laboratory engine tests and vehicle field trial runs using straight vegetable oils as fuel in diesel engines generally gives satisfactory operation. However, long-term operation of straight vegetable oil-fueled engines creates problems in the engine. Higher viscosity and low vaporization characteristics of the vegetable oil leads to combustion chamber deposits, more smoke, oil ring sticking and thickening of the lubricating oil by the vegetable oil contamination. Higher viscosity of the vegetable oil affects its atomization and spray pattern characteristics. Reduction in viscosity of the vegetable oil improves its atomization and combustion characteristics. Blending of vegetable oils with diesel, microemulsion, cracking of oils, and transesterification reduce the viscosity. However, the transesterification process is the preferred method for reducing the viscosity of vegetable oil for commercial purposes. The various feedstock characteristics, biodiesel production technologies, process parameters, biodiesel properties, testing methods, and comparison of various biodiesel production technologies are discussed in the following sections.

13.3 ALKALINE CATALYST TRANSESTERIFICATION

Transesterification is a chemical process of transforming large, branched, triglyceride molecules of vegetable oils and fats into smaller, straight chain molecules, almost similar in size to the molecules of the species present in diesel fuel. Alkaline-catalyzed transesterification is a commercially well-developed biodiesel production process. Alkaline catalysts (NaOH, KOH) are used to improve the reaction rate and to increase the yield of the process. Since the transesterification reaction is reversible, excess alcohol is required to shift the reaction equilibrium to the products side. Alcohols such as methanol, ethanol, or butanol are used in transesterification. The transesterified vegetable oils, that is biodiesel/esters have reduced viscosity and increased volatility relative to the triglycerides present in vegetable oils. A dark, viscous liquid (rich in glycerol) is the by-product of the transesterification process.

$$Triglycerides(TG) + R'OH \overset{catalyst}{\Leftrightarrow} Diglycerides(DG) + R'COOR_1$$

$$Diglycerides(DG) + R'OH \overset{catalyst}{\Leftrightarrow} Monoglycerides(MG) + R'COOR_2$$

$$Monoglycerides(MG) + R'OH \overset{catalyst}{\Leftrightarrow} Glycerol + R'COOR_3$$

The first step is the conversion of the triglycerides to diglycerides, followed by the conversion of the diglycerides into monoglycerides, and finally monoglycerides into glycerol, yielding one methyl ester molecule from each glyceride at each step. Figure 13.1 shows the transesterification reaction of triglycerides to esters.

The reactor is charged with the vegetable oil and heated to about 60 to 70°C with moderate stirring. Meanwhile, about 0.5 to 1.0% (w/w) of anhydrous

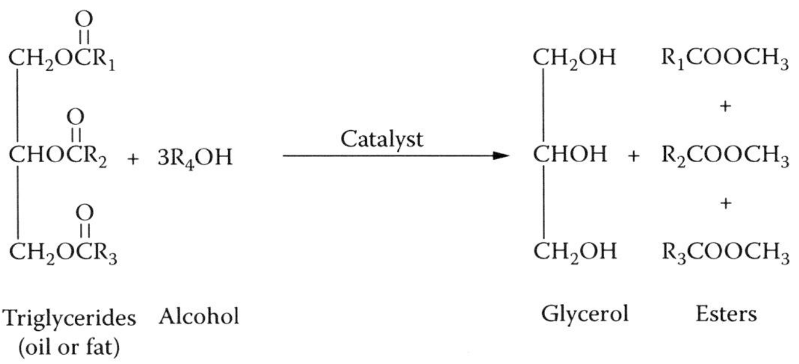

$$\begin{array}{l} CH_2OCOR_1 \\ | \\ CHOCOR_2 \\ | \\ CH_2OCOR_3 \end{array} + 3R_4OH \xrightarrow{\text{Catalyst}} \begin{array}{l} CH_2OH \\ | \\ CHOH \\ | \\ CH_2OH \end{array} + \begin{array}{c} R_1COOCH_3 \\ + \\ R_2COOCH_3 \\ + \\ R_3COOCH_3 \end{array}$$

Triglycerides (oil or fat) Alcohol Glycerol Esters

FIGURE 13.1 Transesterification of triglycerides to esters.

alkaline catalyst (NaOH or KOH) is dissolved in 10 to 15% (w/w) of methanol. This sodium hydroxide–alcohol solution is mixed with the oil and heating and stirring is continued. After 30 to 45 minutes, the reaction is stopped and the products are allowed to settle into two phases. The upper phase consists of esters and the lower phase consists of glycerol and impurities. The ester layer is washed with water several times until the washing becomes clear. Traces of the methanol, catalyst, and free fatty acids in the glycerol phase can be processed in one or two stages depending on the level of purity required. A distillation column recovers the excess alcohol, which can be recycled. The important process parameters, which affect the yield of the transesterification process, are discussed below (Pilar et al. 2004; Antolin et al. 2002).

13.3.1 Alcohol to Oil Molar Ratio

The stoichiometric transesterification requires 3 mol of the alcohol per mole of the triglyceride to yield 3 mol of the fatty esters and 1 mol of the glycerol. However, the transesterification reaction is an equilibrium reaction in which a large excess of alcohol is required to drive the reaction close to completion in a forward direction. The molar ratio of 6:1 or higher generally gives the maximum yield (higher than 98% by weight). Lower molar ratios require a longer time to complete the reaction. Excess molar ratios increase the conversion rate but leads to difficulties in the separation of the glycerol. At optimum molar ratio only the process gives higher yield and easier separation of the glycerol. The optimum molar ratios depend on the type and quality of the vegetable oil used.

13.3.2 Catalyst

The alkaline catalysts such as sodium hydroxide and potassium hydroxide are most widely used. These catalysts increase the reaction rate several times faster than acid catalysts. Alkaline catalyst concentration in the range of 0.5 to 1% by weight yields 94 to 99% conversion efficiency. Further increase in catalyst concentration does not increase the yield, but it adds to the cost and makes the separation process more complicated.

13.3.3 Reaction Temperature

The rate of the transesterification reaction is strongly influenced by the reaction temperature. Generally, the reaction is carried out close to the boiling point of methanol (60 to 70°C) at atmospheric pressure. With further increase in temperature there is more chance of loss of methanol.

13.3.4 Mixing Intensity

The mixing effect is more significant during the slow rate region of the transesterification reaction and when the single phase is established, mixing becomes insignificant. Understanding the mixing effects on the kinetics of the transesterification process is a valuable tool in the process scale-up and design. Generally, after adding the methanol and catalyst to the oil, stirring for 5 to 10 minutes promotes a higher rate of conversion and recovery.

13.3.5 Purity of Reactants

Impurities present in the vegetable oil also affect ester conversion levels significantly. The vegetable oil (refined or crude oil) is filtered before the transesterification reaction. The oil settled at the bottom of the tank during storage would give lower yield because of deposition of impurities such as wax.

13.4 ACID CATALYST TRANSESTERIFICATION

Nonedible oils, crude vegetable oils, and used cooking oils typically contain more than 2% free fatty acids (FFA), and animal fats contain from 5 to 30% FFA. Some very low quality feedstock, such as trap grease, can contain 100% FFA. Moisture or water present in the vegetable oils increases the acid value or the FFA of the oil. It has been reported that FFA content of rice bran rapidly increased within a few hours, showing 5% increase in FFA content per day. The heating of the bran immediately after milling inactivates the lipase and prohibits the formation of the FFA.

The alkaline catalyst reacts with the high-FFA feedstock to produce soap and water. Von Gerpen (2005) advocates that up to 5% FFA, alkaline catalyst can be used for the reaction; however, additional catalyst must be added to compensate for the catalyst lost to the soap. When the FFA value of the vegetable oil is more than 5%, the formation of soap inhibits the separation of the methyl esters from the glycerol and contributes to emulsion formation during the water wash. For these cases, an acid catalyst, such as sulfuric acid, is used to esterify the free fatty acids to methyl esters. Figure 13.2 shows the acid esterification reaction of vegetable oil with methanol.

Canakci and Von Gerpen (2000) and Von Gerpen (2005) report that the standard conditions for the reaction are a reaction temperature of 60°C, 3% sulfuric acid, 6:1 molar ratio of methanol to oil, and a reaction time of 48 h. The ester conversion increased from 87.8 to 95.1% when the reaction time was increased from 48 to 96 h. The drawbacks with acid esterification are water formation and longer reaction duration. The specific gravity of the ester decreases with increase in the reaction tem-

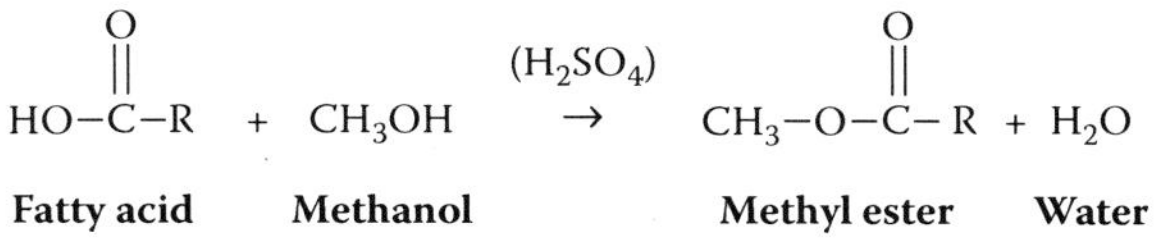

$$\underset{\textbf{Fatty acid}}{HO{-}\overset{\overset{O}{\|}}{C}{-}R} + \underset{\textbf{Methanol}}{CH_3OH} \xrightarrow{(H_2SO_4)} \underset{\textbf{Methyl ester}}{CH_3{-}O{-}\overset{\overset{O}{\|}}{C}{-}R} + \underset{\textbf{Water}}{H_2O}$$

FIGURE 13.2 Acid esterification reaction.

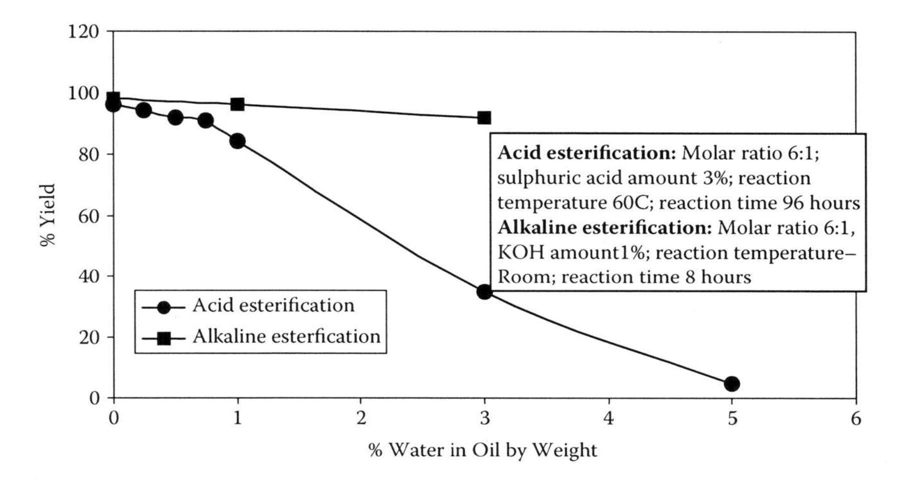

FIGURE 13.3 Effect of water content in the oil on yield of the process. (Reprinted from Canakei, M., and J. Von Gerpen, (2000), Biodiesel production via acid catalysis, *Transactions of ASAE*, 42 (5): 1203–1210, ASAE with permission.)

perature. Figure 13.3 shows the esterification conversion efficiency with respect to water content in the oil. A very small percentage addition of water (0.1%) reduced the ester yield. When more water was added to the vegetable oil, the amount of methyl esters formed was significantly reduced. They also report that more than 0.5% water in the oil decreases the ester conversion to below 90%.

13.5 ALKALINE-ACID CATALYST TWO-STEP ESTERIFICATION PROCESS

The alkaline-acid catalyst two-step esterification process is preferred for oils with FFA about 20 to 50%. The complete conversion of the free fatty acids to esters or the triglycerides to esters is not possible in a single process. Ramadhas, Jayaraj, and Muraleedharan (2005b) developed a two-step esterification process for producing biodiesel from crude rubber seed oil. The two-step esterification process converts low-cost crude vegetable oil into its esters. The first step, the acid-catalyzed esterification process, converts the free fatty acids to esters, reducing the acid value of the oil to about 4. This first step takes less time (about 10 to 30 minutes) compared to acid esterification. The products of the first step (a mixture of triglycerides and esters) are transesterified in the second step using an alkaline-catalyzed transesterification process.

Ghadge and Raheman (2005) developed a two-step esterification process for producing biodiesel from high FFA mauha oil. The high FFA (19%) level of the crude mahua oil was reduced to less than 1% in a first step, acid-catalyzed esterification (1% v/v H_2SO_4) with methanol (0.30 to 0.35 v/v) at 60°C for 1 h reaction time. In the second step, the triglyceride-ester mixture having acid value less than 2 mg KOH/g, was transesterified using alkaline catalyst (0.7% w/w KOH) with methanol (0.25 v/v) to produce biodiesel. The process gave a yield of 98% mauha biodiesel and had comparable fuel properties with that of diesel.

13.6 SUPERCRITICAL ALCOHOL TRANSESTERIFICATION

The transesterification of vegetable oil with the help of catalysts reduces the reaction time but promotes complications in purification of the biodiesel from the catalyst and the saponified products. The purification of the biodiesel and the separation of the glycerol from the catalyst are necessary but increase the cost of the production process. The supercritical alcohol transesterification process is a catalyst-free transesterification process, which is completed in a very short time, about a few minutes. Because of the noncatalytic process, purification of the products of the transesterification reaction is much simpler and environmentally friendly compared to the conventional process.

Saka and Kusdiana (2005) conducted extensive research on the production of biodiesel from vegetable oils and optimization of the process without the aid of catalysts. The process consists of heating a rapeseed oil-methanol mixture (molar ratio up to 42) at its supercritical temperature (350 to 500°C) for different time periods (1 to 4 min). The treated liquid (biodiesel) is removed from the reaction vessel and evaporated at 90°C for about 20 min to remove the excess methanol and water produced during the methyl esterification reaction. The optimized process parameters reported by Saka and Kusdiana (2005) for the transesterification of the rapeseed oil were: molar ratio of 42:1, pressure 430 bar, reaction temperature 350°C for 4 min which yields 95% conversion efficiency. Figure 13.4 describes the yield of the process with respect to the reaction time.

Kusdiana and Saka (2001, 2004b) developed a two-step esterification process, which converted the rapeseed oil to fatty methyl esters in a shorter reaction time under milder reaction conditions than the direct supercritical methanol treatment. The hydrolysis was carried out at a subcritical state of the water to obtain the fatty acids from the triglycerides of the rapeseed oil while methyl esterification of the hydrolyzed products of the triglycerides was carried out near the supercritical methanol condition to achieve fatty acid esters. They studied the kinetics reaction model for the transesterification reaction and reported that at the supercritical temperature below 293°C, the reaction rates are low but much higher at the supercritical state with the rate constant increased by a factor of about 85 at a temperature of 350°C.

Warabi, Kusdiana, and Saka (2004) analyzed the reactivity of the triglyceride and the fatty acids of the rapeseed oil in the supercritical alcohols. In general, with increase in reaction duration, the yield of the alkyl esters was increased. It was also noted that for the same reaction duration treatment, the alcohols with shorter alkyl chains gave better conversion than those with longer alkyl chains. The highest yield of the alkyl esters

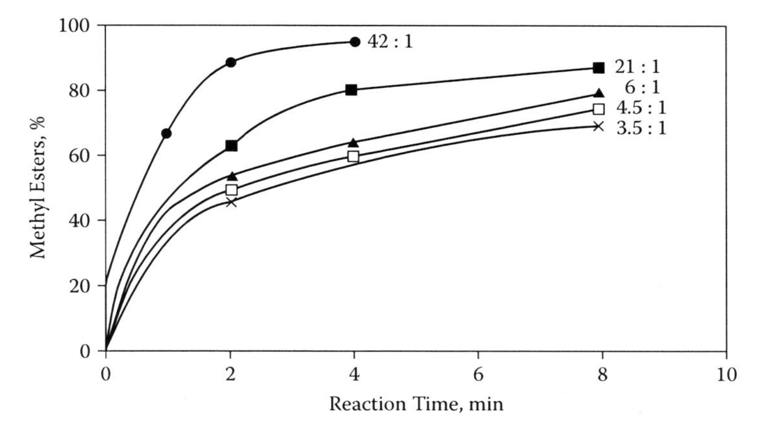

FIGURE 13.4 Yield of the process with respect to reaction time. (Reprinted from Saka, S., and D. Kusdiana. (2005). Biodiesel fuel from rapeseed oil as prepared in supercritical methanol, *International Journal of Fuel*, 80: 225–231, Elsevier Publications, with permission.)

(almost 100%) was obtained with methanol in 15 min, whereas the ethanol and 1-propanol required 45 min. The transesterification reaction temperature influences the reaction rate and yield of the esters and an increase in the reaction temperature, especially at supercritical temperatures, increases the ester conversion. The supercritical temperature of different alcohols at maximum reaction pressure is given in Table 13.2.

Kusdiana and Saka (2004b) analyzed the effect of water on the yield of methyl esters in the transesterification of triglycerides and methyl esterification of fatty acids using the supercritical methanol method. In the case of an alkaline- or acid-catalyzed esterification process, the water had a negative effect, that is, it consumed the catalyst and reduced the efficiency of the catalyst and yield of the process. In catalyst-free supercritical methanolysis, the presence of the water did not affect the yield. They reported that up to 50% water addition did not greatly affect the yield of the methyl esters. The hydrolysis reaction is much faster than transesterification and, hence, the triglycerides are transformed into fatty acids in the presence of water. These are further methyl esterified during the supercritical treatment of the methanol. With the

TABLE 13.2
Critical State and the Maximum Pressure of Various Alcohols

Alcohol	Critical Temperature(°C)	Critical Pressure(MPa)	Pressure at 3000C(MPa)
Methanol	239	8.09	20
Ethanol	243	6.38	15
1-propanol	264	5.06	10
1-butanol	287	4.90	9
1-octanol	385	2.86	6

(Reprinted from Warabi, Y., D. Kusdiana, S. Saka. (2004). Reactivity of triglycerides and fatty acids of rapeseed oil in superciritcal alcohols, Bioresource Technology, 91(3): 283–287, Elsevier Publications, with permission.)

addition of water in the supercritical methanol process, the separation of the methyl esters and glycerol from the reaction mixture becomes much easier. The glycerol is more soluble in water than methanol, which moves to the lower portion and the biodiesel in the upper portion. All the crude vegetable oils and the waste cooking oils can be easily converted to biodiesel by the supercritical methanol method.

13.7 LIPASE-BASED TRANSESTERIFICATION

The commercial biodiesel production industry generally uses alkaline or acid catalysis to produce biodiesel. However, the removal of the catalyst is through the neutralization and eventual separation of the salt from the product esters, which is difficult to achieve. The physiochemical synthesis schemes often result in poor reaction selectivity and may lead to undesirable side reactions. The enzymatic conversion of the triglycerides has been suggested as a realistic alternative to conventional physiochemical methods. The utilization of lipase as the catalyst for biodiesel fuel production has great potential compared with that of chemical methods using alkaline or acid catalysis because no complex operations are needed not only for the recovery of the glycerol but also in the elimination of the catalyst and salt. The key step in the enzymatic processes lies in the successful immobilization of the enzyme, which would allow for its recovery and reuse (Noureddini, Gao, and Philkana 2006; Du et al. 2004).

A typical biodiesel production method using a lipase catalyst developed by Noureddini, Gao, and Philkana (2006) was as follows. The initial conditions were 10 g soybean oil, 3 g methanol (methanol to oil molar ratio of 8.2), 0.5 g water, 3 g immobilized lipase phyllosilicate sol-gel matrix (PS), 40°C, 700 rpm, and 1 h reaction duration. In reactions with ethanol, 0.3 g of water and 5 g of ethanol (ethanol to oil molar ratio of 9.5) were used under identical conditions. The immobilized enzyme was washed with water and after filtration about 90 ± 5 ml of the supernatant was collected. This supernatant may potentially contain free enzyme, partially hydrolyzed precursors, methanol, and soluble oligomers. It has been reported that using methyl acetate as acyl acceptor for biodiesel production from crude soybean oil gave methyl ester yield of 92%, just as high as that of the refined soybean oil. It might be due to more methyl acetate present in the reaction medium resulting in a dilution effect of the lipids in the crude oil sources. Less concentration of the lipids could contribute to less negative effect of the lipids on enzymatic activity. Figure 13.5 describes the product concentration of the soybean esters using lipase.

Modi et al. (2007) used propan-2-ol as an acyl acceptor for the immobilized lipase-catalyzed preparation of biodiesel. The optimum conditions for the transesterification of the crude jatropha (*Jatropha curcas*), karanj (*Pongamia pinnata*), and sunflower (*Helianthus annuus*) oils were 10% Novozym-435 (immobilized *Candida antarctica* lipase B) based on the oil weight, alcohol to oil molar ratio of 4:1 at 50°C for 8 h. Excess methanol leads to the inactivation of the enzyme and glycerol as a major by-product and can also block the immobilized enzyme, resulting in low enzymatic activity. These problems could be limitations for the industrial production of biodiesel with enzymes as catalyst.

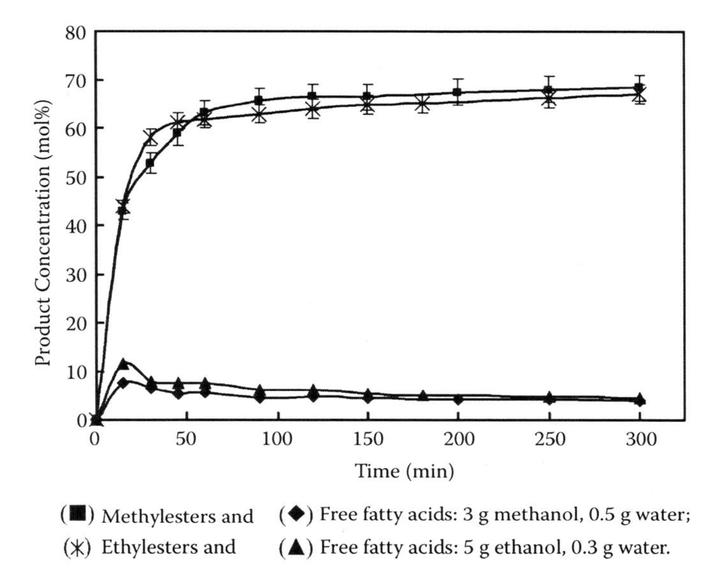

FIGURE 13.5 Time course of the lipase-catalyzed transesterification. (Reprinted from Noureddini, H., Gao, X., Philkans, R. S. [2006], Immobilized pseudomonas cepacia lipase for biodiesel fuel production from soybean oil, *Bioresource Technology*, 96, 769–777. Elsevier Publications, with permission.)

13.8 ULTRASONIC TRANSESTERIFICATION

Low frequency ultrasonic irradiation is considered a useful tool for the emulsification of the immiscible liquids. Stavarache et al. (2005) used the ultrasonic method for the preparation of the emulsion the alkaline-catalyzed esterification process. The collapse of the cavitation bubbles disrupts the phase boundary and causes emulsification by ultrasonic jets that are impinging one liquid to another. It has been reported that with increasing chain length, the miscibility between the oil and alcohol increases, decreasing the reaction time (10 to 20 min) but also making the separation of the esters difficult. The normal chain alcohols react quite rapidly under ultrasonic irradiation. This behavior is due to the increased mass transfer in the presence of the ultrasound. The velocity of an ultrasonic wave through a material depends on its physical properties and, hence, the ultrasonic velocity decreases with increasing density. The ultrasonic properties of an emulsion vary significantly. The droplets of more dense oil move upwards and form a cream layer, while the alcohol moves downwards, facilitating mixing and increasing the contact surface between alcohol and oil. At 40 kHz, the reaction time is shorter, but the transesterification yields are slightly lower than at 28 kHz. The differences in the product yield are mainly due to difficulties in washing. At 40 kHz, the soap is formed in higher amounts and acts as a phase transfer catalyst leading to formation of the esters more rapidly than at 28 kHz. But during the washing, the soap hinders the separation and some ester is trapped into the soap micelles and thus the yield in the isolated product is decreased.

Table 13.3 shows a comparison of the yield of methyl esters with mechanical stirring and ultrasonic irradiation (Stavarache et al. 2005). At higher frequencies, the collapse of cavitation bubbles is not strong enough to impinge one liquid to the other. Thus, the mixing between the two immiscible layers (alcohol and oil) is very poor and emulsification does not occur. The transesterification takes place mainly at the boundary between the two layers. Low frequency ultrasound is efficient, time saving, and economically functional, offering advantages over the conventional procedure. The ultrasonic biodiesel production method can be a valuable tool for the transesterification of fatty acids, aiming to prepare biodiesel fuel on an industrial scale. Table 13.4 describes the comparative profile of various biodiesel production technologies.

13.9 PROPERTIES REQUIREMENT OF THE BIODIESEL

Biodiesel is produced from vegetable oils of varying origin and quality, and hence, it is necessary to establish a standardization of the fuel quality to guarantee engine performance. The fatty acid composition of the oil and the processing technology

TABLE 13.3
Yield of Isolated Methyl Esters

	0.5% (w/w) NaOH		1.0% (w/w)NaOH		1.5% (w/w) NaOH	
Method	Time (min)	Yield (%)	Time (min)	Yield (%)	Time (min)	Yield (%)
Mechanical stirring	60	80	10	91	10	35
Ultrasonic irradiation 28 kHz	40	98	10	95	10	75
Ultrasonic irradiation 40 kHz	20	98	10	91	10	68

(Reprinted from Stavarache, C., Vinatoru, M., Nishimura, R., Maeda, Y. [2005], Fatty acids methyl esters from vegetable oil by means of ultrasonic energy, *Ultrasonics Sonochemistry,* 12(5), 367–372, Elsevier Publications, with permission.)

TABLE 13.4
Comparison of Various Biodiesel Production Technologies

Variable	Alkaline	Acid	Two-Step	Ultrasonic	Lipase	Supercritical
Reaction temperature (°C)	40–70	55–80	40–70	30–40	30–40	240–385
Yield	Normal	Normal	Good	Higher	Higher	Good
Glycerol recovery	Difficult	Difficult	Difficult	Difficult	Easy	–
Purification of ester	Washing	Washing	Washing	Washing	–	–
Production cost	Cheap	Cheap	Cheap	Medium	Expensive	Medium

affects the fuel properties of the biodiesel. The fatty acid composition of oils depends on the climatic condition and the oil extraction method. The flash point of a fuel is the temperature at which it will ignite when exposed to a flame or spark. The flash point of biodiesel is higher than that of petroleum diesel, and hence it is safe for storage. Moreover, the flash point for biodiesel is used as a mechanism to limit the level of unreacted alcohol remaining in the fuel. The flash point of biodiesel is generally around 160°C, but it can be reduced drastically if residual alcohol is present in the biodiesel. The presence of a high level of alcohol in the biodiesel can cause accelerated deterioration of the natural rubber seals and gaskets. Therefore, control of excess alcohol content in the biodiesel on transesterification is required.

The maximum allowable viscosity is limited by considerations related to the engine design and size, and the characteristics of the injection system. The upper limit of the biodiesel viscosity is higher than that of petroleum diesel. However, lower blends of biodiesel with diesel matches the diesel specification. The cold filter plugging point (CFPP) of a fuel reflects its cold weather performance. At low operating temperature, the fuel might thicken in the fuel line and could affect the performance of the fuel pumps and the injectors. The distillation characteristics of biodiesel are quite different from that of diesel. Biodiesel does not contain any highly volatile components and the fuel evaporates only at higher temperature. Biodiesel has molecular chains of 16 to 18 carbons, which have close boiling points. The boiling point of biodiesel is generally between 330 and 357°C.

The cetane number of diesel/biodiesel defines its ignition quality and affects engine performance parameters such as the combustion, stability, drivability, white smoke, noise, and emissions of CO and HC. A higher cetane number of the fuel is an indication of its better ignition properties. Biodiesel has a higher cetane number than conventional diesel fuel, which results in higher combustion efficiency.

The acid number is an indication of the presence of fatty acids in the biodiesel and its degradation due to thermal effects. During the injection process, more fuel is returned than that injected into the combustion chamber of the engine. The high temperature (about 90°C) returned fuel accelerates the degradation of the biodiesel. Thus, a high acid number can cause damage to the injector and result in deposits in the fuel system and affect the life of the pumps and filters.

The neutralization number is the number of milligrams of KOH required to neutralize 1 mol of triglyceride. It is specified to ensure the proper ageing properties of the fuel and reflects the presence of free fatty acids or acids used in the manufacture of the biodiesel and also the degradation of the biodiesel due to thermal effects. The iodine number refers to the amount of iodine required to convert the unsaturated oil into saturated oil. It does not refer to the amount of iodine in the oil but to the presence of the unsaturated fatty acids in the fuel. The iodine number indicates the tendency of a fuel to be unstable as it measures the presence of C=C bonds that are prone to oxidation. Generally, instability of biodiesel would increase by a factor of 1 for every C=C bond on the fatty acid chain. Thus, C18:3 fatty acids are three times more unstable than C18:0 fatty acids. The oxidation stability of biodiesel varies greatly depending on the feedstock used. Poor oxidation stability can cause fuel thickening, formation of gums and sediments, which, in turn, can cause filter clogging and injector fouling (Planning Commission, 2003).

The recommended duration of storage of biodiesel and its blends should not be more than 6 months. The antioxidants must be properly mixed with the fuel for its good effectiveness. In diesel engines, the methyl esters are prone to cause engine crank case oil dilution. The dilution of lubricating case oil by the fuel decreases its viscosity. But high content of unsaturated fatty acids in the ester (indicated by the high iodine number) increases the danger of polymerization in the engine oil. The sudden increase in the lubricating oil viscosity, as encountered in several engine tests, is attributed to the oxidation and polymerization of the unsaturated fuel parts entering into oil through the dilution.

Free glycerol refers to the amount of glycerol that is left in the finished biodiesel. The glycerol is insoluble in biodiesel, hence it is removed by settling. Free glycerol may remain as suspended droplets or dissolved in the biodiesel but can be removed by washing with water. Excessive free glycerol in the biodiesel creates a viscous mixture that can plug the fuel filters and cause combustion problems in the engine. If the reaction is incomplete, the molecules of glycerides left in the reaction mixture are added to the free glycerol that is known as total glycerol. Low levels of the total glycerine ensure the high conversion of oil to mono-alkyl esters.

Ash forming materials may be present in the biodiesel in three forms: abrasive solids, soluble metallic solids, and unremoved catalysts. Abrasive solids and unremoved catalysts can affect the injector, fuel pump, piston, and ring wear and also contribute to engine deposits. The soluble metallic soaps have little effect on the wear but may contribute to filter plugging and engine deposits. The carbon residue of the fuel is indicative of the carbon depositing tendencies of a fuel. The Conradsons Carbon Residue (CCR) test for biodiesel is more important than that for diesel fuel because it shows a high correlation with the presence of free fatty acids, glycerides, soaps, polymers, higher unsaturated fatty acids, and inorganic impurities. The properties that influence the fuel quality for use in engines is specified in most biodiesel standards. The biodiesel specification provided in ASTM D 6751 is shown in Table 13.5.

13.10 CONCLUSIONS

Biodiesel has become more attractive as an alternative fuel to petroleum diesel fuel. Most of the transesterification studies have been done on edible oils such as rapeseed, soybean, sunflower, canola, etc., by using methanol and NaOH/KOH as catalyst. However, there are studies at an advanced stage using nonedible oils that are produced in the wastelands or wild species such as *Pongamia pinnata, Jatropha curcas, Simarouba glauca*, etc. Alkaline-catalyzed transesterification is a promising method for the production of biodiesel from low-FFA vegetable oils. For high-FFA nonedible vegetable oils, acid esterification is the method of choice, and for the medium fatty acid vegetable oils (20 to 50% FFA), the two-step esterification process is preferable. The lipase-catalyzed esterification process is suitable for all types of vegetable oils or fats but it is an expensive process. The supercritical alcohol esterification process is a catalyst-free esterification process that is suitable for any type of vegetable oils. Moreover, the supercritical treatment process is highly advantageous for vegetable oils with more water content such as waste cooking oils or crude vegetable oil. The

TABLE 13.5
ASTM D 6751 Biodiesel Specification

Property	Test Method ASTM	Limits
Flash point (°C)	D 93	Min. 130
Water and sediment (% v)	D 2709	Max. 0.05
Sulfated ash (mass %)	D 874	Max. 0.02
Kinematic viscosity at 40°C (cSt)	D 445	1.9–6.0
Total sulfur (mass %)	D 5453	Max. 0.0015
Carbon residue (mass %)	D 4530	Max. 0.05
Cetane number	D 613	Min. 47
Acid no. (mg KOH/g)	D 664	Max. 0.8
Copper strip corrosion	D 130	Max. No.3
Free glycerin (mass %)	D 6584	0.02
Total glycerin (mass %)	D 6584	0.240
Phosphorus content (% mass)	D 4951	Max. 0.001
Distillation temperature (90% recovered; °C)	D 1160	Max. 360

From ASTM. 2006. Annual Book of ASTM Standards 2006, Vol. 5.04, ASTM D 6751-03.

ultrasonic esterification process is less time consuming and gives higher yield. The suitable biodiesel production process should be selected depending on the availability and quality of the feedstock and the production capacity.

REFERENCES

Antolin, G., F. V. Tinaut, Y. Briceno, V. Castano, C. Perez, and A. I. Ramirez. 2002. Optimisation of biodiesel production by sunflower oil transesterification. *Bioresource Technology* 83: 111–114.

ASTM. 2006. *Annual Book of ASTM Standards, Vol. 5.04, ASTM D 6751-03.* ASTM International.

Barnwal, B. K. and M. P. Sharma. 2004. Prospects of biodiesel production from vegetable oils in India. *Renewable and Sustainable Energy Reviews* 20: 170–186.

Canakci, M. and J. Von Gerpen. 2000. Biodiesel production via acid catalysis. *Transactions of ASAE* 42 (5): 1203–1210.

Du, W., X. Yuanyuan, L. Dehua, and Z. Jing. 2004. Comparative study on lipase-catalyzed transformation of soybean oil for biodiesel production with different acyl acceptors. *Journal of Molecular Catalysis B: Enzymatic* 30: 125–129.

Demirbas, A. 2003. Biodiesel fuels from vegetable oils via catalytic and non-catalytic supercritical alcohol transesterification and other methods: A survey. *Energy Conversion and Management* 44: 2093–2109.

Ghadge, S. V. and H. Raheman. 2005. Biodiesel production from mahua oil having high free fatty acids. *Biomass and Bioenergy* 28: 601–605.

Kusdiana, D. and S. Saka. 2001. Kinetics of transesterification in rapeseed oil to biodiesel fuel as treated in supercritical methanol. *International Journal of Fuel* 80: 693–698.

Kusdiana, D. and S. Saka. 2004a. Two-step preparation for catalyst free biodiesel fuel production. *Applied Biochemistry and Biotechnology* 113: 781–791.

Kusdiana, D. and S. Saka. 2004b. Effects of water on biodiesel fuel production by supercritical methanol treatment. *Bioresource Technology* 91: 289–295.

Modi, M. K., J. R. C. Reddy, B. V. S. K. Rao, and R. B. N. Prasad. 2007. Lipase mediated conversion of vegetable oils into biodiesel using ethyl acetate as acyl acceptor. *Bioresource Technology* 98 (6): 1260–1264.

Noureddini, H., X. Gao, and R. S. Philkana. 2006. Immobilized *Pseudomonas cepacia* lipase for bodiesel fuel production from soybean oil. *Bioresource Technology* 96: 769–777.

Pilar, D. M., B. Evaristo, J. L. Franscisco, and M. Martin. 2004. Optimization of alkali-catalyzed transesterification of *Brassica carinata* oil for biodiesel production. *Energy & Fuel* 18: 77–83.

Planning Commission. 2003. *Report of the Committee on Development of Biofuel*. New Delhi: Government of India.

Ramadhas, A. S., S. Jayaraj, and C. Muraleedharan. 2005a. Use of vegetable oils as I.C. engine fuels: A review. *International Journal of Renewable Energy* 29 (5): 727–742.

Ramadhas, A.S., S. Jayaraj, and C. Muraleedharan. 2005b. Biodiesel production from high FFA rubber seed oils. *International Journal of Fuel* 84 (4): 335–340.

Saka, S. and D. Kusdiana. 2005. Biodiesel fuel from rapeseed oil as prepared in supercritical methanol. *International Journal of Fuel* 80: 225–231.

Shaheed, A. and E. Swain. 1998. Performance and exhaust emission evaluation of a small diesel engine fuelled with coconut oil methyl esters. *Society of Automotive Engineers* 981156.

Stavarache, C., M. Vinatoru, R. Nishimura, and Y. Maeda. 2005. Fatty acids methyl esters from vegetable oil by means of ultrasonic energy. *Ultrasonics Sonochemistry* 12 (5): 367–372.

Von Gerpen, J. 2005. Biodiesel processing and production. *Fuel Processing Technology* 86: 1097–1107.

Warabi, Y., D. Kusdiana, and S. Saka. 2004. Reactivity of triglycerides and fatty acids of rapeseed oil in supercritical alcohols. *Bioresource Technology* 91 (3): 283–287.

14 Lipase-Catalyzed Preparation of Biodiesel

Rachapudi Badari Narayana Prasad and Bhamidipati Venkata Surya Koppeswara Rao

CONTENTS

ABSTRACT

The drawbacks associated with conventional processes for the preparation of biodiesel can be overcome by using lipases as alternate catalysts. Lipases exhibit the ability to esterify and transesterify both fatty acids and acyl glycerols. Different approaches reported in the literature for the preparation of biodiesel from various feedstocks are described in the chapter. The major problem for lipase-catalyzed preparation of biodiesel is deactivation of the lipase in the presence of short chain alcohols like methanol and ethanol. Methodologies reported for stabilizing the lipases and the use of alternate acyl donors are also reviewed.

14.1 INTRODUCTION

The disadvantages caused by physicochemical methods to produce biodiesel can be overcome by using the lipases as alternate catalysts (Haas and Foglia 2005). Lipases are generally effective biocatalysts for having substrate specificity, functional group specificity, and stereo specificity and hence industrial applications of lipases in the oleochemical industry have become more attractive. The advantages for lipase catalysis over chemical methods in the production of biodiesel from vegetable oils include the ability to esterify and transesterify both the free fatty acids (FFA) and acyl glycerols; the production of glycerol as a by-product with minimal water content and very little or no inorganic contamination; and the lipase catalyst can be reused several times. However, the use of enzymatic catalysts has some restrictions due to the high cost of the lipases compared to inorganic catalysts and inactivation of the lipase by the contaminants in the feedstocks, polar short chain alcohols and the by-product glycerol.

Although the enzymatic process for the production of biodiesel is still not commercially feasible, a number of studies have shown that the lipases hold promise as an alternative catalyst to the traditional alkali. These studies mainly describe optimizing the reaction conditions, such as type of enzyme, effect of immobilization of the lipase on reaction, lipase to substrate ratio, oil to alcohol molar ratio, use of the solvent and different acyl donors, temperature, time, etc. (Table 14.1). This chapter describes the work carried out so far for the preparation of biodiesel employing enzymatic approaches.

14.2 PRETREATMENT OF OIL AND LIPASE FOR EFFICIENT ALCOHOLYSIS

The crude vegetable oils contain several components such as lecithin, FFA, waxes, unsaponifiable matter, and pigments. However, it is necessary to pretreat the oil for the removal of the lecithin and pigments by employing degumming and bleaching for efficient conversions during the enzymatic reaction. It is not necessary to remove the FFA during pretreatment as the lipase has the capability of converting the fatty acid into methyl esters along with triacylglycerol.

The crude vegetable oils do not undergo enzymatic alcoholysis, or yields are very low due to the presence of higher amounts of the phospholipids. The inhibition may be due to the interference of the interaction of the lipase molecule with the substrates by the phospholipids bound on the immobilized preparation. Crude and refined soybean oils were subjected to transesterification with methanol using immobilized *Candida antarctica* (Novozym 435) as the biocatalyst (Du et al. 2004a); the yield of methyl esters produced from the crude soybean oil was lower than that from refined soybean oil. The higher the phospholipid content, the lower was the yield of methyl esters. These findings clearly indicate that degummed oil is a better substrate for enzymatic methanolysis.

Several studies were reported for the stabilization or regeneration of the lipase for reusability. The pretreatment of the lipase by immersing in oils may influence the reaction rate of the alcoholysis by improving its activity. A study reported by Samukawa et al. (2000) involves the incubation of the lipase in methyl oleate for 0.5 h and subsequently in soybean oil for 12 h. The methyl ester content reached 97% within 3.5 h by step-wise addition of methanol by employing the pretreated lipase. It was observed that the short chain acyl acceptor glycerol, one of the major by-products during the transesterification reaction, has serious negative effects on the performance of the lipases. The glycerol forms a coating over the enzyme and blocks the active sites, which in turn reduces the activity of the lipase. The treatment of the lipases intermittently with isopropanol to remove the glycerol from the reaction system proved to be effective in retaining its 95% activity (Du, Xu, and Liu 2003) even after 10 to 15 batches of the reaction with more than 94% yield of the methyl esters. In another work, 2-butanol and *t*-butanol were employed to restore the activity of the deactivated enzyme to an extent of 56 and 75%, respectively.

The dialysis method using a flat sheet membrane module was reported to continuously remove the glycerol from the reaction system to reduce the inhibitory effect of the glycerol on the lipase during methanolysis by immobilized *C. antarctica* employing step-wise as well as continuous methanol feeding (Belafi-Bako et al. 2002). Ultrasound pretreatment was also effective in stabilizing the lipase activity and in turn accelerated the transesterification rate of waste oil with methanol (Hong and Min-Hua 2005).

Ethanolysis of sunflower oil with *Mucor mehi* lipase did not yield more than 85% even under optimized conditions but the yields were improved by the addition of silica gel to the reaction medium (Selmi and Thomas 1998). This was due to the adsorption of the glycerol by the silica gel, which reduced the glycerol deactivation of the enzyme. The addition of the silica gel was also useful for the promotion of acyl migration in the transesterification reaction to increase the yield of the biodiesel when 1,3 specific lipase such as lipozyme TL was used (Du et al. 2005). The biodiesel yield was only 66% when 4% lipozyme TL was used, while about 90% biodiesel yield could be achieved when combining 6% silica gel with 4% lipozyme TL, almost as high as that of 10% immobilized lipase used for the reaction.

14.3 LIPASE-CATALYZED ALCOHOLYSIS OF OIL WITH AND WITHOUT THE SOLVENT MEDIUM

The lipase-catalyzed alcoholysis of oil can be achieved with and without the solvent medium. Early work on the application of lipases from *Pseudomonas fluorescens, M. miehei,* and *Candida* sp. for biodiesel preparation was reported using sunflower oil in petroleum ether medium (Mittelbach 1990). Of these, *Pseudomonas* lipase gave almost quantitative yields of the biodiesel. When the reaction was carried out without the organic solvent, only 3% of product was formed during the methanolysis whereas absolute ethanol, 96% ethanol, and 1-butanol produced 70, 82, and 76% of the respective esters. The reaction rates with the homologous alcohols revealed that the reaction rates increased with higher chain length of the alcohol, with or without the addition of water.

The ability of the lipases in the transesterification of several oils such as the soybean, rapeseed, olive, etc. with short chain alcohols was studied using *M. miehei* and *C. antarctica* lipases. *M. miehei* was the most efficient for transesterifying the triglycerides to their alkyl esters with primary alcohols, whereas *C. antarctica* was the most efficient for the branched chain alcohols (Nelson, Foglia, and Marmer 1996). In the presence of hexane medium, 94.8 to 98.5% conversions were achieved for the primary alcohols and 61.2 to 83.8% for the secondary alcohols. However, in solvent-free medium, the yields with methanol and ethanol were lower; in particular the yield with methanol was only 19.4%. *Chromobacterium viscosum, C. rugosa,* and porcine pancreas were screened for transesterification reaction of jatropha oil

TABLE 14.1
Lipase-Mediated Preparation of Biodiesel Using Various Types of Oils, Alcohols, and Lipases

Oil	Alcohol	Lipase(s)	Conditions	Conversion (%)	Remarks	Ref.
Crude and refined soybean	Methanol	*C. antarctica (Novozym 435)*	3 Step methanolysis	94	Pretreatment of enzyme with crude oil for 120 h resulted in good yields	Du et al. (2004a)
Soybean	Methanol	*C. Antarctica (Novozym 435)*	Oil:methanol, 1:8; lipase: 4 wt% of oil; 30°C, 3.5 h	97	Step-wise addition of methanol, with and without water; with and without preincubation of lipase in methyl oleate and soybean oil	Samukawa et al. (2000)
Soybean	Methanol	*Thermomyces lanuginosus (Lipozyme TL IM)*	Oil:methanol, 1:4; lipase: 30 wt% of oil; 30–50°C; 12 h	92	Isopropanol helped to recover glycerol and to stabilize lipase	Du, Xu, and Liu (2003)
Sunflower	Methanol	*C. antarctica (Novozyme 435)*	Oil:methanol, 1:4; lipase: 7 wt% of oil; water: 400 ppm; 50°C; 16 h	97	Step-wise and continuous addition of methanol; glycerol removed by dialysis using a flat sheet membrane module	Belafi-Bako et al. (2002)
Sunflower	Ethanol	*Mucor miehei (Lipozyme)*	Oil:ethanol, 1:3; lipase: 10 wt% of oil; 50°C; 5 h	83	Addition of silica gel improved the yields of ethyl esters	Selmi and Thomas (1998)
Soybean	Methanol	*T. lanuginosus (Lipozyme TL)*	Oil:methanol, 1:3; 6 wt% silica gel and 4 wt% immobilized lipase of oil; 40°C; 48 h	90	Three step-wise additions of methanol; silica gel promoted acyl migration and helped to increase the yield of methyl ester	Du et al. (2005)

Sunflower	Methanol, ethanol (abs), ethanol (96%), 1-propanol, 1-butanol	*Pseudomonas fluorescensMucor miehei (Lipozyme)Candida sp. (SP 382)*	Oil:alcohol,1:3 to 1:12; lipase:10–20 wt% of oil; solvent medium: with/without pet. ether; 25–65°C; 5–14 h	3–82 (without solvent)9–99 (with solvent)	Reaction conducted with or without addition of water and solvent	Mittelbach (1990)
Yellow grease, tallow fat, rapeseed, soybean, olive	Methanol, ethanol, isopropanol, 2-butanol	*M. miehei (Lipozyme IM 60) C. antarctica (SP 435)P. cepaci, Rhizopus delemar, Geotricum candidum*	Oil:alcohol, 1:3; lipase: 10 wt% of oil; hexane as reaction medium; 45°C; 5 h	Up to 98	*M. miehei and C. antarctica were most efficient for transesterifying triglycerides with primary and secondary alcohols, respectively*	Nelson, Foglia, and Marmer (1996)
Jatropha	Methanol, ethanol	*Chromobacterium viscosum, C. rugosa, porcine pancreas*	Oil:alcohol, 1:4; lipase: 10 wt% of oil; 40°C; 8 h	62–92	Reaction carried out with and without immobilization of lipase on celite and with and without water	Shah, Sharma, and Gupta (2004)
Castor	Ethanol	*C. antarctica (Novozym 435), T. lanuginosus (Lipozyme IM)*	Oil:ethanol, 1:10; lipase: 20% of oil; 65°C Oil:ethanol, 1:3; lipase: 20 wt% of oil: 65°C	81.498	Taguchi experimental design was adopted; both reactions were carried out in n-hexane	De Debora et al. (2004)
Cottonseed	Methanol	*C. antarctica (Novozym 435)*	Oil:methanol, 1:4; lipase: 30 wt% of oil; 50°C; 7 h	72–94	Free fatty acid content increased in the product with increasing enzyme quantity	Köse, Tüter, and Aksoy (2002)
Mixture of soybean and rapeseed	Methanol	*C. Antarctica (Novozym 435), Rhizomucor miehei, Rhizopus delemar, Fusarium heterosporum, Aspergillus niger*	Oil:methanol, 1:3; lipase: 4 wt% of oil; 30°C; 48 h	93–98	Step-wise addition of methanol with batch and continuous reaction system	Shimada et al. (1999) Watanabe et al. (2000)

-continued

TABLE 14.1 (continued)
Lipase-Mediated Preparation of Biodiesel Using Various Types of Oils, Alcohols, and Lipases

Oil	Alcohol	Lipase(s)	Conditions	Conversion (%)	Remarks	Ref.
Sunflower Triolein	Methanol	*Rhizomucor mieheiP. fluorescens (Amano AK)*	Oil:methanol, 1:4.5; lipase: 10 wt% of oil; solvent medium: with and without hexane; 40°C; 24 h	80–90	Step-wise addition of 1 M equivalent of methanol at 5 h intervals	Soumanou and Bornscheuer (2003)
Soybean	Methanol	*C. rugosa, P. cepacia, P. fluorescens*	Oil:methanol, 1:3; lipase: 5 wt% of oil; water: 0–20%; 35°C; 80–100 h	13–90	With and without water, presence of water prevents the inactivation of lipase	Kaieda et al. (2001)
Olive, rapeseed, rice bran, soybean	Methanol	*Cryptococcus sp. S-2 (Strain CS2)*	Oil:methanol, 1:4; water: 80 wt% of the substrate contains 2000 U of crude lipase; 30°C; 120 h	80	Single-stage addition of methanol	Kamini and Iefuji (2001)
Soybean	Methanol, ethanol	*Immobilized P. cepacia*	Oil:alcohol, 1:7.5; lipase: 5 wt% of oil; water: 5%, 35°C; 1 hOil:alcohol, 1:15.2; lipase: 5 wt% of oil; water: 3%; 35°C; 1 h	6765	Immobilized lipase was consistently more active than the free enzyme	Noureddini, Gao, and Philkana (2005)
Restaurant grease	Methanol, ethanol	*P. cepacia (PS-30)*	Grease:alcohol, 1;4; lipase: 10 wt% of grease; 40–70°C; 48 h	60–97	Single-step addition of alcohol	Hsu et al. (2003)
Soybean	Methanol	Immobilized Rhizopus oryzae cells within biomass support particles (BSPs)	Oil:methanol, 1:3; lipase: 50 BSPs, 0.1 M acetate buffer 1.5 ml (15% of oil); 35°C; 72 h	91.1	Step-wise addition of methanol in presence of acetate buffer	Ban et al. (2001)

in a solvent-free system with 10% of the lipase based on the oil (Shah, Sharma, and Gupta 2004). Among three lipases used, *C. viscosum* gave good yields (62%) and the yields were further enhanced (71%) when the enzyme was immobilized on Celite 545. The addition of 1% water into the free enzyme preparation and 0.5% water into the immobilized enzyme preparation enhanced the yields of biodiesel to 73% and 92%, respectively. The ethanolysis of castor oil was carried out in *n*-hexane medium using Novozym 435 and Lipozyme IM. The reactions were carried out in the presence and absence of water. Under optimum reaction conditions, a conversion of 81.4% was achieved with Novozym 435 and 98% with Lipozyme IM (De Debora et al., 2004). The yields of biodiesel could be improved using higher dosage of the enzyme, that is, up to 30% based on the oil, when methanolysis was carried out with cottonseed oil using immobilized *C. antarctica* in a solvent-free medium (Köse, Tüter, and Aksoy 2002). However, the FFA content increased in the product with increasing enzyme quantity due to the moisture present in the immobilized lipase.

Enzymes are unstable in short chain alcohols in general and the lower yields of methanolysis could be due to the inactivation of lipase caused by the contact between the lipase and the insoluble methanol that exists as drops in the oil. Indeed, when methanolysis of vegetable oils was conducted with immobilized *C. antarctica,* the lipase was inactivated irreversibly in the presence of half molar equivalent of methanol to oil (Shimada et al. 1999). These findings led to a step-wise incremental addition of the alcohol to safeguard the lipase from the short chain alcohols. A three-step addition of one molar equivalent of methanol under optimized conditions yielded about 98% of biodiesel after 12 h of reaction. Similar results were reported (Watanabe et al. 2000) for the methanolysis of a mixture of soybean and rapeseed oils using immobilized *C. antarctica* by adding the methanol three successive times in a span of 48 h to get the conversions up to 98.4%. This approach maintained more than 95% of the ester conversion even after 50 cycles of the reaction.

The methanolysis of sunflower oil using immobilized *P. fluorescens and Rhizomucor miehei* in the solvent and solvent-free system was investigated by Soumanou and Bornscheuer (2003). About 80% conversions were observed when the reaction was conducted in the presence of *n*-hexane and petroleum ether. A three-step protocol with the step-wise addition of one molar equivalent of the methanol at 5 h intervals reduced the inactivation of the commercial immobilized lipases by the methanol to obtain better yields.

Kaieda et al. (2001) reported the methanolysis of soybean oil with both 1-3-specific and nonspecific lipases in a water-containing system without an organic solvent. Among the nonspecific lipases, *C. rugosa, P. cepacia,* and *P. fluorescens* exhibited significantly high catalytic ability and *P. cepacia* yielded higher contents of the methyl ester in a reaction mixture with 3 molar equivalents of the methanol to the oil. Despite the use of 1,3-specific lipase, the methyl ester yields reached 80 to 90% by step-wise addition of the methanol to the reaction mixture. This was due to the acyl migration from the *sn*-2 position to the *sn*-1 or *sn*-3 position in partial glycerides, which occurred spontaneously.

14.4 NOVEL IMMOBILIZATION TECHNIQUES FOR THE STABILIZATION OF LIPASE

The key step in the enzymatic processes lies in the successful immobilization of the enzyme, which will allow for its recovery and reuse. Immobilization is the most widely used method for enhancing stability in the lipases and to make them more attractive for industrial use. In addition to the commercial approaches of immobilization, some innovative methods have been reported for biodiesel preparation. In one such process, phyllosilicate clay saturated with sodium ions was suspended in water and then exchanged with alkylammonium ions by the addition of cetyltrimethyl ammonium chloride; this mixture was then used in the entrapment of *P. cepacia* with tetramethoxysilane (TMOS) as the polymerization precursor (Hsu et al. 2001). The resultant phyllosilicate sol-gel matrix-based immobilized enzyme (IM PS-30) was then used in the transesterification of tallow and grease and the conversions were more than 95%. In another study, the lipase PS from *P. cepacia* was entrapped within a sol-gel polymer matrix, prepared by the polycondensation of hydrolyzed TMOS and iso-butyltrimethoxysilane (iso-BTMS), and the immobilized lipase was consistently more active than the free lipase for the transesterification of soybean oil; it lost very little activity even after repeated uses (Noureddini et al. 2005).

Thermomyces lanuginose and *C. antarctica* were immobilized on a macroporous acrylic resin and IM PS-30 and employed for the preparation of the methyl and ethyl esters of restaurant grease in solvent-free media employing a one-step addition of the alcohol. The IM PS-30 was the most effective compared to the other lipases even though initially the rate of the reaction was slow. The yield of biodiesel was about 95% after 48 h of reaction. The addition of molecular sieves also improved the methyl ester yields by 20% in a transesterification reaction catalyzed by IM PS-30. In another study, Hsu et al. (2003) defined the reaction variables, such as temperature, effect of solvent, enzyme amount, and mole ratio of reactants, to optimize conditions for alkyl ester production from restaurant grease using IM PS-30. The immobilized lipase was active from 40 to 70°C and the ester yields (60 to 97%) were highest using a grease to alcohol ratio of 1:4 with 10% lipase in the presence of the molecular sieves.

Rhizopus oryzae cells immobilized in biomass support particles (BSPs) were utilized as a whole cell biocatalyst for the methanolysis of soybean oil (Ban et al. 2001). The methanolysis was carried out with step-wise addition of methanol in the presence of 10 to 20% water and the methyl ester yield reached 80 to 90%. In another study, *R. oryzae* cells producing 1,3 positional specificity lipase were cultured with polyurethane foam-based BSPs in an air-lift bioreactor, and the cells immobilized in the BSPs were used as a whole-cell biocatalyst in a repeated batch-cycle methanolysis reaction of soybean oil (Oda et al. 2005). The whole-cell biocatalyst had a higher durability in the methanolysis reaction when obtained from the air-lift bioreactor cultivation than from the shake-flask cultivation. The whole-cell biocatalyst promoted the acyl migration of the partial glycerides and the facilitatory effect was increased by increase in the water content of the reaction mixture, which enhanced the yield of biodiesel, but it was lost gradually with the increasing number of reaction cycles. Cross-linking treatment with the glutaraldehyde to *R. oryzae* cells immobilized in the BSPs as a whole-cell biocatalyst for biodiesel production improved the reusability of the enzyme. The methyl ester content reached 70 to 83% in each cycle using glutaraldehyde-treated lipase, compared to 50% at the sixth batch cycle without glutaraldehyde treatment (Ban et al. 2002).

The ability of a commercial immobilized lipase from *R. miehei* to catalyze the transesterification of soybean oil and methanol was investigated by employing the Response Surface methodology (RSM) and the five-level five-factor Central Composite Rotatable Design (CCRD). The parameters evaluated during this study were the reaction time, temperature, enzyme amount, molar ratio of methanol to soybean oil, and added water content; the biodiesel yield was 92.2% at optimum conditions (Shieh, Liao, and Lee 2003).

14.5 LIPASE-MEDIATED ESTERIFICATION AND TRANSESTERIFICATION OF VEGETABLE OILS CONTAINING FREE FATTY ACIDS AND OTHER LOW-GRADE OILS ISOLATED FROM BLEACHING EARTH

Acid oil is a by-product of the vegetable oil process industry, which contains both FFAs and triglycerides. It could be a cheaper source for the preparation of biodiesel. The acid oils of corn and sunflower contain 75.3% and 55.6% of FFAs and 8.6% and 24.7% triacylglycerols, respectively. The fatty acids of the acid oil were esterified with straight and branched chain alcohols, such as methanol, *n*-propanol, *n*-butanol, *i*-butanol, *n*-amylalcohol, *i*-amylalcohol, and *n*-octanol (Tüter et al. 2004) using immobilized *C. antarctica* in hexane medium. Under optimum reaction conditions, the esterified product of the corn acid oil showed 50% methyl ester content and that of sunflower acid showed 63.6% methyl ester content. However, the acid oil of rapeseed oil was simultaneously esterified and transesterified to fatty acid methyl esters in quantitative yields using immobilized *C. antarctica* lipase (Watanabe et al. 2005). The enzyme was quite stable in both the reaction steps and could be used for more than 100 cycles without significant loss of activity. A similar approach was adopted for both esterification and transesterification of high-FFA (20 to 60%)-containing rice bran oil using Novozyme 435 and IM 60 (Lai et al. 2005).

The spent bleaching earth produced during the bleaching of vegetable oils contains about 40% of oil and may also be used as a good feedstock for the preparation of biodiesel. The residual oil present in the spent bleaching earth obtained from refining of soya, rapeseed, and palm oils was extracted with an organic solvent and the extracted oils were subjected to methanolysis by *R. oryzae* in the presence of 75% water content (by weight of substrate), with a single-step addition of the methanol. The conversion to methyl esters was 55% with palm oil after 96 h reaction (Pizarro and Park 2003). In another study, waste activated bleaching earth was effectively converted to the fatty acid methyl esters using lipase from *C. antartica* in the presence of diesel oil or kerosene or *n*-hexane as the organic solvent (Kojima et al. 2004). The lipase showed highest stability in the diesel oil. Under optimum reaction conditions, nearly quantitative yields of the fatty acid methyl esters were obtained using the diesel oil medium.

14.6 CONTINUOUS PRODUCTION OF ALKYL ESTERS USING PACKED-BED REACTORS

Owing to the high cost of lipases, the continuous process of producing simple alkyl esters using immobilized lipases packed in a fixed-bed reactor has been looked into as a feasible process. Three-step flow methanolysis was conducted (Watanabe et al. 2001) using three columns each packed with 3 g immobilized *C. antarctica* lipase (15 × 80 mm) and a mixture of the waste oil; 1/3 molar equivalent of the required methanol was fed into the first reactor. The eluate was left to stand overnight to allow the glycerol to separate and a mixture of the resulting first-step eluate and another 1/3 molar equivalent of the methanol was then fed into the second reactor. The third-step methanolysis was similarly performed by feeding another 1/3 molar equivalent of the methanol to the third reactor. The flow rates in the three reactors were maintained at 6, 6, and 4 ml/h, respectively. The water (1980 ppm) and FFAs (2.5%) present in the waste oil had little influence on the production of biodiesel. The reaction was carried out with a mixture of rapeseed and soybean oils also. The yield of methyl ester from the vegetable oil mixture and the waste oil was 95.9 and 90.4%, respectively.

The use of a recirculating packed-column reactor has also been reported for the transesterification reaction using a phyllosilicate sol-gel immobilized lipase from *Burkholderia cepacia* (IM BS-30) as a stationary phase (Hsu et al. 2004). Using this packed column reactor, grease was transesterified with ethanol with a flow rate of 30 ml/min in continuous mode without solvent. Under the optimum conditions, more than 96% yield of the ester was achieved. The enzyme was reused in the reactor for continuous production. The reactor, enzyme bed, and the substrate reservoir were washed with *n*-hexane between cycles and the enzyme bed was air dried before reuse. The ester conversions after five cycles of enzyme use were normalized, with the conversion for the first cycle being set at 100%. The conversion to the esters for the second cycle decreased to about 90% and then remained constant for the next three cycles.

Supercritical carbon dioxide as a nonconventional solvent in lipase-catalyzed reactions has received considerable attention in recent years as it is readily separable from the reaction medium by post-reaction step-wise depressurization. The esterification of oleic acid and ethanol was carried out in a continuous packed-bed reactor using supercritical carbon dioxide as the solvent (Goddard, Bosley, and Al-Duri 2000). The reported system did undergo substrate inhibition by the ethanol due to the formation of a dead-end complex between the short chain alcohol (ethanol) and the enzyme, causing enzyme deactivation. The plug flow reaction design equation succeeded in describing the performance of the system under the experimental range investigated.

14.7 ALTERNATE ACYL DONORS FOR THE STABILIZATION OF LIPASES

Short chain alcohols such as methanol and ethanol are commonly used as acyl acceptors for biodiesel production. However, the use of excess alcohol leads to inactivation of the enzyme. In addition, the major by-product glycerol blocks the active sites of the enzyme, resulting in low enzyme activity. Novel acyl acceptors such as methyl acetate, ethyl acetate, and propan-2-ol were reported recently for the interesterification of various oils into biodiesel. A comparative study was reported recently on the Novozym 435-catalyzed transesterification of soybean oil with methanol and interesterification with methyl acetate for the production of biodiesel, with a yield of 92% (Du et al. 2004b). The by-product triacetin obtained during the interesterification of soybean oil with methyl acetate did not deactivate the enzyme and the enzyme could be reused continuously for 100 batches. In another study, the activity of different lipases, such as Novozyme 435, Lipozyme RM IM, and Lipozyme TL IM, were compared for the transesterification of soybean oil using methyl acetate as the acyl acceptor (Xu et al. 2003). The Novozyme 435 gave maximum yield (92%) compared to others. The same authors later reported the kinetics of this reaction and developed a three-consecutive reversible reaction model.

Ethyl acetate has also been used as the acyl acceptor for the production of biodiesel from the crude oils of jatropha, karanja, and sunflower using Novozym 435 and the yields of ethyl esters were 91.3%, 90.0%, and 92.7%, respectively (Modi et al. 2006). The relative activity of the lipase could be well maintained over 12 repeated cycles, whereas it reached zero by the sixth cycle when ethanol was used as an acyl acceptor. Modi et al. (2007) also reported the use of propan-2-ol as an acyl acceptor for the Novozyme 435-catalyzed transesterification reaction for the production of biodiesel from crude jatropha, karanja, and sunflower oils, with good yields. Reusability of the lipase was maintained over 12 repeated cycles with propan-2-ol as the acyl acceptor. Similarly, lipase stability was observed during alcoholysis of safflower oil and triolein with 1-propanol and 1-butanol using free and immobilized *P. fluorescens, P. cepacia, M. javanicus, C. rugosa,* and *R. niveus* (using porous kaolinite particle as a carrier) (Iso et al. 2001). The immobilized *P. fluorescens* exhibited highest activity in these reactions. The activity of the immobilized lipase was superior compared to the free lipase. When methanol and ethanol were used, the alcoholysis reaction proceeded only in an appropriate organic solvent such as 1,4-dioxane to protect the lipase.

14.8 CONCLUSIONS AND FUTURE PERSPECTIVES

The disadvantages caused by physicochemical catalysts can be overcome by using lipases for the preparation of biodiesel in simpler and greener methods. Several lipases can be employed either in free form or in immobilized state for this purpose. Immobilized lipases can be used for the preparation of alkyl esters from vegetable oils and free fatty acids and the enzyme can be used in several cycles. However, the reaction time is still longer compared to the alkali-catalyzed processes. Most of the lipase-mediated transesterification reactions reported for biodiesel preparation are with soybean and sunflower oils. Jatropha, karanja, canola, castor, and cottonseed are other potential candidates. As the lipases can simultaneously esterify and transesterify the FFA and triacylglycerols, low-quality oils such as high-FFA oils, acid oils, restaurant greases, and oils isolated from spent bleaching earth can also be converted into biodiesel in reasonably good yields.

The short chain alcohols such as methanol and ethanol as acyl donors lead to inactivation of lipases. The major by-product glycerol also blocks the active sites of the enzymes, resulting in low lipase activity. Novel acyl acceptors such as methyl acetate, ethyl acetate, and propane-2-ol are alternatives to safeguard the activity of the lipases. The cost of lipase production is still the main obstacle to commercializing the lipase-catalyzed process for biodiesel production. However, hopefully in the near future, this can be overcome and biodiesel production based on lipase commercialized.

REFERENCES

Ban, K., M. Kaieda, T. Matsumoto, A. Kondo, and H. Fukuda. 2001. Whole cell biocatalyst for biodiesel fuel production utilizing *Rhizopus oryzae* cells immobilized within biomass support particles. *Biochemical Engineering Journal* 8: 39–43.

Ban, K., S. Hama, K. Nishizuka, M. Kaieda, T. Matsumoto, A. Kondo, H. Noda, and H. Fukuda. 2002. Repeated use of whole-cell biocatalysts immobilized within biomass support particles for biodiesel fuel production. *Journal of Molecular Catalysis B: Enzymatic* 17: 157–165.

Belafi-Bako, K., F. Kovacs, L. Gubicza, and J. Hancsok. 2002. Enzymatic biodiesel production from sunflower oil by *Candida antarctica* lipase in a solvent-free system. *Biocatalysis and Biotransformation* 20: 437–439.

De Debora, O., M. Di Luccio, F. Carina, D. R. Clarissa, B. J. Paulo, L. Nadia, M. Silvana, A. Cristiana, and J. V. de Oliveira. 2004. Optimization of enzymatic production of biodiesel from castor oil in organic solvent medium. *Applied Biochemistry and Biotechnology* 115: 771–780.

Du, W., Y. Xu, and D. Liu. 2003. Lipase-catalyzed transesterification of soya bean oil for biodiesel production during continuous batch operation. *Biotechnology and Applied Biochemistry* 38: 103–106.

Du, W., Y. Xu, J. Zeng, and D. Liu. 2004a. Novozym 435-catalyzed transesterification of crude soya bean oils for biodiesel production in a solvent-free medium. *Biotechnology and Applied Biochemistry* 40: 187–190.

Du, W., Y. Xu, D. Liu, and J. Zeng. 2004b. Comparative study on lipase-catalyzed transformation of soybean oil for biodiesel production with different acyl acceptors. *Journal of Molecular Catalysis B: Enzymatic* 30: 125–129.

Du, W., Y.-Y. Xu, D. Liu, and Z.-B. Li. 2005. Study on acyl migration in immobilized lipozyme TL-catalyzed transesterification of soybean oil for biodiesel production. *Journal of Molecular Catalysis B: Enzymatic* 37: 68–71.

Goddard, R., J. Bosley, and B. Al-Duri. 2000. Lipase-catalyzed esterification of oleic acid and ethanol in a continuous packed bed reactor, using supercritical CO_2 as solvent: Approximation of system kinetics. *Journal of Chemical Technology and Biotechnology* 75: 715–721.

Haas, M. J. and T. A. Foglia. 2005. Alternate feedstocks and technologies for biodiesel production. In *The Biodiesel Handbook,* ed. G. Knothe, J. Gerpen, and J. Krahl, 42–61. Champaign, IL: AOCS Press.

Hong, W. and Z. Min-Hua. 2005. Effect of ultrasonic irradiation on enzymatic transesterification of waste oil to biodiesel. Preprints of Symposia – American Chemical Society, Division of Fuel Chemistry, 50: 773–774.

Hsu, A.-F., K. C. Jones, W. N. Marmer, and T. A. Foglia. 2001. Production of alkyl esters from tallow and grease using lipases immobilized in a phyllosilicate sol-gel. *Journal of American Oil Chemists' Society* 78: 585–588.

Hsu, A.-F., K. C. Jones, T. A. Foglia, and W. N. Marmer. 2003. Optimization of alkyl esters from grease using phyllosilicate sol-gel immobilized lipase. *Biotechnology Letters* 25: 1713–1716.

Hsu, A.-F., K. C. Jones, T. A. Foglia, and W. N. Marmer. 2004. Continuous production of ethyl esters of grease using an immobilized lipase. *Journal of American Oil Chemists' Society* 81: 749–752.

Iso, M., B. Chen, M. Eguchi, T. Kudo, and S. Shrestha. 2001. Production of biodiesel fuel from triglycerides and alcohol using immobilized lipase. *Journal of Molecular Catalysis B: Enzymatic* 16: 53–58.

Kaieda, M., T. Samukawa, A. Kondo, and H. Fukuda. 2001. Effect of methanol and water contents on production of biodiesel fuel from plant oil catalyzed by various lipases in a solvent-free system. *Journal of Bioscience and Bioengineering* 91: 12–15.

Kamini, N. R. and H. Iefuji. 2001. Lipase-catalyzed methanolysis of vegetable oils in aqueous medium by *Cryptococcus* sp. S-2, *Process Biochemistry* 37: 405–410.

Kojima, S., D. Du, M. Sato, and E. Y. Park. 2004. Efficient production of fatty acid methyl ester from waste activated bleaching earth using diesel oil as organic solvent. *Journal of Bioscience and Bioengineering* 98: 420–424.

Köse, Ö., M. Tüter, and H. A. Aksoy. 2002. Immobilized *Candida antarctica* lipase-catalyzed alcoholysis of cotton seed oil in a solvent-free medium. *Bioresource Technology* 83: 125–129.

Lai, C.-C., S. Zullaikah, S. R. Vali, and Y. H. Juyl. 2005. Lipase-catalyzed production of biodiesel from rice bran oil. *Journal of Chemical Technology and Biotechnology* 80: 331–337.

Mittelbach, M. 1990. Lipase catalyzed alcoholysis of sunflower oil. *Journal of American Oil Chemists' Society* 67: 168–170.

Modi, M. K., J. R. C. Reddy, B. V. S. K. Rao, and R. B. N. Prasad. 2006. Lipase-mediated transformation of vegetable oils into biodiesel using propan-2-ol as acyl acceptor. *Biotechnology Letters* 28: 637–640.

Modi, M. K., J. R. C. Reddy, B. V. S. K. Rao, and R. B. N. Prasad. 2007. Lipase-mediated conversion of vegetable oils into biodiesel using ethyl acetate as acyl acceptor. *Bioresource Technology* 98: 1260–1264.

Nelson, L. A., T. A. Foglia, and W. N. Marmer. 1996. Lipase-catalyzed production of biodiesel. *Journal of American Oil Chemists' Society* 73: 1191–1195.

Noureddini, H., X. Gao, and R. S. Philkana. 2005. Immobilized *Pseudomonas cepacia* lipase for biodiesel fuel production from soybean oil. *Bioresource Technology* 96: 769–777.

Oda, M., K. Masaru, H. Shinji, Y. Hideki, K. Akihiko, I. Eiji, and F. Hideki. 2005. Facilitatory effect of immobilized lipase-producing *Rhizopus oryzae* cells on acyl migration in biodiesel-fuel production. *Biochemical Engineering Journal* 23: 45–51.

Pizarro, A. V. L. and E. Y. Park. 2003. Lipase-catalyzed production of biodiesel fuel from vegetable oils contained in waste activated bleaching earth. *Process Biochemistry* 38: 1077–1082.

Samukawa, T., M. Kaieda, T. Matsumoto, K. Ban, A. Kondo, Y. Shimada, H. Noda, and H. Fukuda. 2000. Pretreatment of immobilized *Candida antarctica* lipase for biodiesel fuel production from plant oil. *Journal of Bioscience and Bioengineering* 90: 180–183.

Selmi, B. and D. Thomas. 1998. Immobilized lipase-catalyzed ethanolysis of sunflower oil in a solvent-free medium. *Journal of American Oil Chemists' Society* 75: 691–695.

Shah, S., S. Sharma, and M. N. Gupta. 2004. Biodiesel preparation by lipase-catalyzed transesterification of jatropha oil. *Energy Fuels* 18: 154–159.

Shieh, C.-J., H.-F. Liao, C.-C. Lee. 2003. Optimization of lipase-catalyzed biodiesel by response surface methodology. *Bioresource Technology* 88: 103–106.

Shimada, Y., Y. Watanabe, T. Samukawa, A. Sugihara, H. Noda, H. Fukuda, and Y. Tominaga. 1999. Conversion of vegetable oil to biodiesel using immobilized *Candida antarctica* lipase. *Journal of American Oil Chemists' Society* 76: 789–793.

Soumanou, M. M. and U. T. Bornscheuer. 2003. Improvement in lipase-catalyzed synthesis of fatty acid methyl esters from sunflower oil. *Enzyme and Microbial Technology* 33: 97–103.

Tüter, M., H. A. Aksoy, E. E. Gilbaz, and E. Kurşun. 2004. Synthesis of fatty acid esters from acid oils using lipase B from *Candida Antarctica*. *European Journal of Lipid Science and Technology* 106: 513–517.

Watanabe, Y., Y. Shimada, A. Sugihara, H. Noda, H. Fukuda, and Y. Tominaga. 2000. Continuous production of biodiesel fuel from vegetable oil using immobilized *Candida antarctica* lipase. *Journal of American Oil Chemists' Society* 77: 355–360.

Watanabe, Y., Y. Shimada, A. Sugihara, and Y. Tominaga. 2001. Enzymatic conversion of waste edible oil to biodiesel fuel in a fixed-bed bioreactor. *Journal of American Oil Chemists' Society* 78: 703–707.

Watanabe, Y., P. Pinsirodom, T. Nagao, T. Kobayashi, Y. Nishida, Y. Takagi, and Y. Shimada. 2005. Production of FAME from acid oil model using immobilized *Candida antarctica* lipase. *Journal of American Oil Chemists' Society* 82: 825–831.

Xu, Y., W. Du, D. Liu, and J. Zeng. 2003. A novel enzymatic route for biodiesel production from renewable oils in a solvent-free medium. *Biotechnology Letters* 25: 1239–1241.

15 Biodiesel Production With Supercritical Fluid Technologies

Shiro Saka and Eiji Minami

CONTENTS

ABSTRACT

At present, the alkaline catalyst method is applied commercially to produce biodiesel. However, the process is not simple and not applicable to wastes of oils and fats. Therefore, a one-step supercritical methanol method, the Saka process, was developed as a noncatalytic process. In this process, even wastes of oils and fats that are high in water and free fatty acids can be converted to biodiesel. However, the reaction conditions are drastic (350°C, >20 MPa), thus a special alloy such as hastelloy is required for the reaction vessel. Additionally, the biodiesel produced is thermally deteriorated. Therefore, to realize milder reaction conditions, a two-step supercritical methanol method, the Saka-Dadan process, was developed, which consisted of the hydrolysis of oils and fats in subcritical water and subsequent methyl esterification of the fatty acids produced in supercritical methanol. In this process, milder reaction conditions (270°C, <10 MPa) can be realized using ordinary stainless steel instead of a special alloy. Moreover, due to the removal of the glycerol after the hydrolysis process, the biodiesel satisfies most of the requirements of the EU and U.S. standards.

15.1 INTRODUCTION

TG + MeOH ⇌ DG + FAME

DG + MeOH ⇌ MG + FAME

MG + MeOH ⇌ G + FAME

FIGURE 15.1 Three step-wise transesterification reactions of triglyceride.

Biodiesel fuel, which is defined as fatty acid methyl ester (FAME), is one of the most promising bioenergies used as a substitute for fossil diesel and can be produced commercially with methanol by transesterification of triglyceride, which is a major component of oils and fats in vegetables and animals. In the transesterification reaction (Figure 15.1), the triglyceride (TG) is converted step-wise to diglyceride (DG), monoglyceride (MG), and finally glycerol (G). At each step, one molecule of FAME is produced, consuming one molecule of the methanol. These reactions are reversible, although the equilibrium lies towards the production of FAME.

Most methods for biodiesel production involve the use of an alkali catalyst, although acid catalysts and a combination of acid and alkali catalysts can also be used. However, each of these methods has disadvantages as well. Supercritical fluids have recently received attention as a new reaction field due to their unique properties and noncatalytic effects. In this chapter, current progress in biodiesel production by supercritical fluid technologies is introduced and discussed.

15.2 SUPERCRITICAL FLUID

A pure substance changes its form to be solid, liquid, or gas, depending on conditions of temperature and pressure. However, when the temperature and pressure go beyond the critical point, the substance becomes a supercritical fluid. In the supercritical state, the molecules in the substance have high kinetic energy like a gas and high density like a liquid. It is, therefore, expected that the chemical reactivity can be enhanced, particularly when a protic solvent becomes supercritical. In addition, the dielectric constant of its supercritical fluid is lower than that of liquid due to a cleavage of the hydrogen bonds in a protic solvent. For example, the dielectric constant of supercritical methanol (critical temperature T_c = 239°C, critical pressure P_c = 8.09 MPa) becomes 7 at the critical point, while that of liquid methanol is about 32 at ambient temperature (Franck and Deul 1978). The former value is equivalent to that of the nonpolar organic solvent, and it can dissolve well many kinds of nonpolar organic substances, such as oils and fats. In supercritical methanol, therefore, a homogeneous (one-phase) reaction between the oils/fats and methanol can be realized. Furthermore, the ionic product of a protic solvent such as water (T_c = 374°C, P_c = 22.1 MPa) and methanol is increased in the supercritical state. Therefore, the solvolysis reaction field can be achieved, thus resulting in hydrolysis in the water and methanolysis in the methanol (Holzapfel 1969).

By taking these interesting properties into consideration, noncatalytic biodiesel production methods have been developed during the last decade using supercritical methanol. One such method is the one-step supercritical methanol method (Saka process); another is the two-step supercritical methanol method (Saka-Dadan process).

15.3 ONE-STEP SUPERCRITICAL METHANOL METHOD (SAKA PROCESS)

In the supercritical methanol, TG in oils/fats is converted into the fatty acid methyl ester (FAME) by transesterification without catalyst due to its methanolysis ability (Figure 15.2) (Saka and Dadan 2001). At 300°C (20MPa), a relatively poor conversion to the FAME is observed. Under temperatures over 350°C, however, the reaction rate increases remarkably, resulting in a good conversion (Figure 15.3). This transesterification follows a typical second-order reaction, in which the reaction equations for TG, DG, and MG can be described as follows (Diasakou, Louloudi, and Papayannakos 1998):

$$\frac{dC_{TG}}{dt} = -k_{TG}C_{TG}C_M + k'_{TG}C_{DG}C_{FAME} \tag{15.1}$$

$$\frac{dC_{DG}}{dt} = -k_{DG}C_{DG}C_M + k'_{DG}C_{MG}C_{FAME} + k_{TG}C_{TG}C_M - k'_{TG}C_{DG}C_{FAME} \tag{15.2}$$

$$\frac{dC_{MG}}{dt} = -k_{MG}C_{MG}C_M + k'_{MG}C_GC_{FAME} + k_{DG}C_{DG}C_M - k'_{DG}C_{MG}C_{FAME} \tag{15.3}$$

where C_{TG}, C_{DG}, C_{MG}, C_G, C_{FAME}, and C_M refer to the molar concentrations of TG, DG, MG, glycerol, FAME, and methanol in the reaction system, respectively. Similarly, when the reaction rate constants of TG, DG, and MG are equal to each other, the rate of FAME formation can be described as below:

$$\frac{dC_{FAME}}{dt} = kC_OC_M - k'C_O'C_{FAME} \tag{15.4}$$

$$(C_O = C_{TG} + C_{DG} + C_{MG},\ C_O' = C_{DG} + C_{MG} + C_G)$$

Because of the backward reaction shown in these equations, a larger amount of methanol must be added in the reaction system to achieve a higher yield of FAME.

With regard to the interaction between the methanol and the oils/fats, the reaction system initially forms a two-phase liquid system at ambient temperature and pressure because the solvent properties of the methanol are significantly different from those of the oils/fats, such as the dielectric constant. As the reaction temperature increases, the dielectric constant of the methanol decreases to be closer to that of the oils/fats, allowing the reaction system to form one phase between the methanol and the oils/fats so that the homogeneous reaction takes place (Saka and Minami 2005). Therefore, there are no limitations of mass transfer on the reaction, allowing it to proceed

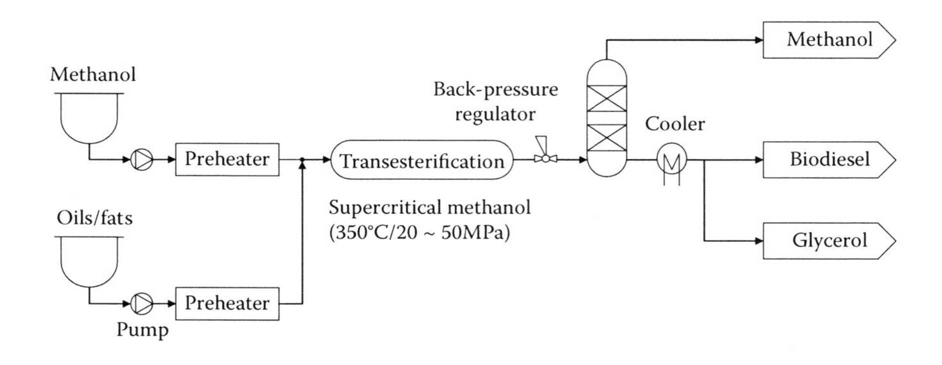

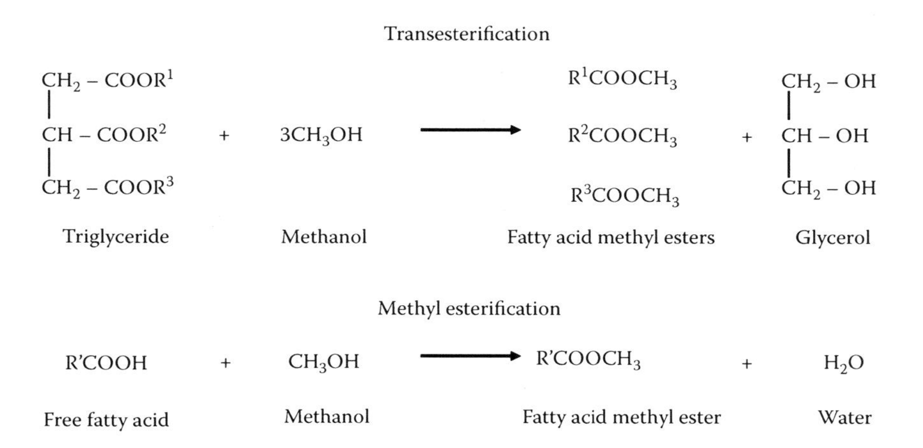

FIGURE 15.2 Scheme of the one-step supercritical methanol method (Saka process) and reactions of oils and fats involved in biodiesel production (R^1, R^2, R^3, R': hydrocarbon groups). (From Saka, S. and K. Dadan. 2001. *Fuel* 80: 225–231. With permission.)

in a very short time. Compared to the alkali-catalyzed method, in which the stirring effect is significant in a heterogeneous two-phase system, stirring is not necessary in the supercritical methanol because the reaction system is already homogeneous.

Another important achievement in the one-step supercritical methanol method is that the FFA can be converted to FAME by methyl esterification (Figure 15.2) (Dadan and Saka 2001), while in the case of the alkali-catalyzed method, they are converted to the saponified products, which must be removed after the reaction. Therefore, the one-step method can produce a higher yield of FAME than the alkali-catalyzed method, especially for low-quality feedstock containing FFA.

Based on these lines of evidence, the superiority of the one-step supercritical methanol method can be summarized as follows: (1) the production process becomes simple, (2) the reaction is fast, (3) the FFA can be converted simultaneously to FAME through methyl esterification, and (4) the yield of FAME is high.

Although this process has many advantages to produce a high yield of biodiesel, it requires restrictive reaction conditions of, for example, 350°C and 20 MPa. Under

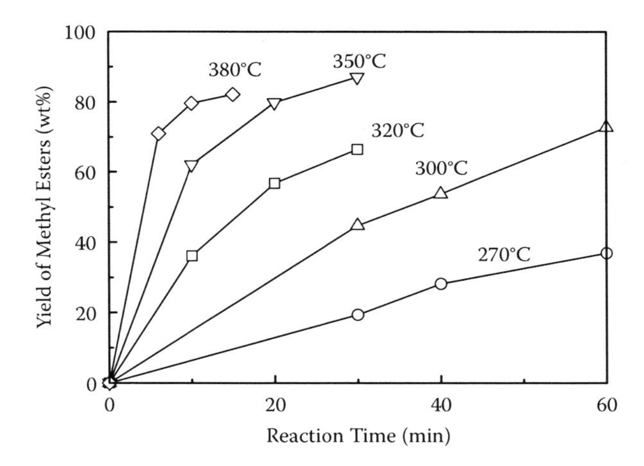

FIGURE 15.3 Transesterification of rapeseed oil to fatty acid methyl esters in supercritical methanol at various temperatures (reaction pressure, 20 MPa; molar ratio of methanol to triglyceride, 42). (From Minami, E. and S. Saka. 2006. *Fuel* 85: 2479–2483. With permission.)

these conditions, a special alloy (e.g., Inconel and Hustelloy) is required for the reaction tube to avoid its corrosion. In addition, the methyl esters, particularly from polyunsaturated fatty acids such as methyl linolenate, are partly denatured under these severe conditions (Tabe et al. 2004).

15.4 TWO-STEP SUPERCRITICAL METHANOL METHOD (SAKA-DADAN PROCESS)

To realize more moderate reaction conditions, the two-step supercritical methanol method was developed (Figure 15.4) (Dadan and Saka 2004). In this method, the oils and fats are first treated in subcritical water for the hydrolysis reaction to produce fatty acids (FA). After the hydrolysis, the reaction mixture is separated into the oil phase and water phase by decantation. The oil phase (upper portion) contains FA, while the water phase (lower portion) contains glycerol. The separated oil phase is then mixed with methanol and treated under supercritical conditions for the methyl esterification. After removing the unreacted methanol and water produced in the reaction, the FAME can be obtained as biodiesel.

The hydrolysis of the oils and fats consists of three step-wise reactions similar to transesterification (Figure 15.1): one molecule of the TG is hydrolyzed to the DG producing one molecule of the FA, and the DG is repeatedly hydrolyzed to the MG, which is further hydrolyzed to glycerol, producing all together three molecules of the FA. As a backward reaction, however, the glycerol reacts with the FA to produce the MG. In a similar manner, the DG and MG also return to the TG and DG, respectively, consuming one molecule of the FA. In subcritical water, the hydrolysis reaction occurs without catalyst (Dadan and Saka 2004). A good conversion of oils and fats to the FA can be achieved at low temperatures, between 270 and 290°C (20

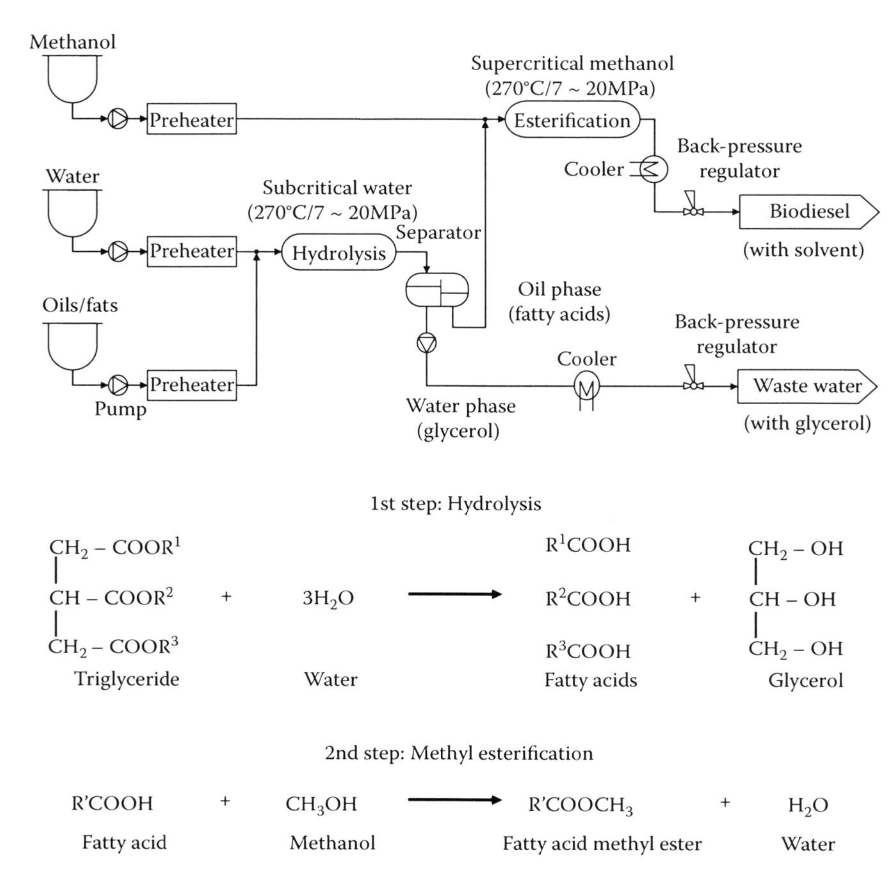

FIGURE 15.4 Scheme of the two-step supercritical methanol method (Saka-Dadan process) and reactions of oils and fats involved in biodiesel production (R^1, R^2, R^3, R': hydrocarbon groups). (From Dadan, K. and S. Saka. 2004. *Appl. Biochem. Biotechnol.* 115: 781–791. With permission.)

MPa), compared with one-step transesterification, but higher temperature results in faster hydrolysis (Figure 15.5).

In the hydrolysis reaction of the oils and fats, the yield of FA is very slowly increased in the early stage of the reaction, especially at the lower temperatures of 250 and 270°C (Figure 15.5). The rate of FA formation, then, becomes faster when the treatment is prolonged. This phenomenon can be explained by the reaction equation:

$$\frac{dC_{FA}}{dt} = \left(kC_O C_W - k'C_O{}'C_{FA}\right) \times C_{FA} \tag{15.5}$$

where C_{FA} and C_W refer to the concentrations of FA and water, respectively. In this equation (15.5), the formula in parenthesis depicts a typical second-order reaction, while the factor C_{FA} describes the effect of autocatalytic reaction by the FA. The

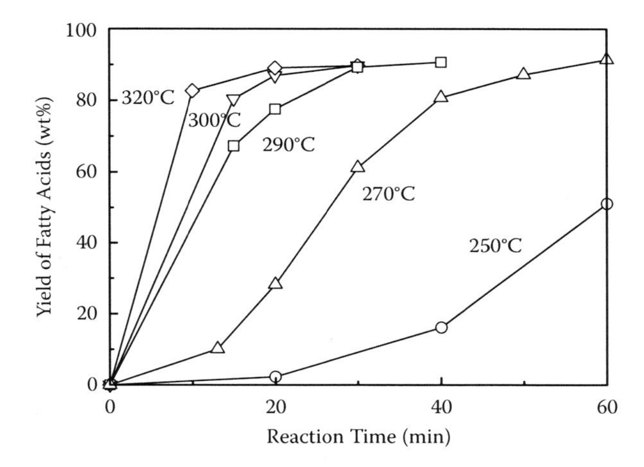

FIGURE 15.5 Hydrolysis of rapeseed oil to fatty acids in subcritical water at various temperatures (reaction pressure, 20 MPa; molar ratio of water to triglyceride, 54). (From Minami, E. and S. Saka. 2006. *Fuel* 85: 2479–2483. With permission.)

equation is based on the assumption that the FA produced by hydrolysis acts as the acid catalyst in subcritical water. Therefore, the hydrolysis of the oils and fats in subcritical water is proved successfully by Equation (15.5) (Minami and Saka 2006). For more efficient hydrolysis reaction, therefore, the addition of FA to the oils and fats can be expected to enhance hydrolysis in subcritical water due to its acidic character. In a similar manner, the back-feeding of the FA produced to the reaction system can be expected to enhance the hydrolysis reaction.

The second part of the two-step supercritical methanol method deals with the methyl esterification of the FA, the hydrolyzed products of the oils and fats, by the supercritical methanol treatment. Similar to the hydrolysis reaction, the esterification of the FA is almost completely performed at between 270 and 290°C and 20 MPa (Figure 15.6). In the case of methyl esterification, the yield of FAME tends to increase quickly in the early stage of the reaction, whereas the rate of FAME formation becomes slower as the reaction proceeds. This is because the FA itself acts as an acid catalyst in the methyl esterification as well as hydrolysis (Minami and Saka 2006). Therefore, the autocatalytic mechanism by the FA can be applied for the methyl esterification as in the following equation:

$$\frac{dC_{FAME}}{dt} = \left(kC_{FA}C_M - k'C_{FAME}C_W\right) \times C_{FA} \tag{15.6}$$

The autocatalytic methyl esterification offers a unique effect of the methanol concentration on the FAME yield. In Figure 15.7, a higher yield is achieved when less methanol is added to the reaction system. For example, about 94% of the FAME is obtained with a molar ratio of 8/1 (methanol/FA), whereas only 87% is obtained in 42/1 methanol ratio when treated at 290°C and 20 MPa for 30 min.

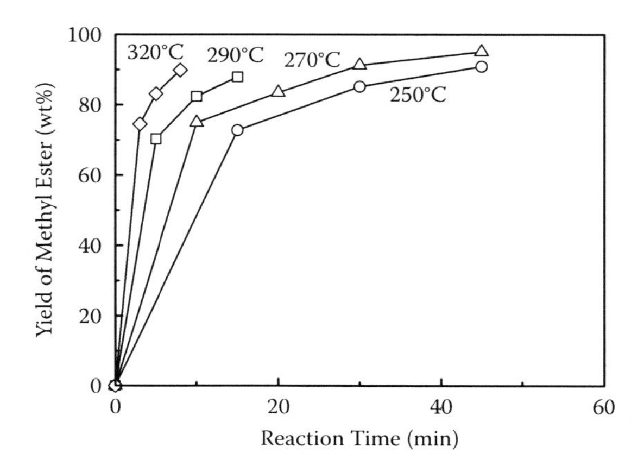

FIGURE 15.6 Methyl esterification of oleic acid to its methyl ester in supercritical methanol at various temperatures (reaction pressure, 20 MPa; molar ratio of methanol to oleic acid, 14). (From Minami, E. and S. Saka. 2006. *Fuel* 85: 2479–2483. With permission.)

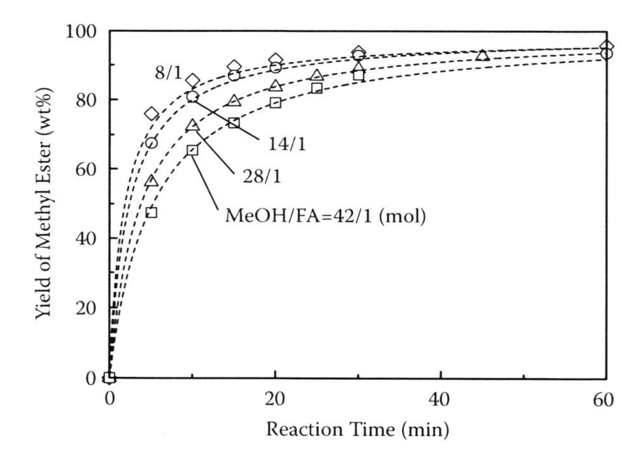

FIGURE 15.7 Effect of methanol concentration on methyl ester yield from oleic acid as treated in supercritical methanol at 290°C and 20 MPa. (From Minami, E. and S. Saka. 2006. *Fuel* 85: 2479–2483. With permission.)

In the autocatalytic reaction by the FA, less methanol makes the FA concentration higher in the reaction system, thus achieving faster methyl esterification. Based on Equation (15.6), theoretical curves actually fit well with the experimental results, as represented by the dotted lines shown in Figure 15.7. After the equilibrium, however, a large amount of methanol is more preferable to realize a higher yield of the FAME due to suppression of the backward reaction.

Based on these lines of evidence, milder reaction conditions (270~290°C, 7~20 MPa) can be achieved by the two-step supercritical methanol method, compared with the one-step method. In designing a manufacturing plant for the supercritical fluid process, lower temperature and lower pressure are more desirable. The two-step method allows, therefore, the use of common stainless steel instead of special alloys such as Inconel and Hastelloy for reactors.

Coincidentally, the two-step method can produce high-quality biodiesel fuel. In the case of the one-step method, glycerol always exists in the reaction system and reacts with the FAME to reproduce MG as a backward reaction. Similarly, MG and DG are also reversed to DG and TG, respectively, consuming one molecule of the FAME. In the two-step method, on the other hand, glycerol is removed prior to the methyl esterification reaction. Therefore, such a backward reaction can be depressed in the methyl esterification step.

15.5 PROPERTIES OF BIODIESEL

Among the standard specifications for biodiesel, such as EN 14214 (European Commission of Normalization 2003) and ASTM D 6751 (American Society for Testing and Materials 2003), the total glycerol content G_{total} (wt% on the biodiesel) described in Equation (15.7) is one of the most important characteristics because the glycerides significantly affect the biodiesel properties such as viscosity, pour point, carbon residue, and so on.

$$G_{total} = 0.1044W_{TG} + 0.1488W_{DG} + 0.2591W_{MG} + W_G \tag{15.7}$$

where W_{TG}, W_{DG}, W_{MG}, and W_{G} are wt% of TG, DG, MG, and free glycerol on biodiesel, respectively. In EU and U.S. standards, the G_{total} must be less than 0.24 and 0.25 wt%, respectively.

As mentioned previously, low total glycerol content can be expected in the two-step method, because this method can depress the backward reaction of the glycerol. Actually, no glycerides are detected in biodiesel prepared by the two-step method from waste rapeseed oil and dark oil (Table 15.1) (Saka et al. 2005). Concomitantly, other biodiesel properties can also satisfy the specifications in the EU standard.

As shown in Table 15.1, waste rapeseed oil can be a good raw material as it contains only a small amount of FFA. Therefore, it is available even for the alkali-catalyst method as well as the supercritical methanol methods. However, dark oil, which is a by-product from oil/fat manufacturing plants that contains large amounts of FFA (>60%), is not available for the alkali-catalyzed method. In the case of the two-step method, however, the conversion is made successfully (Table 15.1). Thus, the two-step supercritical methanol method can produce high-quality biodiesel from various feedstocks through relatively milder reaction conditions. However, a backward reaction of the FAME to the FA exists due to the water formed by the methyl esterification. For this reason, acid value by the two-step method tends to be rather high. At present, therefore, a re-esterification step is adapted at the pilot plant in Japan to satisfy the specification for the acid value (<0.5 mg/g in the EU standard).

TABLE 15.1
Biodiesel Fuel Evaluation Prepared by the Two-Step Supercritical Methanol Method

Properties	EN 14214	Raw Materials	
		Waste Rapeseed Oil	Dark Oil
Density, g/ml	0.86~0.90	0.883	0.883
Viscosity (40°C), mm²/s	3.5~5.0	4.70	4.41
Pour point, °C	–	-7.5	-2.5
Cloud point, °C	–	-8	-2
CFPP, °C	–	-8	-3
Flash point, °C	>120	173	161
10% carbon residue, wt%	<0.3	0.04	0.04
Cetane number	>51	54	50
Ester content, wt%	>96.5	99.5	96.1
Total glycerol, wt%	<0.25	N.D.	N.D.
Water content, wt%	<0.05	0.04	0.03
MeOH content, wt%	<0.2	–	0.011
Sulfur, mg/kg	<10	<3	14
Oxidation stab., h[a]	>6	>>6	8.8
Acid value, mg KOH/g	<0.5	0.32	0.29
Iodine value, g I_2/100 g	<120	99	107
Gross calorific value, kJ/g	–	39.7	39.7

[a] Antioxidant was added.

From Saka et al. 2005. With permission.

15.6 CONCLUSIONS AND FUTURE PERSPECTIVES

To overcome the various drawbacks in the conventional alkali-catalyzed method, two novel processes have been developed employing noncatalytic supercritical methanol technologies. The one-step method can produce biodiesel through the transesterification of oils and fats in supercritical methanol, with a simpler process and shorter reaction time. In addition, a higher yield of the FAME was achieved due to the simultaneous conversion of the FFA through methyl esterification. The two-step method, on the other hand, realized more moderate reaction conditions than those of the one-step method, keeping the advantages previously obtained. By this method, furthermore, high-quality biodiesel can be obtained because glycerol is removed before the methyl esterification step. These production methods have a tolerance for the FFA and water in the oil/fat feedstocks, especially in the case of the two-step method. Therefore, various low-grade waste oils and fats, such as waste oils from the household sector and rendering plants, can be used as raw materials.

REFERENCES

American Society for Testing and Material. 2003. ASTM D6751-03; Standard Specification for Biodiesel Fuel Blend Stock (B100) for Middle Distillate Fuels, 1-6.

Dadan, K. and S. Saka. 2001. Methyl esterification of free fatty acids of rapeseed oil as treated in supercritical methanol. *J. Chem. Eng. Jpn.* 34: 383–387.

Dadan, K. and S. Saka. 2004. Two-step preparation for catalyst-free biodiesel fuel production. *Appl. Biochem. Biotechnol.* 115: 781–791.

Diasakou, M., A. Louloudi, and N. Papayannakos. 1998. Kinetics of the non-catalytic transesterification of soybean oil. *Fuel* 77: 1297–1302.

European Commission of Normalization. 2003. EN 14214; Automotive Fuels - Fatty Acid Methyl Esters (FAME) for Diesel Engines - Requirements and Test Methods, 1-17.

Franck, E. U. and R. Deul. 1978. Dielectric behavior of methanol and related polar fluids at high pressures and temperatures. *Faraday Disc. Chem. Soc.* 66: 191–198.

Holzapfel, W. 1969. Effect of pressure and temperature on the conductivity and ionic dissociation of water up to 100 kbar and 1000°C. *J. Chem. Phys.* 50: 4424–4428.

Minami, E. and S. Saka. 2006. Kinetics of hydrolysis and methyl esterification for biodiesel production in two-step supercritical methanol process. *Fuel* 85: 2479–2483.

Saka, S. and K. Dadan. 2001. Biodiesel fuel from rapeseed oil as prepared in supercritical methanol. *Fuel* 80: 225–231.

Saka, S. and E. Minami. 2005. A novel non-catalytic biodiesel production process by supercritical methanol as NEDO "High Efficiency Bioenergy Conversion Project." *Proc. of 14th Euro Biomass Conf Exhib on Biomass for Energy, Industry and Climate Protection,* October 17–21, 2005, Paris, France, 1419–1422.

Saka, S., E. Minami, K. Yamashita, H. Wada, N. Okada, Y. Toide, H. Miyauchi, Y. Nagasato, S. Okamura, M. Hattori, H. Murakami, and N. Matsui. 2005. NEDO "High Efficiency Bioenergy Conversion Project": R&D for biodiesel fuel production by two-step supercritical methanol method. *Proc. of 14th Euro Biomass Conf Exhib on Biomass for Energy, Industry and Climate Protection,* Octoberl7–21, 2005, Paris, France, 1056–1059.

Tabe, A., K. Dadan, E. Minami, and S. Saka. 2004. Kinetics in transesterification of rapeseed oil by supercritical methanol treatment. *Proc. 2nd World Conf Technol Exhib on Biomass for Energy, Industry and Climate Protection*, May 10–14, 2004, Rome, Italy, 1553–1556.

16 Palm Oil Diesel Production and Its Experimental Test on a Diesel Engine

Md. Abul Kalam, Masjuki Hj Hassan, Ramang bin Hajar, Muhd Syazly bin Yusuf, Muhammad Redzuan bin Umar, and Indra Mahlia

CONTENTS

ABSTRACT

This chapter presents the status of palm oil diesel (POD) production and its experimental test on a multicylinder diesel engine. The test results obtained are brake power, specific fuel consumption (SFC), exhaust emissions, anti-wear characteristics of fuel-contaminated lubricants, and fuel Rancimat test characteristics. It was found that B20X fuel showed better overall performance such as improved brake power, reduced exhaust emissions and shows better lube oil quality as compared to other tested fuels. The specific objective of this investigation is to improve the performance of B20 fuel using an antioxidant additive.

16.1 INTRODUCTION

With reference to the world energy scenario, some 85 to 90% of world primary energy consumption will continue (until 2030) to be based on fossil fuels (DOE 2007). However, after 2015, usage of renewable energy, natural gas, and nuclear energy will be increased because of stringent emissions regulations and limited fossil fuel reserves. The total renewable energy demand will increase from 2% (2002) to 6% (2030), and fuel from biomass will be one of the major resources, followed by solar and hydropower generation. Fuel from biomass (as well as vegetable oils) conversion, such as biodiesel, is becoming a new alternative, renewable fuel to be used for heating, transportation, and electricity generation.

The biodiesel is produced primarily in some 10 to 15 countries, with four to five types of vegetable oil. The total production of biodiesel from various types of vegetable oil is about 2 to 3 million tones per year. Details regarding production of biodiesel on a country basis can be found in Kalam and Masjuki (2005). Table 16.1 lists vegetable oil production by country and shows the area under plantation for each..

Palm oil is produced mainly in Malaysia and Indonesia. Malaysia is the leader in terms of production and export. It produces about 55% of the world's palm oil and exports 62% of world palm oil in the form of cooking oil and oil products. Palm oil has become one of the most crucial foreign exchange earners for the country. Total export earnings for palm oil products increased by 160% to US$9.50 billion in 2005 from US$3.00 billion in 1996 (Choo et al. 2005). The palm oil production area has increased from 38,000 ha in 1950 to about 4.2 million ha in 2005, occupying more than 60% of agricultural land in the country. The rapid expansion in oil palm

TABLE 16.1
World Vegetable Oil Plantation Areas and Oil Production 2005

Oils	Oil Production (Million Tons)	Leading Countries	Plantation Area (million ha)
Soybean	29.15	United States and Brazil	78.65
Palm	29.6	Malaysia and Indonesia	8.9
Rapeseed	14.7	Europe	27.8
Sunflower	9.2	France and Italy	19.5
Coconut	4.5	The Philippines	10.4

cultivation resulted in a corresponding increase in the palm oil production from less than 100,000 tonnes in 1950 to 16.28 million tonnes in 2005. The oil palm yields on average 3.66 tonne/ha of oil per year. Malaysian palm oil currently goes into food (80%) and in the nonfood sector (20%), which includes making soaps and detergents, toiletries, cosmetics, biodiesel, and other industrial usages.

16.1.1 Biodiesel Production and Marketing Status in Malaysia

Since the 1980s, the Malaysian Palm Oil Board (MPOB), in collaboration with the local oil-producing company Petronas, has carried out transesterification of crude palm oil into palm oil diesel (POD). It is now under design to build a 60,000 tonnes per annum palm oil diesel plant based on a previous pilot plant at the MPOB headquarters with a capacity of 3,000 tonnes per annum. In addition, the Malaysian government is also trying to build a biodiesel plant (at a cost of about US$60 million) to produce biodiesel from palm oil. This plant will produce two types of fuel: (1) a blend of petroleum diesel (95%) with palm oil (5%) for local usage without modifications in the diesel engines; and (2) biodiesel, produced by the conversion of palm oil into methyl ester, which can be used as fuel (B100). In 2005, Malaysia produced over 16 million tonnes of crude palm oil and some 500,000 tonnes were converted into biodiesel. Currently, 10% of the palm oil production has been allocated for the biodiesel project. It will further stabilize the price of palm oil in the international market and subsequently contribute to the Malaysian palm oil industry (Yoo et al. 1998) as well as partial replacement of diesel fuel. The consumption of diesel fuel was 4.84 and 5.34 billion liters in 2004 and 2005, respectively, when the target was set to replace at least 5% of diesel with palm oil by the year 2007. As a trial, more than 150 vehicles (buses, trucks, and lorries) are being run on a palm diesel blend to evaluate engine noise, lube oil, degradation emissions, and performance characteristics. At present, Malaysia exports palm oil to over 100 countries and exports palm oil diesel (POD) to Korea, Germany, and Japan. The local prices of net palm oil and POD production are US$0.39 and US$0.60 per liter, respectively, and the commercial diesel fuel price is US$0.26 per liter. Currently, the government is trying to promote biodiesel production and utilization through incentives and tax exemption.

16.1.2 Biodiesel Standardization

The term biodiesel refers to methyl esters of long chain fatty acids derived from vegetable oils. The Fuel Standards Regulations 2001 under the Fuel Quality Standards Act 2000 define biodiesel as "a diesel fuel substitute obtained by esterification of oil derived from plants or animals" (Fuel Quality Standards Regulations 2001). It also can be used as a fuel in compression ignition engines without any modification.

Germany and the EU have biodiesel standards for rapeseed methyl ester, DIN E51606 and EN 14214, respectively. The United States has produced a biodiesel standard for soybean methyl ester. Japan and Korea have also produced biodiesel standards. The EU standard EN 14214 is often used as the reference for other nations considering adoption of biodiesel standards.

In Malaysia, biodiesel is prepared from palm oil by the methanol transesterification process. Currently, Malaysia produces two types of palm biodiesel, normal palm biodiesel with pour point of 15°C, which can only be used in tropical countries, and low-pour-point biodiesel (-21°C to 0°C), which can be used in temperate countries to meet the seasonal pour point requirements (summer grade, 0°C; spring and autumn grades, -10°C, and winter grade, -20°C). The world biodiesel standard comparisons are summarized in Table 16.2.

Palm oil-based biodiesel has been tested locally (Kalam and Masjuki,2005; Choo et al. 2005) and internationally (Ramadhas, Jayaraj, and Muraleedharan 2006) in B20 and B100 forms. The results showed that B20 produces lower brake power and increases wear after long-term engine operation. The fuel B100 produces higher nitrogen oxide (NOx) emission and lower brake power due to the O_2 and water that it contains, which contribute to oxidation, plugging the fuel filter, and formation of deposits on the piston-cylinder head, and the used lubricant has increased wear debris. However, generally NOx is considered the main problem in biodiesel fuel. The formation of NOx is mainly due to the high combustion temperature of the long chain fatty acid (with oxygen content) in the biodiesel. During combustion, the long chain fatty acids are broken into short chain fatty acids and polarization of combustion products. The short chain fatty acids contain high energy, which results in the oxidation. If the biodiesel is treated with a suitable antioxidant additive, which can absorb the energy of the short chain fatty acids, NOx will be reduced and the fuel thermal conversion energy increased. The U.S. National Biodiesel Board (2007) has presented test results on the effect of fuel-borne catalyst on NOx emissions from soybean oil-based biodiesel blend with diesel fuel No.1 (the commercial pipeline-grade kerosene widely used by the municipalities). The results showed that the fuel-borne catalyst could reduce 5% of the NOx emissions. MPOB has used different types of additive to observe the oxidative stability of the palm oil diesel. It was found that the antioxidant additive was effective in increasing the Rancimat induction period (Liang et al. 2006). However, no information is available on engine tests with palm oil diesel (as B20) using antioxidant additive to investigate the performance, emissions, and wear characteristics.

16.2 EVALUATION OF PALM OIL-BASED BIODIESEL

A schematic diagram of a fuel system with dynamometer engine is shown in Figure 16.1. The specifications of the indirect injection (IDI) diesel engine are shown in Table 16.3. The dynamometer instrumentation used was fully equipped in accordance with SAE recommended practice, J1349 JUN90. A variable speed range from 1000 to 4000 rpm with half-throttle setting was selected for performance test such as to measure the brake power and specific fuel consumption (SFC). The emission test was done with constant 50 Nm load and at constant 2250 rpm engine speed. The same test procedure and practice were followed for all the test fuels. A Bosch gas analyzer model ETT 008.36 was used to measure the HC and CO emissions. A Bacharach model CA300NSX gas analyzer (Standard version, k-type probe) was used to measure the NOx concentration in vppm (parts per million by volume).

TABLE 16.2
Standardization of Biodiesel

Country		Germany[a]	USA[b]	Korea[c]	Malaysia[d]	
Standard/Specification		DIN E 51606	ASTM D6751	B20	B100	LPPP[e]
Date		Sep-97	10-Jan-02	30-Sept-04	Aug-2005	
Application		FAME	FAME	FAME	FAME	FAME
Density 15°C	g/cm3	0.875–0.90	0.80–0.90	0.86–0.90	0.8783	0.87–0.9
Viscosity 40°C	mm2/s	3.5–5.0	1.9–6.0	1.9–5.5	4.415	4–5
Distillation 95%	°C	–	≤360	–	–	–
Flash point	°C	>100	>130	>120	182	150–200
Cloud point	°C	–	–	–	15.2	-18–0
CFPP	°C	0/-10/-20	–	–	15	-18–3
Pour point	°C	–	–	–	15	-21–0
Sulfur	% mass	<0.01	–	<0.001	<0.001	<0.001
CCR 100%	% mass	<0.05	<0.05	–	–	–
10% dist.resid.	% mass	–	–	<0.5	0.02	0.025
Sulfated ash	% mass	<0.03	0.02	<0.02	<0.01	<0.01
(Oxid) Ash	% mass	–	–	<0.02	–	–
Water and sediment	mg/kg	<300	<500	<500	<500	<500
Oxidation stability	h/110°C	–	–	>6	–	–
Total contaminant	mg/kg	<20	–	<24	–	–
Cu Corrosion	3 h/50°C	1	<No. 3	1	1a	1a
Cetane no.	–	>49	>47	–	–	–
Acid value	mg KOH/g	<0.5	<0.80	–	0.08	<0.3
Methanol	% mass	<0.3	–	<0.2	<0.2	<0.2
Ester content	% mass	–	–	>96.5	98.5	98–99.5
Monoglycerides	% mass	<0.8	–	<0.8	<0.4	<0.4
Diglycerides	% mass	<0.4	–	<0.2	<0.2	<0.2
Triglycerides	% mass	<0.4	–	<0.2	<0.1	<0.1
Free glycerol	% mass	<0.02	0.02	<0.02	<0.01	<0.01
Total glycerol	% mass	<0.25	0.24	<0.25	<0.01	<0.01
Iodine no.	–	<115	–	–	58.3	53–59
C18:3 and high unsat. acids	% mass	–	–	<1	<0.1	<0.1
Phosphorous	mg/kg	<10	<10	<10	–	–
Alkaline met. (Na, K)	mg/kg	<5	–	<5	–	–
Linolinec acid	% mass	–	–	<12	<0.5	<0.5
Lubricity 60°C	μm	–	–	<460	–	–

[a] Data from BLT (2000).

[b] Data from U.S. National Biodiesel Board (2007).

[c] Data from Lee and Park (2004).

[d] Data from MPOB (2005).

[e] LPPP, low-pour-point palm oil diesel.

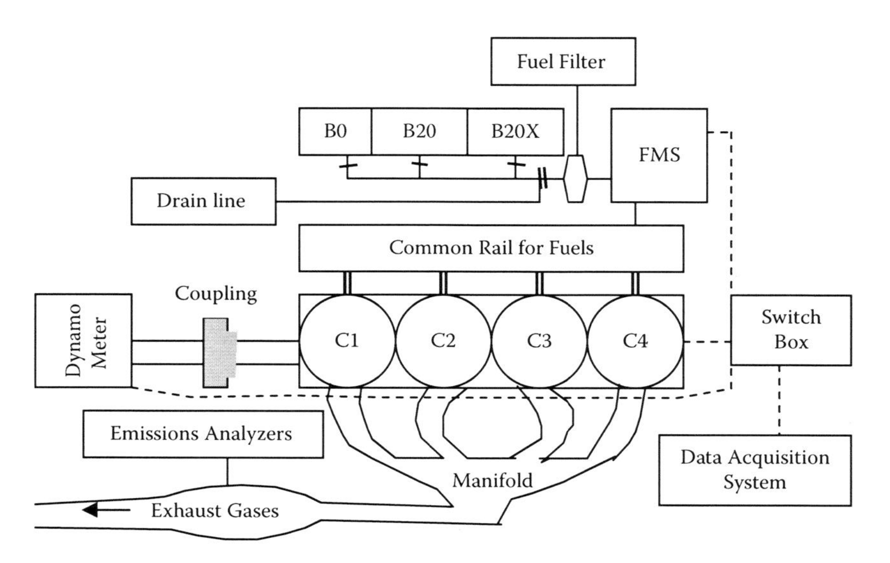

FIGURE 16.1 Schematic diagram of fuel system with dynamometer engine.

TABLE 16.3
Specification of Diesel Engine Being Used

Engine	Isuzu
Model	4FB1
Type	Water-cooled, 4 strokes
Combustion	Indirect injection (IDI)
Number of cylinders	4
Bore × Stroke	84 × 82 mm
Displacement	1817 cc
Compression ratio	21:1
Nominal rated power	39 kW/5000 rpm
Maximum torque speed	1800–3000 rpm
Dimension (L × W × H)	700 × 560 × 635 (mm)
Cooling system	Pressurized circulation

16.2.1 Test Fuels

The analysis and preparation of the test fuels were conducted at the Engine Tribology Laboratory, Department of Mechanical Engineering, University of Malaya. Three test fuels were selected: (1) 100% conventional diesel fuel (B0) supplied by the Malaysian petroleum company Petronas, (2) B20 as 20% POD blended with 80% B0, and (3) B20X as B20 with X% antioxidant additive (in this investigation X was 1% only). The blending process was done using a mechanical homogenizer stirrer at room temperature with stirring speed of 2000 rpm. The major properties of the fuels used are shown in Table 16.4.

TABLE 16.4
Major Properties of Fuels

Property	B0	B20	BOX
High calorific value, MJ/kg	46.80	45.40	45.87
Kinematic viscosity, cSt at 40°C	3.60	4.13	4.22
Cetane number	53	51	51
Specific density, g/cm3	0.832	0.848	0.858

16.2.2 Additive

The fuel B20 was treated with 1% octylated/butylated diphenylamine antioxidant to make the additive-added biodiesel B20X. This antioxidant helped lower the combustion temperature as it absorbed the heat from the short chain fatty acid during the combustion. The properties of the antioxidant were (1) viscosity at 40°C, 280 (mm^2/s), (2) density at 20°C (g/m^3), 0.98, (3) flash point (°C) 185.

16.2.3 Anti-Wear Characteristics

The anti-wear characteristics of the B0-, B20-, and B20X-contaminated lubricants in terms of the coefficient of friction, wear scar diameter of the used balls, and flash temperature parameter (FTP) were obtained using a tribometer such as a four ball wear machine. The four ball wear machine was used as required by the standard IP-239. This is a simple method for testing the anti-wear properties of the used lubricating oils. It consists of a device by means of which a ball bearing is rotated in contact with three fixed ball bearings, which are immersed in the lubricant sample. Different loads are applied on the balls by a load lever that gives a correlative pressure-act as similar as in the piston cylinder frictional zone caused. Hence, the results obtained from the four balls test machine gives an indication of the quality of the fuel-contaminated lube oil that is used in the engine. Table 16.5 shows the compositions of the test lubricant samples. Details of the four ball test method and experimental set up are given in Masjuki and Maleque (1997) and Ichiro et al. (2007).

TABLE 16.5
Lubricant Test Sample Specifications for Testing of Four Ball Machine

No	Sample	Specifications
1.	B0	100% commercial lubricant (SAE 40 grade)
2.	1% B20	1% of fuel B20 and 99% of pure lubricant
3.	2% B20	2% of fuel B20 and 98% of pure lubricant
4.	3% B20	3% of fuel B20 and 97% of pure lubricant
5.	1% B20X	1% of fuel B20X and 99% of pure lubricant
6.	2% B20X	2% of fuel B20X and 98% of pure lubricant
7.	3% B20X	3% of fuel B20X and 97% of pure lubricant

16.3 EVALUATION OF PALM OIL BIODIESEL

16.3.1 Brake Power Output

The results of the brake power output from the diesel engine for every test fuel showed that the fuel B20X produced higher brake power over the entire speed range in comparison to other fuels (Figure 16.2). The B20X produced an average of 11.82 kW brake power over the entire speed range followed by B20 (11.38 kW) and B0 (11.50 kW), which was 2.93% higher brake power than fuel B20. The maximum brake power obtained at 2500 rpm was 12.28 kW from the B20X fuel followed by 11.93 kW (B0) and 11.8 kW (B20). This could be attributed to the effect of the fuel additive in the B20 blend, which influenced the conversion of the thermal energy to work, or increased the fuel conversion efficiency by improving the fuel ignition and combustion quality (complete combustion). A similar effect of additive on increasing diesel fuel conversion efficiency was achieved by Gvidonas and Slavinskas (2005).

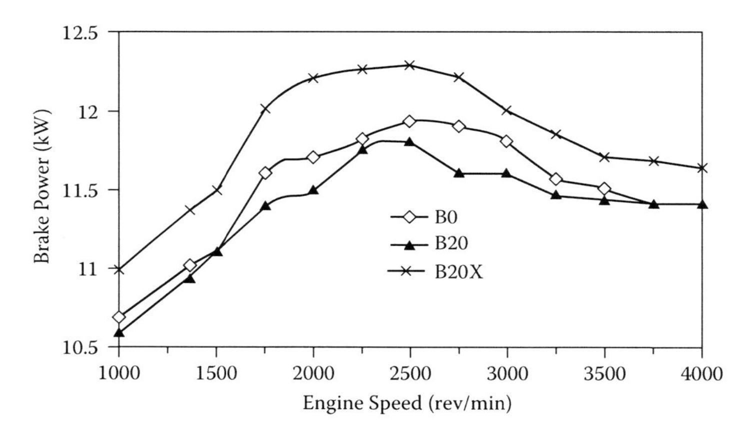

FIGURE 16.2 Brake power output vs. engine speed.

16.3.2 Specific Fuel Consumption

Figure 16.3 shows the SFC for all the fuels. The performance of the B20 and B20X was similar to that of the B0 up to an engine speed of 2250 rpm. After that, the fuel consumption of B20 increased. The B20X showed similar SFC to B0 up to an engine speed of 3500 rpm. This result was due to the presence of 1% antioxidant additive in B20, which produced fuel conversion similar to B0 fuel up to 3500 rpm and then produced higher fuel conversion as compared to B0 fuel at engine speeds higher than 3500 rpm. The lowest SFC was obtained from the B20X fuel, followed by the B0 and B20 fuels. The average SFC values over the speed range were 405 g/kW·h, 426.69 g/kW·h, and 505.38 g/kW·h for B20X, B0, and B20 fuels, respectively.

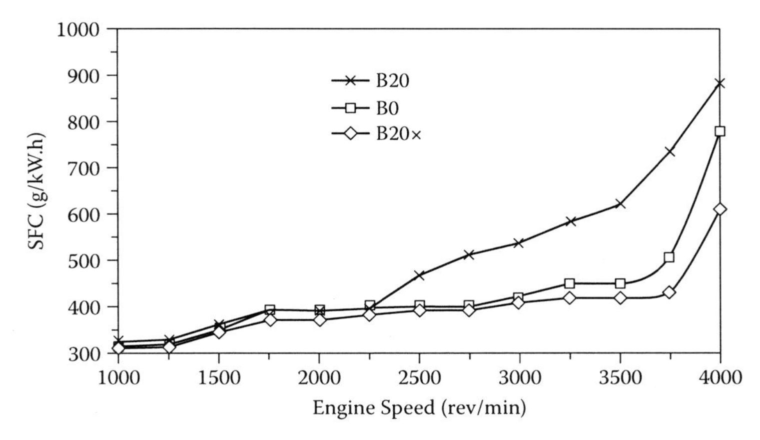

FIGURE 16.3 Specific fuel consumption vs. engine speed.

16.3.3 Oxides of Nitrogen Emission

The effect of the antioxidant additive in the biodiesel blended fuel on NOx emission is shown in Figure 16.4. The NOx concentration decreased with the B20X fuel (92 ppm), which was lower than the B20 (119 ppm) and B0 (115 ppm) fuels. The NOx are produced mainly from the fuel-air high combustion temperature. At high combustion temperature in the cylinder, the long chain hydrocarbons (in the diesel fuel) break into short chain hydrocarbons and long chain fatty acids (in the biodiesel) break into short chain fatty acids. These short chain hydrocarbons and short chain fatty acids contain high energy in the polarized form, which produce oxidation. However, the antioxidant absorbs the energy of the short chain fatty acid, hence the NOx is reduced (Figure 16.4). The difference of the NOx concentration between the B20X and B20 fuels (22% reduction) is the effect of 1% antioxidant additive. This result is contrary to oxygenate additive, which increases the NOx (Gong et al. 2007).

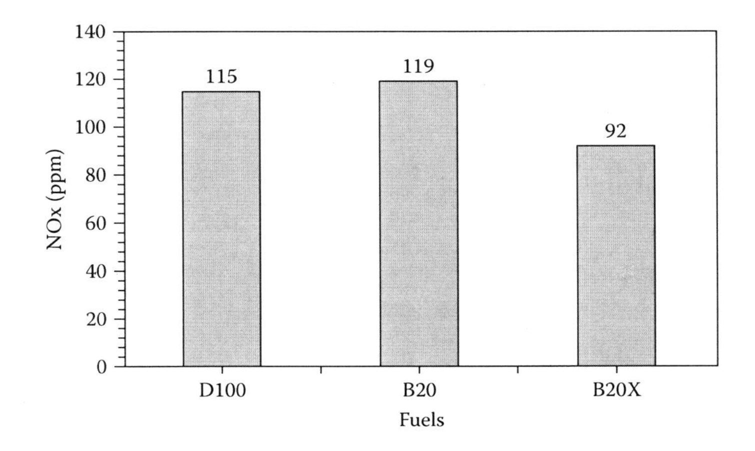

FIGURE 16.4 NOx emission at constant load of 50 Nm and engine speed of 2250 rpm.

16.3.4 Carbon Monoxide Emission

Carbon monoxide is formed during the combustion process with rich air-fuel mixtures when there is insufficient oxygen to fully burn all the carbon in the fuel to CO_2. However, a diesel engine normally uses more oxygen (excessive air) to burn fuel, which has little effect on the CO emissions. Since the operating conditions are exclusively lean (1.8 × the stoichiometric fuel air ratio), the CO concentration value for all the fuels is less than 1% (Figure 16.5). It is found that among all the fuels, the B20X produces the lowest level of CO emissions, 0.1%, followed by the B20 (0.2%) and B0 (0.35%). This is because the 1% additive in the biodiesel blended fuel produces complete combustion through enhancing the vaporization and atomization as compared to the B20 and B0 fuels.

16.3.5 Hydrocarbon Emission

Figure 16.6 shows the hydrocarbon (HC) emissions for all the test fuels. The B20X produced the lowest HC emission (29 ppm), followed by the B20 (34 ppm) and B0 (41 ppm). The difference between the B20 and B20X was 5 ppm, revealing that the B20X produced better combustion than B20 and B0 fuels. Hence, adding the antioxidant with the B20 has a beneficial effect in reducing HC emission. The reduction in HC is mainly the result of complete combustion of the B20X fuel within the combustion period as confirmed by combustion characteristics (for palm oil diesel and other biological fuels) such as net heat release rate and mass burn fraction (Masjuki, Abdulmuin, and Sii 1997; Masjuki, Kalam, and Maleque 2000). Around 60% mass (of each of the test fuels) was burnt within 0 and 20°C. After top dead center (ATDC), the remaining fuel mass was burnt within 20 to 50°C. ATDC. The B20X reduced 30% and B20 17% as compared to the B0 fuel. Hence, it could be stated that the B20 fuel with the antioxidant additive could be effective as an alternative fuel for diesel engines because it reduced the emission levels of NOx, CO, and HC.

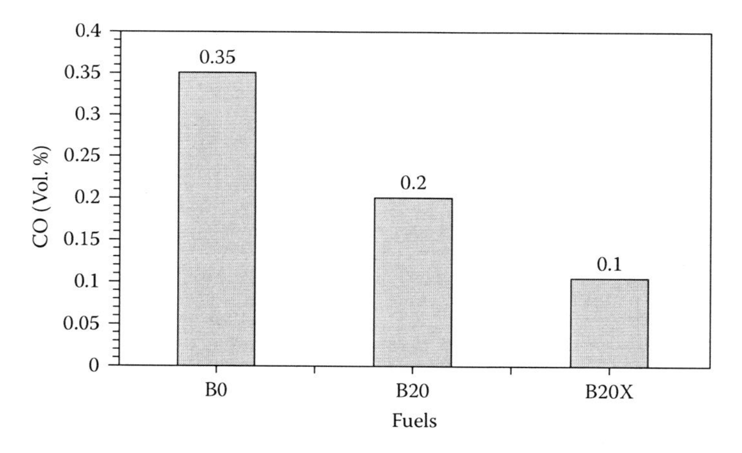

FIGURE 16.5 CO emission at constant load of 50 Nm and engine speed of 2250 rpm.

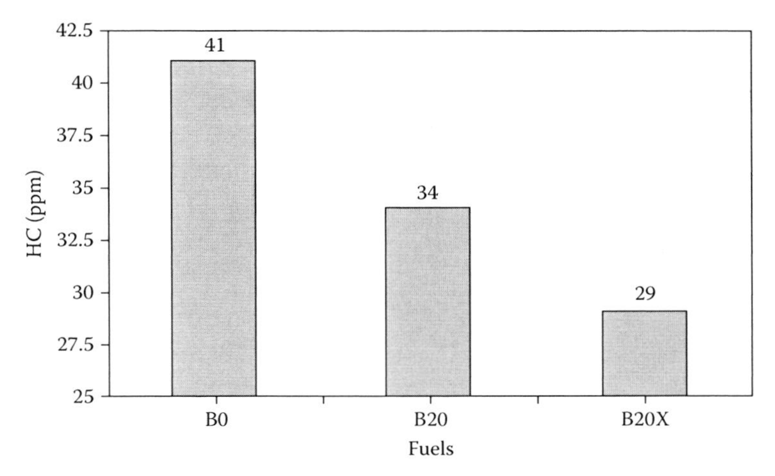

FIGURE 16.6 HC emission at constant load of 50 Nm and engine speed of 2250 rpm.

16.3.6 Wear Scar Diameter

Figure 16.7 shows the wear scar diameter (WSD) of the used ball for all the lubricant samples (Table 16.5) with contaminated fuels. The highest WSD (3.7481 mm) was produced by the pure lubricant as lubricant sample B0. All the B20-contaminated lubricants, for example 1%, 2%, and 3% B20 produced WSD of 3.5253, 3.452, and 3.5147 mm, respectively. The lowest WSD (3.4191 mm) was obtained from 2% B20X. It can be said that the lubricants contaminated with antioxidant additive fuel produced comparatively lower WSD than the lubricants contaminated with B20 and B0, which was the effect of 1% antioxidant additive in the B20.

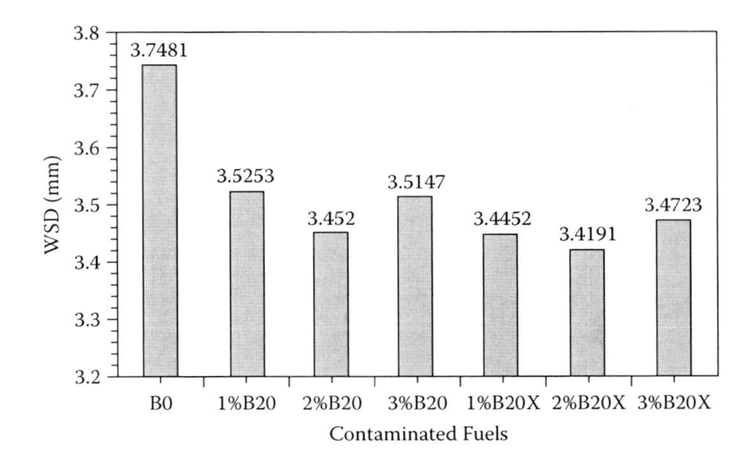

FIGURE 16.7 Wear scar diameter (WSD) of used ball with various contaminated fuels at constant load of 50 Nm.

16.3.7 Flash Temperature Parameter

Figure 16.8 shows the FTP for all the contaminated lubricants. The maximum and minimum FTP were obtained from the 2% B20X- and B0-contaminated lubricants, respectively. The maximum FTP value means that good lubricating performance occurred, indicating less possibility of the lubricant film breakdown. This phenomenon has also been observed by other workers (Husnawan et al. 2005; Masjuki et al. 2005), which apparently indicated that the additive in the fuel acted as anti-wear additive for lubricating oil. For 1 to 3% of the B20X-contaminated lubricants, better FTP was observed as compared to the B20- and B0-contaminated lubricants.

16.3.8 Friction Properties

Figure 16.9 shows the friction torque that is developed by various lubricant samples. It was found that the lowest level of friction torque was developed by the BOX-contaminated fuels. The maximum friction torque was produced by the pure lubricant (B0) as 53.05 kg-m. The low friction torque means good lubricity as well as lower coefficient of friction. Hence, it can be said that the antioxidant additive with the B20 was effective as a lubricant additive.

Figure 16.10 shows the variation of the friction coefficient for all the fuel-contaminated lubricants. The lowest coefficient of friction was obtained from the B20X-contaminated lubricants. The lower coefficient friction means developing low friction torque by the lubricants within the frictional surfaces. The maximum coefficient of friction was produced by the B0- and 3% B20-contaminated lubricants. The lowest coefficient of friction was achieved by the 1 to 3% B20X-contaminated lubricants. Hence, the antioxidant additive in B20 fuel was effective in reducing the coefficient of friction.

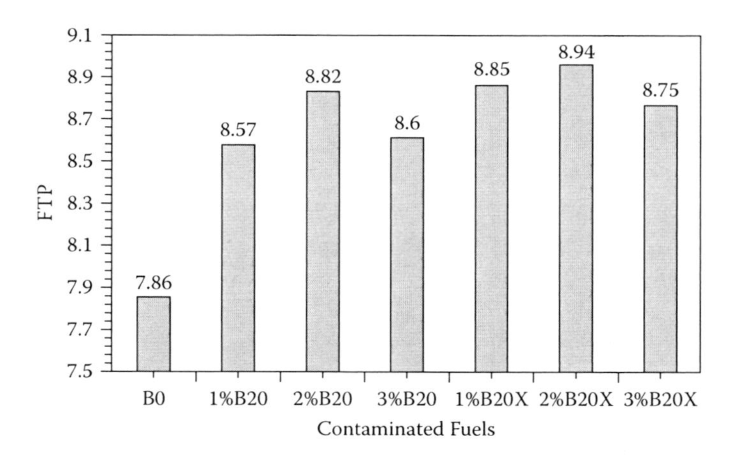

FIGURE 16.8 Flash temperature parameter (FTP) of used lubricants vs. contaminated fuels at constant load of 50 Nm.

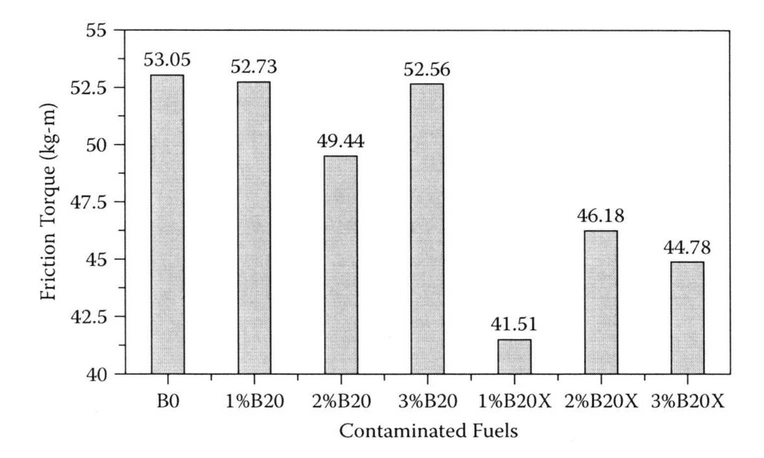

FIGURE 16.9 Friction torque for lubricants contaminated with various fuel at constant load of 50 Nm.

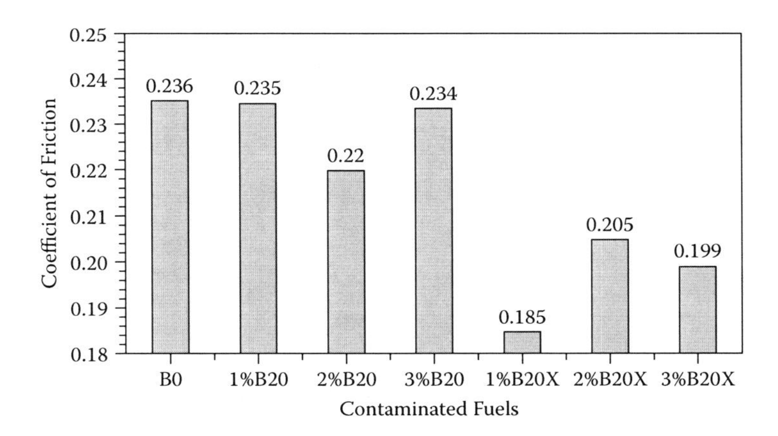

FIGURE16.10 Coefficient of friction for lubricants contaminated with various fuel at constant load of 50 Nm.

The tribometer test showed that a certain level of the biodiesel (as the B20) and the biodiesel with antioxidant additive (as B20X) showed good performance as compared to the pure lubricant. This was mainly due to reducing the lubricant viscosity to a level that reduced the frictional forces, which affected the fuel conversion efficiency as well as enhanced the fuel economy. Tribometer tests with higher load (greater than 50 Nm, such as 60 to 100 Nm) and higher percentage of fuels (greater than 3%, such as 4% and 5%) contaminating the lube oil were also conducted. It was found that above 4%, all the contaminated lubricants showed adverse results as compared to the pure lubricant. The higher percentage of the fuel in the lubricant reduces the lubricant film strength quality.

16.3.9 Oxidative Stability

The Rancimat test is a standard method for testing the oxidative stability of biodiesel samples in accordance with EN 14214. Figure 16.11 shows the variation in viscosity for B20, B20X, and B100 (100% palm oil diesel) at 40°C. The viscosity of the B20X was consistent over the period. The viscosity of the B100 increased after the fourteenth week mainly due to oxidation. The viscosity of the B20X was slightly reduced, from 4.25 cSt to 4.15 cSt after the eighth week due to oxygen molecules absorbed by the antioxidant additive, and then increased to its original level. But the B100 and B20 showed an increasing trend, for example, after 18 weeks, the B100 increased in viscosity from 4.40 cSt to 4.60 cSt mainly due to oxidation.

The effect of fuel storage duration on total base number (TBN) is shown in Figure 16.12. Total base number is a measure of oil alkalinity, which is an indication of its ability to counter the corrosive effects of oxidation. Higher TBN values mean more stability of the lubricating oil. A positive TBN value indicates the absence of free strong acids (Toms 1994).

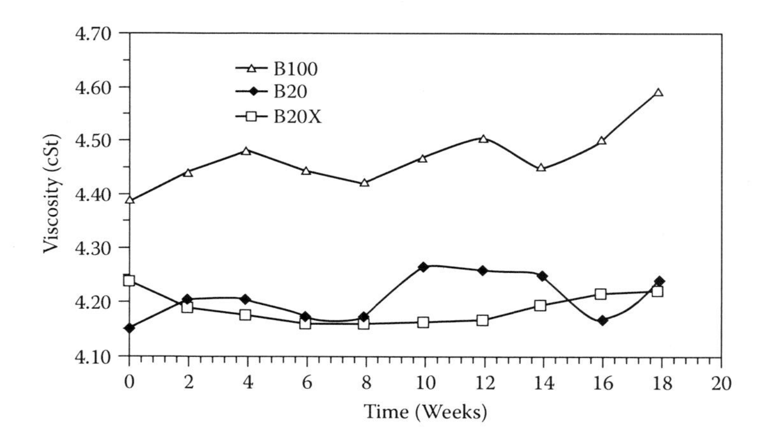

FIGURE 16.11 Variation of viscosity (at 40°C) vs. time in weeks.

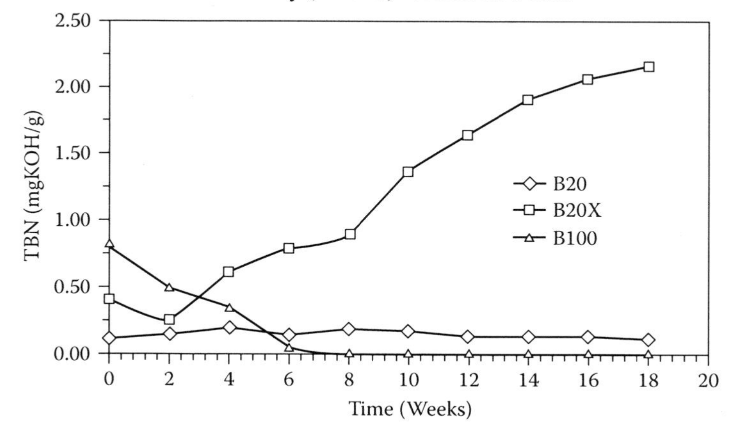

FIGURE 16.12 Variation of total base number (TBN) vs. time in weeks.

16.4 CONCLUSIONS

From the above, it can be concluded that palm oil diesel properties are comparable to other biodiesels such as those from rapeseed and soybean. The palm oil biodiesel with additive (B20X) produced higher brake power and lower SFC as compared to B0 and B20 fuels, and reduces NOx, CO, and HC emissions. It also showed desirable properties on the lubricant test and the oxidative stability test. Thus, palm oil biodiesel can be effectively used as transportation fuel.

ACKNOWLEDGMENTS

The authors wish to thank the Malaysian Palm Oil Board for supplying biodiesel, the Ministry of Science, Technology and Innovation of Malaysia for a supporting IRPA Grant, Mr. Sulaiman bin Arifin for technical assistance provided, and University of Malaya, which made this study possible.

REFERENCES

BLT. 2000. *Standardization of Biodiesel*. Wieselburg, Austria: BLT. Retrieved June 22, 2007 from http://www.senternovem.nl/mmfiles/149997_tcm24-124377.pdf.

Choo, Y. M., M. A. Ngan, H. Ahmad, K. Y. Cheah, A. M. Rusnani, Y. K. C. Andrew, L. L. N. Harrison, S. F. Cheng, C. L. Yung, and B. Yusof. 2005. Recent development in palm biofuel. *Proceedings of the 2005 PIPOC International Palm Oil Congress: Technological Breakthroughs and Commercialization - The Way Forward*, September 25–29, Selangor, Malaysia, 311–319.

DOE (Department of Energy). 2007. *International Energy Outlook 2007*. Washington, DC: DOE/EIA. Retrieved June 22, 2007 from http://www.eia.doe.gov/oiaf/ieo/index.html.

Gong, Y., S. Liu, H. Guo, T. Hu, and L. Zhou. 2007. A new diesel oxygenate additive and its effects on engine combustion and emissions. *Applied Thermal Engineering* 27: 202–207.

Gvidonas, L. and S. Slavinskas. 2005. Influence of fuel additives on performance of direct-injection diesel engine and exhaust emissions when operating on shale oil. *Energy Conversion and Management* 46: 1731–1744.

Husnawan, M., H. H. Masjuki, T. M. I. Mahlia, M. G. Saifullah, and M. Varman. 2005. The effect of oxidized and non-oxidized palm oil methyl ester on the stability properties during time of storage. Paper No. JSAE 20056050. *Proceedings of the 18th Internal Combustion Engine Symposium*, December 20–21, Jeju, Korea.

Ichiro, M., H. Keiji, M. Michimasa, and M. Shigeyuki. 2007. Investigation of anti-wear additives for low viscous synthetic esters: Hydroxyalkyl phosphonates. *Tribology International* 40: 626–631.

Kalam, M. A. and H. H. Masjuki. 2005. Recent development on biodiesel in Malaysia. *Journal of Scientific & Industrial Research* 64: 920–927.

Lee, J.S. and S. C. Park. 2004. Recent developments on biofuels for transport in Korea. *Proceedings of the 2nd International Symposium on Sustainable Energy Systems*, December 17–18, Kyoto University, Japan, 93–99.

Liang, Y. C., Y. M. Choo, C. S. Foon, M. H. Ngan, C. C. Hock, and B. Yusof. 2006. The effect of natural and synthetic antioxidants on the oxidative stability of palm diesel. *Fuel* 85: 867–870.

Masjuki, H. H. and Maleque, M. A. 1997. Investigation of anti-wear characteristics of palm oil methyl ester using a four ball tribometer test. *Wear* 206: 179–186.

Masjuki, H. H., M. Z. Abdulmuin, and H. S. Sii. 1997. Indirect injection diesel engine operation on palm oil methyl esters and its emulsions. *Journal of Automobile Engineering (Part D), I. Mech. E. London* 211: 291–299.

Masjuki, H. H., M. A. Kalam, and M. A. Maleque. 2000. Combustion characteristics of biological fuel in diesel engine. paper no-2000-01-0689. Presented at SAE 2000 World Congress, Detroit, Michigan.

Masjuki, H. H., M. G. Saifullah, M. Husnawan, M. S. Faizul, and M. G. Shaaban. 2005. Flash temperature parameter number prediction model by design of tribological experiments for basestock mineral oil containing palm olein and aminephosphate additives. Paper No. WTC2005-63193. Proceedings of World Tribology Congress (WTC III), September 12–16, Washington, DC.

MPOB (Malaysian Palm Oil Board). 2005. *Palm Biofuel and Palm Biodiesel Fuels for the Future*. Bangi, Malaysia: MPOB.

Ramadhas, A. S., S. Jayaraj, and C. Muraleedharan. 2006. Theoretical modeling and experimental studies on biodiesel-fueled engine. *Renewable Energy* 31: 1813–1826.

Toms, L. A. 1994. *Machinery Oil Analysis: Methods, Automation and Benefits*. Pensacola, FL: Larry A. Toms Technical Services.

U.S. National Diesel Board. 2007. *Specification for Biodiesel*. Jefferson City, MO: National Biodiesel Board. Retrieved June 22, 2007 from http://biodiesel.org/pdf_files/fuelfactsheets/BDSpec.PDF.

17 Biodiesel from Rice Bran Oil

Yi-Hsu Ju and Andrea C. M. E. Rayat

CONTENTS

ABSTRACT

There is a growing interest in the development and utilization of alternative fuels. This is driven by several factors, which include environmental concerns regarding the further use of petroleum fuels, energy security and independence, growth, and commitment to international accords such as the Kyoto Protocol. Currently, the widespread production and use of such alternative fuels as biodiesel is hindered by its uncompetitive price against petroleum-based diesel fuel. The high cost of raw material, usually refined vegetable oils, largely contributes to the expensive cost of biodiesel. There is now an intensifying search for a cheaper raw material for biodiesel production. One of these is rice bran oil. Rice, which is the staple food of more than half of the world's population, is produced at a rate of about 600 megatons per year. Rice bran is a by-product of rice milling. Given the magnitude of annual rice production, an enormous amount of bran is available. Unfortunately, bran is considered a low-value material and mostly treated as an agricultural waste. This chapter shows how oil from rice bran can be used as a feedstock for biodiesel production. Rice bran also contains protein, and other important bioactive compounds, if harvested from the bran, can be sold as high-value by-products. In this regard, this chapter also highlights conditions for the retention and subsequent recovery of important bioactive compounds in rice bran after biodiesel production. When integrated into the process economics, the sale of these valuable co-products from rice bran processing to biodiesel is one possible way of reducing the price of biodiesel.

17.1 INTRODUCTION

World annual rice production is about 600 million tons. More than 85% of this comes from Asia, of which 90% is from China, India, Indonesia, Bangladesh, Vietnam, Thailand, Myanmar, and the Philippines. An increase in energy consumption is expected in these countries. Such energy supply is normally in the form of petroleum oil. The importation of oil for energy uses up a country's important financial reserves, which otherwise can be used to finance essential infrastructures. It is estimated that Asian countries import about 60% of their requirement for oil. The reliance on imported oil poses a threat to a country's energy security.

The International Energy Agency (IEA) emphasizes the likelihood of better living standards in developing countries with an increased access to energy services. Reducing energy poverty is deemed as an urgent need to sustain a country's development. There is also the looming environmental damage that is reportedly caused by the increasing use of energy. At least 30% of air pollution emissions are attributed to the transport sector. In line with this is the global understanding to reduce the cumulative emissions of greenhouse gases, as stipulated in the Kyoto Protocol which was entered into force in February 2005. The protocol stipulates that involved parties will have to reduce their overall emissions of greenhouse gases by at least 5% below 1990 levels in the commitment period 2008 to 2012. As of early 2006, there are 158 countries that ratified, accepted, or acceded to the Kyoto Protocol. This international agreement has been one of the driving forces in the search for and promotion and development of renewable energy and other innovative environment-friendly technologies worldwide. Most developing countries in Asia are parties to this protocol.

Thus, the search for alternative fuels is inevitable worldwide and so for the Asian region. The development and utilization of alternative transport fuels is important due to the prominent environmental concerns, and more precisely because of its consequence for a country's energy independence, growth, and international reputation. These alternative fuels should not only be technically plausible for application, but should also be readily available, economically viable, environment friendly, and should be produced preferably with sustainability. In this respect, biodiesel is being developed and promoted as an alternative to petroleum diesel fuel. Biodiesel, which is composed of monoalkyl esters of fatty acids, is receiving intensified interest as a renewable fuel that is nontoxic and biodegradable. A major concern with this biofuel is its high price. Without tax holidays or government subsidies, current biodiesel production is economically unattractive. More research, development, and technological advancements are still required with the current production. The cost of feedstock oil contributes at least 70% to the biodiesel price. Hence, the use of inexpensive, nonedible feedstock and the recovery of high-value co-products during its production may considerably lower the cost of biodiesel.

The prospect of producing biodiesel from rice bran, an underutilized by-product from rice milling, is discussed in the succeeding sections. The annual world rice cultivation yields an estimated 47 million tons of rice bran, from which about 9 million tons of rice bran oil (RBO) could be available for the production of biodiesel. The prospective biodiesel production from RBO in Asian countries is about 10 billion liters. On average, this amounts to at least 10% of these countries' diesel requirements. Figure 17.1 illustrates the potential biodiesel production against the estimated diesel requirements for the world's top ten rice producing countries.

17.2 RICE BRAN OIL PROCESSING

Rice is the staple food of about 55% of the world's human population. It is grown in most countries, with the largest production in Asia, where 135 million hectares of rice area is cultivated. On a dry weight basis, the whole grain rice contains about 3% embryo, 70% endosperm, 20% hull, and 8% rice bran. This composition differs to some extent among rice varieties, and with specific cultivation methods and conditions (Palipane and Swarnasiri 1985; Goffman, Pinson, and Bergman 2003). Rice bran is obtained as a by-product of rice milling. It is particularly obtained during the second stage of rice milling, after the rice has been dehulled. Typical compositions of rice bran, its oil, and meal cake are presented in Table 17.1. Until recently, rice bran has been used primarily as animal feeds. Only about 2 million metric tons of bran are converted to oil annually. Thailand and Japan are the top exporting and importing countries, respectively, of refined RBO. In 2003, Japan imported about 18,900 tons of refined RBO while the amount of refined RBO that Thailand exported was roughly 18,800 tons.

At present, increasing interest is centered on the nutritive quality of RBO, which has been reported to contain biologically active compounds and antioxidants such as γ-oryzanol, fatty acid steryl esters, phytosterols, tocopherols, tocotrienols, lecithin (phospholipids), and wax esters. RBO has physicochemical properties that are distinct from other edible oils and which render conventional vegetable oil refining

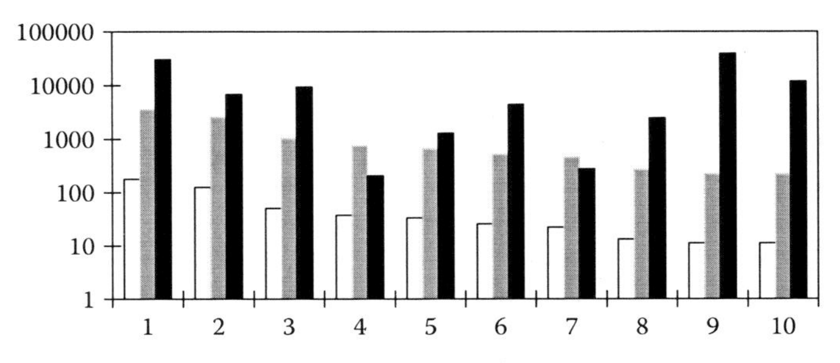

(□) Rice production [10^6 tons] (▩) Biodiesel potential from RBO [10^6 L]
(■) Estimated diesel requirement [10^6 L]

Country: (1) China; (2) India; (3) Indonesia; (4) Bangladesh; (5) Vietnam; (6) Thailand; (7) Myanmar; (8) Philippines; (9) Japan; (10) Brazil

FIGURE 17.1 Estimated biodiesel yield from rice bran oil among top rice-producing countries.

TABLE 17.1
Typical Composition of Rice Bran, Meal Cake, and Oil

Rice Bran (%, Wet Weight)		Rice Bran Meal Cake (%, Dry Weight)	
Water	9–12	Protein	14–16
Oil	15–27	Dietary fiber	12–15
Crude protein	4–17	Phytic acid	5–9
Crude rice bran oil[a] (%, Dry Weight)			
Triacylglycerides	60–86	Polar lipids	9–12
Partial glycerides	7–14	Sterols	3–5
Free fatty acids	2–8	Tocols	0.1–0.2
Waxes	1–4	T-Poryzanol or γ-oryzanol	1–3

[a] The free fatty acid (FFA) composition of crude rice bran oil may actually reach as much as 80% depending on the storing age and condition of the rice bran.

Combined from Ignacio and Juliano (1968), Goffman, Pinson, and Bergman (2003), Juliano (1985), Wang et al. (1999), Zullaikah et al. (2005), and Vali et al. (2005).

processes unsuitable. The high content of partial glycerides, polar lipids, and wax esters in RBO makes its viscosity approximately two times that of other vegetable oils. A considerable amount of gum is also present in crude RBO due to the unsaturated nature of its fatty acids. Hence, the oil needs to be degummed and dewaxed prior to refining. However, complete wax removal is often difficult. The residual wax imparts haziness to the oil and is the main reason for the darker color of refined RBO (Ju and Vali 2005). This dark color is the main cause of the low appeal of edible RBO to consumers.

Another difficulty encountered in the refining of crude RBO is caused by its high free fatty acid (FFA) content. Crude RBO from fresh rice bran contains 6 to 8% FFA. However, due to the presence of active enzymes in rice bran, FFA content of rice bran during storage increases by 1 to 7% per day (Zullaikah et al. 2005). Depending on the storage conditions, 35 to 70% FFA may be obtained in a month. It is possible to reduce this FFA increase by immediately extracting the oil from rice bran right after milling. However, rice mills are often scattered in locations, thereby making the collection of large quantities of fresh rice bran for immediate oil extraction impractical. As a result, most crude RBO contains 40 to 50% FFA (Kosugi, Kuneida, and Azuma 1994). Figure 17.2 illustrates a typical curve of FFA and triglycerides (TG) content in rice bran under different storage conditions. It was reported that at temperatures higher than 50°C, lipase activity is significantly reduced, resulting in less FFA formation in rice bran (Zullaikah et al. 2005).

Generally, it is not practical to refine crude RBO with FFA content higher than 10%. In particular, less than 5% FFA is desired for the cost-effective production of refined RBO. The high FFA content also results in a large loss of neutral oil because during refining, oil recovery is inversely proportional to the FFA content

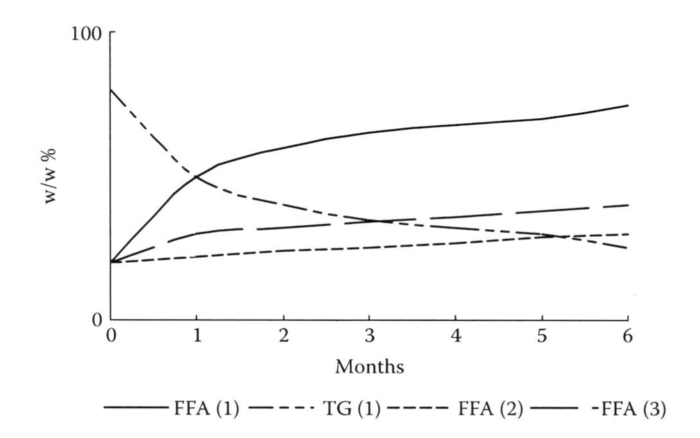

FIGURE 17.2 Free fatty acid formation in rice bran during storage: (1) stored at 25°C; (2) dried at 95°C under vacuum for 1 h and stored at 25°C; (3) stored at 5°C.

and the loss is normally two to three times the FFA content (Lai et al. 2005). Therefore, inactivation of enzymes is important prior to oil extraction if refined RBO is desired.

The process of enzyme inactivation, called bran stabilization, further increases the product cost of refined oil. Furthermore, most bioactive compounds in crude RBO are lost during refining. For example, as much as 90% of the oryzanol content in crude RBO was lost during refining (Krishna et al. 2001). Recently, a process employing simultaneous degumming and dewaxing in the physical refining of RBO has been reported in which only a minimal oryzanol content (<10%) from the original crude RBO was lost in the process (Rajam et al. 2005). However, about 20% of the tocols was lost in the refining, although the amount left was still higher than tocols content in commercially refined RBO. The process appears to be promising but complete process analysis, including process economics, still needs to be established. On the whole, processing of crude RBO into edible oil is not economically attractive as of this moment.

Aside from processing into edible oil, other efforts are being made to harness the nutrients in RBO. Some processes are currently being developed to recover and purify the bioactive compounds such as γ-oryzanol (Xu and Godber 1999; Saska and Rossiter 1998). Because the total bioactive compounds content of RBO is about 5%, the direct recovery of these compounds from RBO will result in a complex procedure with a lot of oil components as by-products. Indeed, acid oil or soap-stock from crude oil refining comes out as the more suitable feedstock for oryzanol (3.3 to 7.4 w/w% on a dry basis) (Krishna et al. 2001).

Another process option, which is the main focus of this chapter, is to convert the FFA and TG in RBO into fatty acid methyl esters (FAME), or biodiesel before separating from it and purifying the remaining bioactive compounds.

17.3 RBO FOR BIODIESEL PRODUCTION

The environmental and other benefits of using biodiesel compared to petroleum diesel are well known. It is recognized, however, that the current expensive price of biodiesel prevents the full utilization of such biofuel. The prices of oil feedstock and by-product meal cake were cited as the two most important factors in the economics of biodiesel production (Ju and Vali 2005). In this respect, RBO is one of the most valuable resources among the nonconventional oils investigated for biodiesel production. Compared to traditional oils derived from cereal or seed sources, crude RBO is an inexpensive feedstock for biofuel production. The current price of rice bran in the United States is about US$55/ton. The price of degummed and dewaxed RBO is estimated to be around US$0.18/lb (US$396/ton). This is about 40% cheaper than the prices of refined vegetable oils, which are presently the feedstocks for commercial biodiesel production. In other places, where there is less trade of rice bran, the price may even be lower because the bran is normally considered as an agricultural waste.

The high nutritional quality of soybean meal cake makes it possible to sell it at higher market price than other meals. As a result, the price of biodiesel from soybean is normally lower than that from other vegetable oils (Ju and Vali 2005). Like soybean meal cake, defatted rice bran is a rich source of protein, other carbohydrates, and phytochemicals, which have high commercial value. The essential amino acids profile in rice bran protein isolate was found to be similar to that prescribed for children and similar to that of soy protein isolate and casein (Wang et al. 1999). The lysine content in rice bran protein is reportedly higher than in rice endosperm protein or any other cereal bran proteins (Juliano 1985). Hence, rice bran meal cake has the potential of gaining a high commercial market value like the meal cake of soybean. Furthermore, crude rice bran wax is also available as a potentially important co-product. Vali et al. (2005) reported the production of food-grade wax from degummed and dewaxed crude RBO. The utilization of these co-products may significantly lower the product cost of biodiesel from RBO. Figure 17.3 illustrates a possible flow diagram for the production of biodiesel and its co-products from rice bran.

17.4 TECHNICAL ASPECTS OF RBO PROCESSING TO BIODIESEL

The processing of RBO to biofuel appears to be less complicated than RBO refining. In biofuel processing, there is relatively less concern with residual solvents than in RBO refining. In addition, the presence of high initial amounts of FFA in RBO is not problematic to the biodiesel process. Hence, RBO stabilization or special storage infrastructure to minimize the increase of FFA in rice bran is not required. However, certain considerations remain if an optimal recovery of different products in the processing of RBO to biodiesel is desired. First, the choice of process conditions should favor the retention of most bioactive compounds, which are mostly heat-sensitive and may degrade at certain high temperatures (Zullaikah et al. 2005). Furthermore, most of the bioactive compounds present in RBO are susceptible to alkaline treatment due to their phenolic character. It was shown that treating RBO with a base decreased the oryzanol content by as much as 90% and temperatures above 240°C

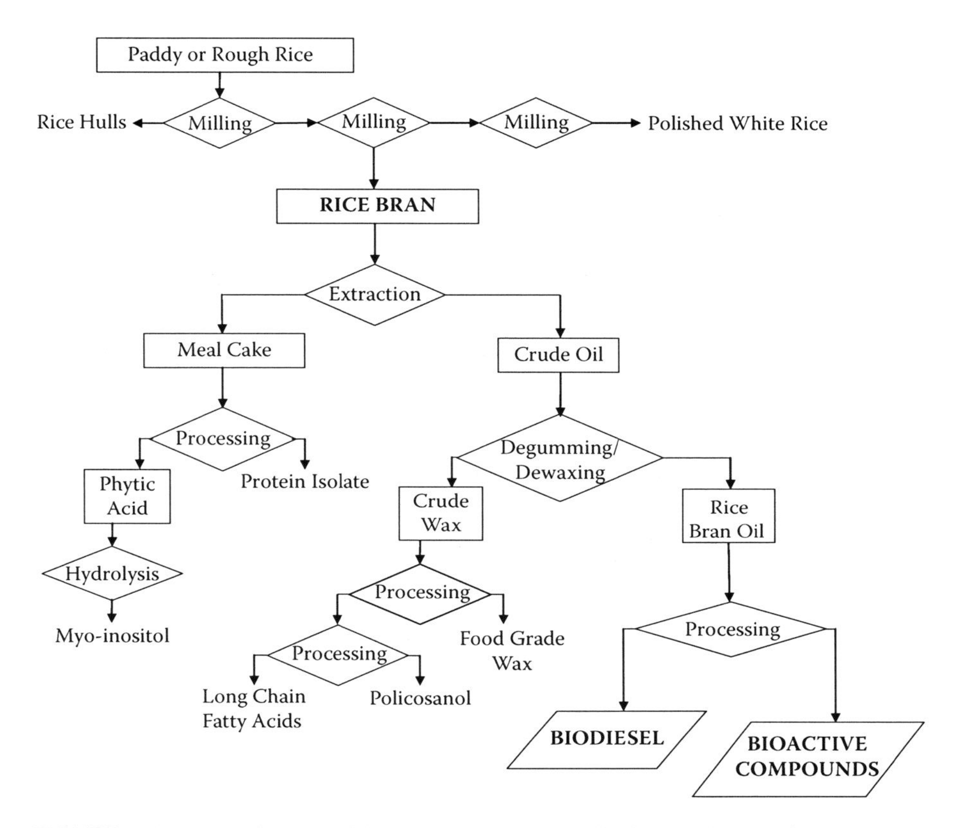

FIGURE 17.3 Flow diagram of rice bran processing to biodiesel and co-products.

resulted in considerable loss of the tocols (Krishna et al. 2001). Second, it is desirable to remove most, if not all, wax esters, phospholipids, and gums as these components interfere with the conversion of RBO to biodiesel. It was reported that small amounts of phospholipids resulted in the partial deactivation of lipase during the enzymatic conversion of RBO to biodiesel (Lai et al. 2005). Hence, degumming and dewaxing of crude RBO prior to further processing is essential. However, refining of crude RBO is not required for biodiesel production because important minor components with nutritional value may be lost during refining.

17.4.1 Extraction of RBO

Solvent extraction is the most commonly used method in commercial oil extraction, with hexane as the most widely used solvent. Commercial hexane, which contains 50 to 85% *n*-hexane and some isomers, has been cleared by the FDA as an extraction solvent. Hexane recoveries in commercial oil mills are usually higher than 96%. There is no need to pelletize the bran prior to extraction. Rice bran in flake form is enough to result in efficient extraction (Ju and Vali 2005). It is also possible to recover RBO from rice bran by using supercritical carbon dioxide. However, such methods are often less economical than conventional hexane extraction and requires further investigation to attain commercial viability.

17.4.2 Degumming and Dewaxing of Crude RBO

The purpose of degumming and dewaxing is to remove fat-soluble impurities in the oil. Dewaxing is especially required for RBO because of its high content of wax esters. Degumming is usually done by adding polar solvents to the oil under adequate mixing to allow polar lipids to be extracted into the polar phase. The mixture is then cooled and centrifuged whereby wet gum is removed with the water phase. Water degumming is the preferred method if minimal loss of bioactive compounds is desired. It was found that a processing temperature of about 70°C and an addition of 4% water (based on the oil weight) was enough to substantially remove the gums (Indira et al. 2000). A novel degumming process employed the use of 1% (v/w) $CaCl_2$ solution, which achieved simultaneous degumming and dewaxing (Rajam et al. 2005).

17.4.3 Acid-Catalyzed Biodiesel Production from RBO

The acid-catalyzed conversion of RBO to fatty acid methyl esters (FAME) involves two steps. FFA is first transformed into FAME with water as the by-product. Some of the acylglycerides may be converted during the first phase of the reaction. However, it has been shown that acid catalysis of acylglycerides is slow. To completely convert the remaining acylglycerides, a second step in the conversion was proposed (Zullaikah et al. 2005). The water and glycerol produced during the first step were removed before subjecting the mixture to a second conversion step. The water-soluble components were extracted by washing the mixture with water. In this extraction process, methanol and catalyst were also removed. Another batch of acidic methanol was added for the second conversion step of the remaining acylglycerides to FAME. The amount of methanol added was four to six times the required stoichiometric amount for the total conversion of acylglyceride. Methanol from the first step can easily be recovered by distillation and subsequently reused in the process.

Among acid catalysts that may be used are sulfuric acid, nitric acid, and hydrochloric acid. The esterification process may proceed using 1 to 5% (w/w, based on oil) H_2SO_4, and about five to six times the stoichiometric amount of methanol is required for the total conversion of FFA. Sufficient agitation should be provided. The temperature of the reaction is usually between 60 and 65°C. Although high-temperature operation has its advantages, such conditions will incur relatively higher expenses for the required high-pressure vessels.

The first step normally takes about 2 h for the FFA content to drop below 5%, while the second step may take 6 to 8 h to fully convert the remaining acylglyceride to FAME. In a typical run, at the end of the first step, the reaction mixture contains about 3% FFA, 35% acylgycerides, and 62% FAME. At the end of the second step, the FAME content in the product is more than 96% (Zullaikah et al. 2005).

Distillation may be employed for the separation of FAME from the reaction mixture. However, since boiling points of FAME from RBO are generally higher than 300°C, atmospheric distillation is not recommended if the bioactive compounds are to be recovered. Vacuum distillation with lower distillation temperature can be used. Laboratory-scale vacuum (about 5 mmHg) distillation of the FAME up to 220°C resulted in 99% pure FAME in the distillate (Zullaikah et al. 2005). The recovery of FAME in the distillate was 96%. The residue contained about 18% γ-oryzanol and 20% mixture of sterols, steryl esters, and tocols.

17.4.4 Lipase-Catalyzed Biodiesel Production from RBO

The use of lipase (triacylglycerol acylhydrolyses, E.C. 3.1.1.3) to catalyze the reaction of oils to FAME has been studied extensively. Lipases can catalyze the hydrolytic reactions of acylglycerols and the synthetic reactions of their corresponding esters. Some considerations of the lipase-catalyzed production of FAME include operation under a certain (minimum) amount of water and not too high methanol concentration. It is detrimental to use too much excess methanol as opposed to alkaline- or acid-catalyzed processes in producing biodiesel.

The lipase-catalyzed conversion of RBO to FAME is carried out in a two-step reaction. This is because at a certain period after the start of the reaction, the water and glycerol produced result in the deactivation of lipase (Lai et al. 2005). Therefore, a second step is necessary, in which the enzyme in the first step is reused after regaining its activity, which results from incubating the enzyme in *tert*-butanol for at least 1 h after washing with hexane. The preferred mode of addition of methanol in the lipase-catalyzed production of biodiesel is the intermittent or repetitive batch mode. A one-time addition of methanol leads to too high concentrations that the lipase cannot tolerate. Sufficient mixing is also required. However, excess shear may inactivate the enzyme.

After the first-step and prior to the second step, the reaction mixture is washed with water to extract water-soluble components. The second step of the enzymatic process proceeds faster than that of the acid-catalyzed second-step reaction. The lipase-catalyzed transesterification reduces the triacylglycerol content to about 2% in 2 h and in about 3 h more than 98% FAME can be obtained in the product. Since enzymatic processes are usually considered expensive, a biodiesel production process from RBO may employ an acid-catalyzed reaction as the first-step and lipase-catalyzed reaction as the second step. In the second step, only a small amount of enzyme is required because the amount of remaining triacylglycerol will be relatively smaller compared to when the original RBO is subjected to lipase-catalyzed reaction. Figure 17.4 illustrates a generic function diagram of biodiesel production from rice bran.

17.4.5 *In Situ* Esterification/Transesterification

In the method described in the previous sections, the production of biodiesel started from the extraction of oil from rice bran and subsequently using the oil for the conversion of FFA and acylglycerides to FAME (biodiesel). Another method of producing biodiesel is using rice bran directly as the substrate for esterification without the oil being extracted first. Rice bran was subjected to a mixture of sulfuric acid (catalyst) and methanol (Özgül-Yücel and Türkay 2002). Rice bran was prepared such that the size was about 0.6 mm. Forty milliliters of methanol (with 2.5% v/v acid)

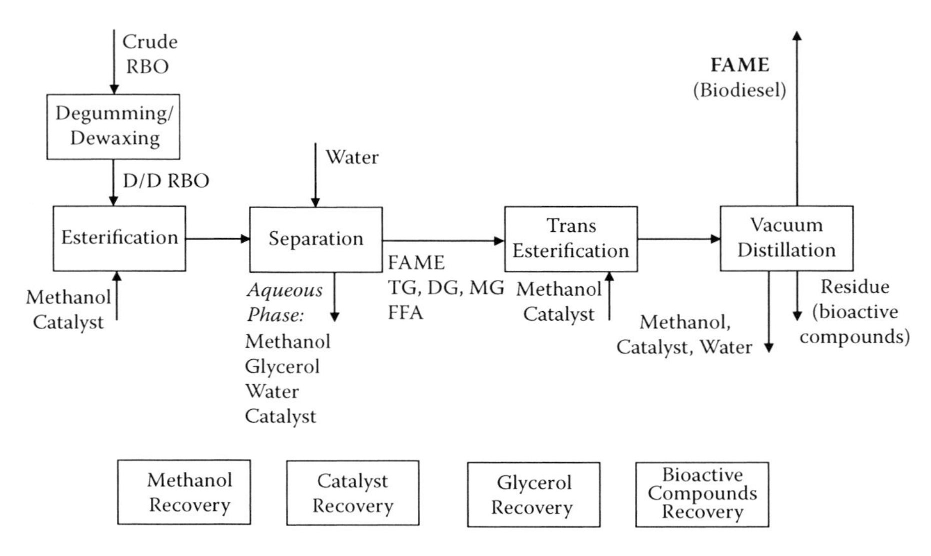

FIGURE 17.4 Function diagram of biodiesel production from rice bran oil.

per gram rice bran were added to the bran. The acidified methanol was refluxed for 1 h at 65°C. After *in situ* reaction, the bran was filtered and the reaction mixture was washed with methanol and water. Organic components in the mixture were extracted with hexane. Although the process reduces the FFA content in the residual oil in the bran better than the extraction process, FAME content by this method is quite lower than the previously mentioned processes. Using rice bran with high FFA content (approximately 75%), the FAME content only increased from 80% in 30 min to about 87% in 5 h. Adding more acid or methanol did not significantly affect the methyl esters content. *In situ* production of biodiesel from rice bran remains a challenge for rice bran with low to medium FFA content.

The *in situ* process may be modified by incubating rice bran in acidified methanol instead of having the reaction proceed in refluxed condition. Although not yet reported in the case of RBO, this *in situ* process may yield more methyl esters than the one previously described. Nevertheless, *in situ* transesterification of oil in oilseeds was still found to be less efficient than the normal process of transesterification of oil (Haas et al. 2004). Improvement in this process still needs to be addressed, especially with the large amount of methanol and catalyst used as compared to other processes where oil reacts with methanol to form methyl esters.

17.4.6 Choice of Alcohols for the Alcoholysis of RBO

In a lipase-catalyzed process, it appears that methanol is the best among alcohols tested, including ethanol, propanol, butanol, and isobutanol. However, it is recognized that the optimal conditions for each alcohol in an enzymatic reaction may differ. It was also reported that branched alcohols may produce fatty acid alkyl esters that have better fuel properties than FAME (Knothe 2005). However, the cheaper price of methanol among the other alcohols remains the driving force behind its more popular use.

17.5 PROSPECTS OF BIODIESEL FROM RBO

The basic properties of fatty acids and FAME from RBO are provided in Tables 17.2 and 17.3. The World-Wide Fuel Charter (WWFC), which is made by association members of international car and engine manufacturers, provides technical background on fuel properties. The WWFC acknowledges that FAME ensures lubricity of injection equipment and reduces exhaust particulate matter. The excessive rise in viscosity and the loss of fluidity at low temperatures were observed for some FAME. This seems not to be the case for FAME from RBO because RBO contains a high percentage of unsaturated fatty acids (more than 70% are oleic and linoleic acids). It was reported that feedstocks with relatively low content of saturated long chain fatty acids generally yield biodiesel with lower cloud point (CP) and pour point (PP) (Dunn 2005). CP is the temperature at which crystals become visible, while the temperature at which excessive crystal formation prevents free pouring of the fluid is the PP. The lower the CP and PP, the better is the fuel's performance at cold temperatures.

Other important properties include cetane number (CN), density, and viscosity. A high CN results in less hydrocarbon and CO emissions, and reduces fuel consumption and combustion noise. The CN generally increases with increasing chain length and decreases with increasing unsaturation. It has been reported that higher CN correlates with lower nitrogen oxides (NO_x) exhaust emission (Knothe 2005). An inspection of the CN of the constituent FAME from RBO reveals that these FAMEs have, on average, a cetane rating above 40. Most biodiesel from plant sources, including soybean and rapeseed oil, have CN higher than 47. The minimum CN of biodiesel for automotives is 47 (ASTM D6751) or 51 (European biodiesel standard, EN 14214)

TABLE 17.2
Fatty Acids of Degummed and Dewaxed Rice Bran Oil

Common Name of Fatty Acid	%[a]	Chemical Formula	Melting Point[b](°C)	Boiling Point[c](°C)	Density[d] (g/cm³)
Myristic	0.1–0.3	$C_{14}H_{28}O_2$	53.9	250^{100}	0.8622^{54}
Palmitic	14.2–15.2	$C_{16}H_{32}O_2$	63.1	351.5	0.8527^{62}
Palmitoleic	0.1–0.4	$C_{16}H_{30}O_2$	-0.1	$131^{0.06}$	
Stearic	1.6–2.1	$C_{18}H_{36}O_2$	68.8	232^{15}	0.9408^{20}
Oleic	41.5–42.9	$C_{18}H_{34}O_2$	13.4	286^{100}	
Linoleic	37.3–38.3	$C_{18}H_{32}O_2$	-12	$229\text{-}30^{16}$	0.9022^{20}
Linolenic	2.2–2.5	$C_{18}H_{30}O_2$	-11	231^{17} $129^{0.05}$	0.9164^{20}
Behenic	0.1–0.3	$C_{22}H_{44}O_2$	81	306^{60}	0.8223^{90}
Lignoceric	0.1–0.4	$C_{24}H_{48}O_2$			

[a] Degummed and dewaxed rice bran oil. Data from Zullaikah et al. (2005).

[b] Data from Lide (2002).

[c] Data from Lide (2002). The superscript is the pressure in mmHg at which the boiling point was measured.

[d] Data from Lide (2002). The superscript is the temperature in °C at which the density was measured.

TABLE 17.3
Properties of Fatty Acid Methyl Esters (FAME) From Rice Bran Oil

Methyl Ester of	Chemical Formula	Melting Point[a] (°C)	Boiling Point[b] (°C)	Density[c]	Viscosity[d] (cSt)	Cetane Number[e]
Myristic acid	$C_{15}H_{30}O_2$	19	295, 155[7]	0.8671[20]	3.23	66.2
Palmitic acid	$C_{17}H_{34}O_2$	30	417, 148[2]	0.8247[75]	4.32–4.38	74.5
Stearic acid	$C_{19}H_{38}O_2$	39.1	443, 215[15]	0.8498[40]	5.51	86.9
Oleic acid	$C_{19}H_{36}O_2$	-20	218.5[20]		4.45–4.51	55–59.3
Linoleic acid	$C_{19}H_{34}O_2$	-35	215[20]	0.8886[10]		38.2–42.2
Linolenic acid	$C_{19}H_{32}O_2$		182[3]	0.8960[25]		
Behenic acid	$C_{23}H_{46}O_2$	54				

[a] Data from Lide (2002).
[b] Data from Lide (2002). The numbers in parenthesis are the pressures (in mmHg) at which the measurements were made.
[c] Data from Lide (2002). The superscript is the temperature in °C at which the density was measured.
[d] Data from Knothe (2005).
[e] Data from Knothe (2005).

(Knothe, Van Gerpen, and Krahl 2005). It is, therefore, presumed that biodiesel from RBO, given its component FAME, will also have a CN greater than or about 47.

As mentioned above, the use of FAME improves the lubricity of petroleum diesel fuel. Most developed countries require low or even zero sulfur in petroleum diesel fuels. During the process of desulfurization, nonsulfur polar compounds are also removed (Knothe 2005). These compounds are reported to be responsible for the lubricity in petroleum diesel fuels. Hence, the lubricity of such fuels is negatively affected after desulfurization. In this aspect, biodiesel has an advantage over petroleum diesel because biodiesel from plant sources has natural components that impart sufficient fuel lubricity.

Biofuel from plant sources may contain antioxidants that can augment the oxidative stability of the fuel. The oxidative stability of biodiesel is important during extended storage (Knothe 2005). The presence of air, elevated temperatures, and trace metals have been reported to aid the oxidation process. Further, the degree of unsaturation of fatty esters has large effects on the autoxidation of FAME. Linoleic and linolenic acids and their esters are generally more susceptible to autoxidation (Knothe 2005). RBO, like soybean oil and rapeseed oil, contains large amounts of oleic and linoleic acids. This would seem to be a disadvantage as the oxidation process will result in the deterioration of biofuel. However, it was shown recently that crude palm oil methyl esters that contained about 644 ppm vitamin E (a mixture of tocols) and 711 ppm β-carotene showed better oxidative stability than distilled palm oil methyl esters, which barely contained such antioxidants (Liang et al. 2006). It was shown that about 0.1% α-tocopherol in biodiesel was enough to meet the required specification of the EN 14214 in terms of oxidative stability. Note that RBO contains about 0.2% of these potent natural antioxidants. With proper process design, distilled RBO biodiesel may have the required or even better oxidative stability if sufficient amounts of these antioxidants are retained.

17.6 CONCLUSIONS

It is recognized that the nutritive value of rice bran is due to the presence of bioactive compounds in the oil. One way to benefit from this is to process rice bran oil to refined oil for human consumption. However, there are issues in the refining of rice bran oil. This chapter presented such issues and showed an alternative way to obtain nutritive bioactive compounds in rice bran. This alternative is the production of biodiesel from rice bran oil. Oil from rice bran makes a good feedstock for biodiesel production. Because rice bran is generally regarded as an agricultural waste, it is a potential cheap raw material for producing inexpensive biodiesel. Based on a survey of the literature, it was shown that the process of producing biodiesel from rice bran oil can be carried out in such a way that these bioactive compounds are retained. Specifically, these conditions include the use of an acid instead of a base catalyst to avoid damaging the bioactive compounds, and low-temperature (vacuum distillation) purification of biodiesel because some bioactive compounds are heat labile. The recovery of these bioactive compounds occurs during the purification of biodiesel. With proper process design, these compounds can be recovered as co-products with purified biodiesel. If the recovery and sale of these high-value bioactive compounds is integrated in the process economics of biodiesel production from rice bran oil, then the price of biodiesel from rice bran oil may potentially be cheaper than the price of petro-diesel.

REFERENCES

Dunn, R. O. 2005. Cold weather properties and performance of biodiesel. In *The Biodiesel Handbook*. Champaign, IL: AOCS Press, pp. 83–121.

Goffman, F. D., S. Pinson, and C. Bergman. 2003. Genetic diversity for lipid content and fatty acid profile in rice bran. *J. Am. Oil Chem. Soc.* 80(5): 485–490.

Haas, M. J., K. M. Scott, W. N. Marmer, and T. A. Foglia. 2004. In situ alkaline transesterification: An effective method for the production of fatty acid esters from vegetable oils. *J. Am. Oil Chem. Soc.* 81(1): 83–89.

Ignacio, C. C. and B. O. Juliano. 1968. *Physicochemical properties of brown rice from Oryza species and hybrids. J. Agric. Food Chem.* 16(1): 125–127.

Indira, T. N., J. Hemavathy, S. Khatoon, A. G. G. Krishna, and S. Bhattacharya. 2000. Water degumming of rice bran oil: A response surface approach. *J. Food Eng.* 43: 83–90.

Ju, Y. H. and S. R. Vali. 2005. Rice bran oil as a potential resource for biodiesel: A review. *J. Sci. Ind. Res.* 64: 866–882.

Juliano, B. O. 1985. Rice bran. In *Rice Chemistry and Technology*. St. Paul, MN: American Association of Cereal Chemists, pp. 647–687.

Knothe, G. 2005. Dependence of biodiesel fuel properties on the structure of fatty acid alkyl esters. *Fuel Process Technol.* 86: 1059–1070.

Knothe, G., J. Von Gerpen, and J. Krahl (eds.). 2005. *The Biodiesel Handbook*. Champaign, IL: AOCS Press.

Kosugi, Y., T. Kuneida, and N. Azuma. 1994. Continual conversion of free fatty acids in rice bran oil to triacylglycerol by immobilized lipase. *J. Am. Oil Chem. Soc.* 71: 445–448.

Krishna, A. G. G., S. Khatoon, P. M. Shiela, C. V. Sarmnadal, T. N. Indira, and A. Mishra. 2001. Effect of refining crude rice bran oil on the retention of oryzanol in the refined oil. *J. Am. Oil Chem. Soc.* 78(2): 127–131.

Lai, C. C., S. Zullaikah, S. R. Vali, and Y. H. Ju. 2005. Lipase-catalyzed production of biodiesel from rice bran oil. *J. Chem. Technol. Biotechnol.* 80: 331–337.

Liang, Y. C., C. Y. May, C. S. Foon, M. A. Ngan, C. C. Hock, and Y. Basiron. 2006. The effect of natural synthetic antioxidants on the oxidative stability of palm diesel. *Fuel* 85: 867–870.

Lide, D. R. (ed.). 2002. *Handbook of Chemistry and Physics,* 83rd ed. Boca Raton, FL: CRC Press.

Özgül-Yücel, S. and S. Türkay. 2002. Variables affecting the yields of methyl esters derived from in situ esterification of rice bran oil. *J. Am. Oil Chem. Soc.* 79(6): 611–614.

Palipane, K. B. and C. D. P. Swarnasiri. 1985. Composition of raw and parboiled rice bran from common Sri Lankan varieties and from different types of rice mills. *J. Agric. Food Chem.* 33: 732–734.

Rajam, L., D. R. S. Kumar, A. Sundaresan, and C. Arumughan. 2005. A novel process for physically refining rice bran oil through simultaneous degumming and dewaxing. *J. Am. Oil Chem. Soc.* 82(3): 213–220.

Saska, M. and G. J. Rossiter. 1998. Recovery of γ-oryzanol from rice bran oil with silica-based continuous chromatography. *J. Am. Oil Chem. Soc.* 75(10): 1421–1427.

Vali, R., Y. H. Ju, T. N. B. Kaimal, and Y. T. Chern. 2005. Process for preparation of food grade rice bran wax and determination of its composition. *J. Am. Oil Chem. Soc.* 82(1): 57–64.

Wang, M., N. S. Hettiarachchy, M. Qi, W. Burks, and T. Siebenmorgen. 1999. Preparation and functional properties of rice bran protein isolate. *J. Agric. Food Chem.* 47: 411–416.

Xu, Z. and S. Godber. 1999. Purification and identification of component of γ-oryzanol in rice bran oil. *J. Agric. Food Chem.* 47: 2724–2728.

Zullaikah, S., C.-C. Lai, S. R. Vali, and Y. H. Ju. 2005. A two-step acid-catalyzed process for the production of biodiesel from rice bran oil. *Bioresource Technol.* 96: 1889–1896.

18 Biodiesel Production Using Karanja *(Pongamia pinnata)* and Jatropha *(Jatropha curcas)* Seed Oil

Lekha Charan Meher, Satya Narayan Naik, Malaya Kumar Naik, and Ajay Kumar Dalai

CONTENTS

ABSTRACT

Biodiesel consists of mono-alkyl esters of long chain fatty acids, produced by transesterification of vegetable oil with methanol or ethanol. In developing countries such as India, the use of edible oils for biodiesel is not economically feasible. The nonedible oils are the potential feedstock for the development of biodiesel fuel. These oils include karanja, jatropha, neem, simarouba, sal, mahua, etc. The nonedible oils contain some toxic components (unsaponifiable matter) and sometimes high free fatty acids that create difficulties during conventional methods of biodiesel preparation. This chapter deals with the characterization of karanja and jatropha oils, and the preparation and fuel quality of biodiesel derived from them.

18.1 INTRODUCTION

Increased industrialization and the growing transport sectors worldwide face major challenges in terms of energy demand as well as increased environmental concerns. The rising demand for fuel and the limited availability of mineral oil provide incentives for the development of alternative fuels from renewable sources with less environmental impact. One of the possible alternatives to petroleum-based fuels is the use of fuels from plant origins (Encinar et al. 1999). The use of biofuel as a renewable resource combines the advantages of almost unlimited availability and ecological benefits such as an integrated closed carbon cycle.

Vegetable oil was used as fuel in the early 1900s (Knothe 2001). However, at that time the ready availability of conventional diesel fuel gave little incentive for the development of alternative fuels from renewable sources. The first use of vegetable oil-based fuel, the ethyl esters of palm oil, as a diesel substitute was reported in a Belgian patent in 1937 (Knothe 2005). Research work on the development of vegetable oil-based alternative diesel fuel gained importance in the 1990s. The major oilseed crops identified for the development of the triglyceride-based fuel include sunflower, safflower, soybean, rapeseed, linseed, cottonseed, peanut, and canola (Peterson 1986).

The use of edible-grade oil as a feedstock for biodiesel seems insignificant for the developing countries such as India, which are importers of edible oils. Various nonedible, tree-borne oils, such as jatropha, karanja, neem, etc., are the potential feedstock for development of the triglyceride-based fuels. The oils derived from these nonedible oilseeds are toxic and do not find use for edible purposes. This chapter describes the oils derived from karanja and jatropha and their use as feedstock for the development of alternative diesel fuel.

Karanja (*Pongamia pinnata*) and jatropha (*Jatropha curcas*) are two oilseed plants that produce nonedible oils and are not exploited widely due to the presence of toxic components in their oils. *Pongamia pinnata* Syn. *P. glabra* trees are widely distributed through the humid lowland tropics commonly found in India and Australia and also in Florida, Hawaii, Malaysia, Oceania, the Philippines, and the Seychelles. The karanja is a medium-sized evergreen tree, which has minor economic importance in India. The fruit or pod is about 1.7 to 2 cm in length, 1.25 to 1.7 cm wide, and weighs about 1.5 to 2 g. The seeds are collected manually and decorticated using a hammer. The hulls are separated by winnowing. The karanja seed kernel contains 27 to 39 wt% oil. The oil is extracted from the kernel by traditional expeller, which yields 24 to 26% oil. The oil contains toxic flavonoids such as karanjin and a di-ketone pongamol as major lipid associates, which make the oil nonedible. The oil has been used chiefly for leather tanning, lighting, and to a smaller extent in soap making, medicine, and lubricants. The main constraints to greater use of karanja oil in soaps is its color and odor, as well as the ineffectiveness of conventional refining, bleaching, and deodorization in improving the quality of the oil (Bringi 1987).

Jatropha curcas is a drought-resistant shrub or tree grown in Central and South America, Southeast Asia, India, and Africa. The plant was propagated from South America to other countries in Africa and Asia by the Portuguese (Gubitz, Mittelbach, and Trabi 1999). Jatropha is easily propagated by cutting; it is planted as a fence to protect fields because it is not browsed by cattle. It is well adapted to arid and semiarid regions and often used for soil erosion control. The seeds of the jatropha resemble castor seeds, somewhat smaller in size (0.5 to 0.7 g) and dark brown in color. The oil content of the seed varies from 30 to 40%. The oil is toxic due to the presence of diterpenes, mainly phorbol esters, responsible for tumor-promoting activity. The flavonoids vietin and isovitexin have been isolated from *J. curcas* grown in India (Iwu 1993). The oil has been used as a purgative, to treat skin diseases, and to soothe pain such as that caused by rheumatism (Gubitz, Mittelbach, and Trabi 1999). Now, these nonedible oilseeds have become important for the preparation of triglyceride-based biodiesel fuel.

18.1.1 Karanja and Jatropha Oils as Feedstock for Biodiesel

The physicochemical properties of karanja and jatropha oils are listed in Table 18.1. Karanja oil is yellowish orange to brown, whereas jatropha oil is pale yellow in color. Karanja and jatropha oils contain 3 to 5% and 0.4 to 1.1%, respectively, of lipid associates (unsaponifiable matter) responsible for the toxicity and development of the dark color on storage. The fatty acid compositions of both oils are listed in Table 18.2. The karanja oil contains 44.5 to 71.3% oleic acid as the major fatty acid. Oleic and linoleic acids are the major fatty acid in jatropha oil. There are slight variations in the composition of the fatty acids depending on the agroclimatic conditions; stearic acid content ranging from 3.9 to 5.25% has been reported in the mature seeds of *J. curcas*, but was not detected in some oilseeds of *J. curcas* (Nagaraj and Mukta 2004). The jatropha oil has a hydroxyl value of 4 to 20 mg KOH/g (see Table 18.1). After conventional refining and bleaching, the hydroxyl value of the oil is reduced to almost 1 mg KOH/g, indicating that the hydroxyl value is not contributed by the fatty acids but due to some of the lipid associates such as curcine and curcasin (Bringi 1987).

TABLE 18.1
Physicochemical Characteristics of Jatropha and Karanja Oils

Characteristics	Jatropha Oil	Karanja Oil
Acid value (mg KOH/g)	3–38	0.4–12
Hydroxyl value (mg KOH/g)	4–20	–
Saponification value (mg KOH/g)	188–196	187
Iodine value (g/100 g)	93–107	86.5
Unsaponifiable matter (% w/w)	0.4–1.1	2.6

TABLE 18.2
Fatty Acid Composition (wt%) of Jatropha and Karanja Oils

Fatty Acids	Jatropha Oil (% by Weight)[a]	Karanja Oil (Results from GC Analysis) (% by Weight)[b]
Palmitic acid ($C_{16:0}$)	12.6	11.6
Stearic acid ($C_{18:0}$)	3.9	7.5
Oieic acid ($C_{18:1}$)	41.8	51.5
Linoleic acid ($C_{18:2}$)	41.8	16.0
Linolenic acid ($C_{18:3}$)	–	2.6
Eicosanoic acid ($C_{20:0}$)	–	1.7
Eicosenoic acid ($C_{20:1}$)	–	1.1
Docosanoic acid ($C_{22:0}$)	–	4.3
Tetracosanoic acid ($C_{24:0}$)	–	1.0
Unaccounted for	–	2.7

[a] Data from Nagaraj and Mukta (2004).
[b] GC, gas chromatography.

18.1.2 Fatty Acid Alkyl Esters as Biodiesel

The plant-based triglycerides usually contain free fatty acids, phospholipids, sterols, water, odorants, and other lipid associates, which make the oil unsuitable for use as fuel directly in existing diesel engines. Karanja and jatropha oils contain large amounts of free fatty acids (FFA) and some lipid associates such as flavonoids or forbol esters. The higher molecular weight, higher viscosities, poor cold flow properties, deposit formation due to poor combustion, and low volatilities are the main constraints in using the vegetable oils directly as fuel. The solution to the viscosity problem has been approached by four routes: dilution, microemulsification, pyrolysis, and transesterification. Among the techniques developed, the conversion of the oil by transesterification with short chain alcohol produces cleaner and more environmentally safe fuel with improved fuel quality.

18.2 PRODUCTION OF BIODIESEL FROM KARANJA OIL

The alkali-catalyzed methanolysis of karanja oil was studied for the preparation of methyl esters (Meher, Vidya Sagar, and Naik 2006). The optimization study of the methanolysis provided the following reaction conditions: catalyst concentration 1% KOH (w/w of oil); MeOH/oil molar ratio 6:1; reaction temperature 65°C and stirring rate 600 rpm for 2 h, which resulted in 97 to 98% methyl esters. The yield of the methyl esters vs. time with the optimized reaction condition is shown in Figure 18.1. Equation (18.3) shows the effect of the reaction variables on the rate of formation of the methyl esters.

Increasing the catalyst concentration up to 1% resulted in more rapid formation of the methyl esters. The presence of excess amounts of the catalyst may lead to saponification of the triglyceride, forming soaps, which increase the viscosity of the

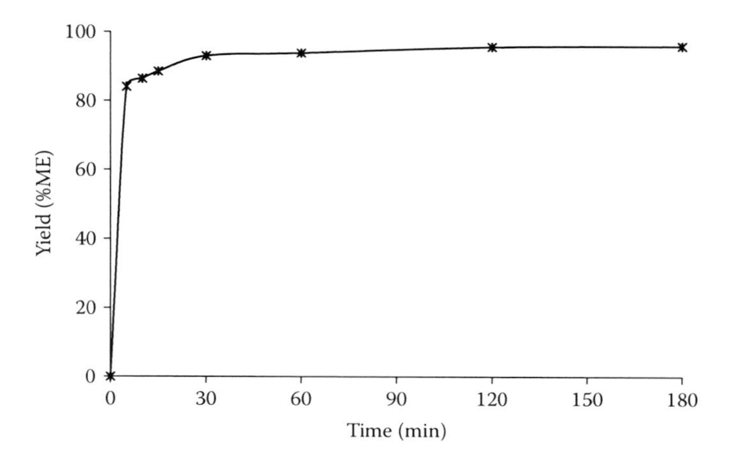

FIGURE 18.1 Formation of methyl esters during KOH-catalyzed transesterification of karanja oil under optimized reaction conditions (catalyst 1 wt% KOH, MeOH/oil molar ratio 6:1, reaction temperature 65°C, rate of stirring 600 rpm).

reaction medium. Increasing the molar ratio of the methanol to oil increases the rate of formation of the methyl esters. The reaction was faster with a high molar ratio of MeOH to oil, whereas longer reaction time was required for the lower molar ratio to get the same conversion. Mixing is very important in triglyceride transesterification, as oils or fats are immiscible with alcoholic methanol solution. Once the two phases are mixed by stirring and the reaction is started, stirring is no longer needed (Ma, Clements, and Hanna 1999). Increasing the reaction temperature up to boiling point of the methanol increases the rate of methyl ester formation. The same yields can be obtained at room temperature by simply extending the reaction time (Freedman, Pryde, and Mounts 1984). A reaction temperature above the boiling point of the alcohol is avoided because at high temperature, it tends to accelerate the saponification of the glycerides by the alkaline catalyst before completion of the alcoholysis (Dorado et al. 2004).

The conversion of karanja oil to methyl esters can be expressed by the following equations:

$$Q = \frac{at}{1+bt} \tag{18.1}$$

$$\left(\frac{dQ}{dt}\right)_{t\to 0} = a \tag{18.2}$$

where Q is conversion, a is the initial rate of formation of methyl esters and b is a constant.

The initial rate a for the formation of methyl ester can be expressed as:

$$a = A \times (\text{moles of MeOH per mole of oil})^{p} \times (\text{percent KOH})^{q} \times (\text{rate of stirring})^{r} \times (\text{temperature in °C})^{s} \quad (18.3)$$

The values of p, q, r, and s are 1.255, 0.38, 0.115, and 0.155, respectively, obtained from optimization of methanolysis of karanja oil, and A is a constant where $A = 0.185$.

The transesterification of karanja oil with ethanol was studied for the preparation of karanja ethyl esters. The yield of ethyl esters was 95% under the optimized reaction conditions. The study of the transesterification of high-FFA karanja oil with methanol and ethanol resulted in lower yield of the methyl/ethyl esters. The acid value of the karanja oil was increased by adding oleic acid to the oil. On increasing the FFA content of the oil from 0.3 to 5.3 for the methanolysis, the methyl ester content in the product decreased from 97 to 6%, as shown in Figure 18.2. Likewise for the ethanolysis, the yield decreased sharply. A process that utilizes high-FFA feedstock needs pretreatment of the raw material to reduce its acid value before the transesterification with the alkaline catalyst (Canakci and Von Gerpen 2001, 1999). The acid-catalyzed esterification can be followed by alkali-catalyzed transesterification for higher conversion of the oil to alkyl esters. The effect of water on the ethanolysis revealed that the formation of the esters decreased linearly with increase in the amount of the water in the reaction medium. The presence of water during transesterification causes the hydrolysis of the ester group of the triglyceride, resulting in FFAs. The presence of water in the alkali-catalyzed reaction leads to saponification.

18.2.1 Effect of Reaction Time on Acid Value during Pretreatment

Pretreatment of karanja oil containing 3.2 to 20% FFA was carried out with sulfuric acid catalyst for methyl esterification. The decrease in the acid value of the karanja oil with time during acid-catalyzed methyl esterification is shown in Figure 18.3. The acid values decreased from 41.9 to 3.8 mg KOH/g during 0.5% H_2SO_4-catalyzed

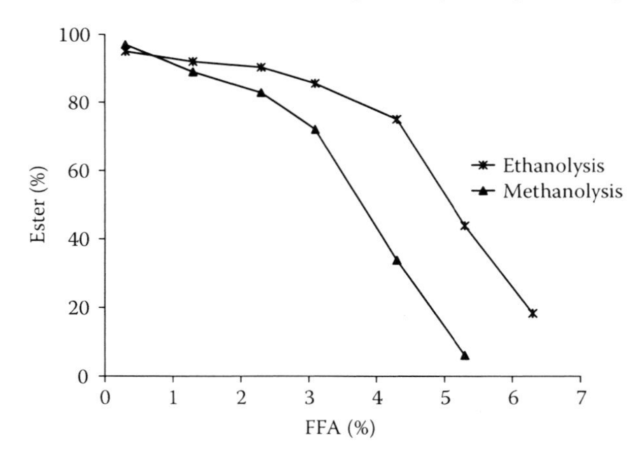

FIGURE 18.2 Effect of free fatty acid during alkali-catalyzed transesterification of karanja oil (catalyst 1 wt% KOH, MeOH/oil molar ratio 6:1, reaction temperature 65°C, reaction time 3 h, rate of stirring 600 rpm).

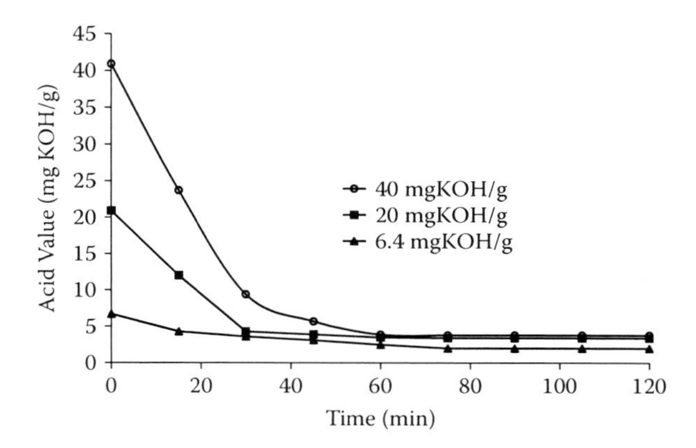

FIGURE 18.3 Effect of reaction time on acid value during pretreatment (catalyst 0.5% H_2SO_4, MeOH/oil molar ratio 6:1, reaction temperature 65°C, rate of stirring 600 rpm).

pretreatment of karanja oil containing 20% FFA in 1 h. The decrease in the acid value during pretreatment is also dependent on the amount of acid catalyst used (Canacki and Von Gerpen 2001).

18.2.2 Effect of Alcohol on the Pretreatment Step

Methanol and ethanol were used for the esterification of FFA during the pretreatment step. The final acid value of 20% FFA karanja oil was higher for ethyl esterification in comparision to methyl esterification. This might be due to the high reactivity of methanol as compared to ethanol. However, the final acid value for 20% FFA karanja oil after ethyl esterification was 4.6 mg KOH/g, after which the transesterification of the pretreated oil with ethanol was feasible using the alkali-catalyzed route.

18.2.3 Alkali-Catalyzed Transesterification

The acid-catalyzed esterification of the FFA in the oil reduces the acid value of the oil to 4–5 mg KOH/g depending on the initial acid value and the type of alcohol used. The pretreated oil can be transesterified with an alkali catalyst. Part of the alkali used for the reaction compensates for the acidity due to H_2SO_4 and the remaining portion acts as a catalyst for the transesterification. The alkali-catalyzed transesterification is accomplished in the same way as in the reaction using low-FFA karanja oil.

Table 18.3 shows the methyl and ethyl ester yield from karanja oil containing FFA up to 20%. The results reveal that there is no significant change in the yield of esters with respect to amounts of the FFA present in the oil.

Heterogeneous catalysis has also been used for the production of biodiesel from karanja oil in which solid acid catalysts such as Hβ-zeolite, montmorillonite K-10, and ZnO were employed by Karmee and Chadha (2005) for the methanolysis. The conversion was low as compared to the alkaline-catalyzed route. Meher et al, (2006) used solid basic catalyst for biodiesel preparation from high-FFA karanja oil. The

TABLE 18.3
Effect of Free Fatty Acids on the Yield of Methyl and Ethyl Esters during the Dual-Step Process

FFA of Karanja Oil (%)	Yield of Karanja Methyl Esters	Yield of Karanja Ethyl Esters
0.3	97[a]	95[a]
3.2	96.7	–
10	96.6	94.6
20	96.6	95.4

[a] Yield of esters by single-step transesterification.

alkali metal (Li, Na, K) doped the CaO catalyst as the strong alkalinity catalyzed the transesterification, resulting in 94.9% methyl esters (using 2% Li-impregnated CaO catalyst, molar ratio of MeOH/oil of 12:1, reaction time of 6 h at 65°C in a batch reactor). Increasing the FFA from 0.48 to 5.75 decreased the methyl ester formation from 94.9 to 90.3%. The decrease in the yield of the methyl esters was due to the formation of the metallic soap (calcium salt of free fatty acids) by the reaction of the calcium with the free fatty acids consuming a part of the catalyst. The biodiesel layer containing the metallic soap was purified and the resulting biodiesel had total methyl ester content of 98.6% and acid value of 0.3 mg KOH/g, which satisfied the ASTM specifications for biodiesel.

18.2.4 Unsaponifiable Matter from Karanja Oil and Biodiesel

The major lipid associates in the karanja oil are karanjin (1.1 to 4.5%) and pongamol (0.2 to 0.7%). The karanjin and pongamol content were determined by using the reverse phase HPLC method described by Gore and Satyamoorthy (2000). The karanjin and pongamol content were 1.6 and 0.7%, respectively, and the unsaponifiable matter in the oil was 2.6% (w/w). After completion of the reaction, these unsaponifiable components get crystallized and distributed at 1.56 and 0.88% concentration in the glycerol and methyl esters layers, respectively. There was no detection of the pongamol but 0.009% of karanjin was detected in the purified methyl esters.

18.3 PRODUCTION OF BIODIESEL FROM JATROPHA OIL

The free fatty acid content is the key parameter for identifying the process of biodiesel preparation. The acid value of jatropha oil ranges from 3 to 38 mg KOH/g (Munch and Kiefer 1986). The jatropha oil with low FFA was transesterified to methyl esters and ethyl esters by using the conventional alkali catalyst method. In a typical biodiesel preparation, 2000 g of the crude jatropha oil was transesterified with a solution of 30 g KOH in 331 g methanol. The reaction was carried out in a batch reactor in two steps at 30°C. The oil was mixed with two parts of the methanolic KOH solution and the reaction mixture was stirred for 30 min and the glycerol layer allowed to separate. The upper organic layer was mixed with one part methanolic KOH and stirred for a further 30 min. After 5 h settling time, the glycerol layer was separated and the ester layer was washed with warm water, passed over Na_2SO_4 which resulted in 92% theoretical yield of the methyl esters. Biodiesel prepared on a pilot scale had 99.5% purity of the methyl esters (Foidl et al. 1996).

The single-step alkali-catalyzed transesterification of the jatropha oil was studied using 1% KOH as catalyst and 6:1 molar ratio of methanol to oil at 65°C with stirring at 600 rpm for 3 h. The esters content in the biodiesel was 98%.

The dual-step process, as described for karanja oil, was also carried out for preparing biodiesel from jatropha oil. The pretreatment step of the jatropha oil needs a longer time for completion of the methyl esterification of FFA compared to the karanja oil. The second step, that is, the alkali-catalyzed transesterification, was carried out according to a procedure similar to that used for karanja oil.

18.4 KINETICS OF TRANSESTERIFICATION

The kinetics of the transesterification of karanja oil with methanol and ethanol were studied with 100% excess of alcohol and 1% KOH as the catalyst. The forward and reverse reactions followed a pseudo-first- and second-order kinetics, respectively, with a good fit obtained at all the temperatures. The activation energies of the forward and reverse reactions are given in Table 18.4. The forward and reverse reactions of the first step had activation energies of 13.579 and 13.251 Kcal/mol, while the activation energies of the third step were 7.363 and 4.592 Kcal/mol, respectively. The low activation of the third step for the conversion of MG (monoglyceride) to GL (glycerol) was due to the diffusion limitation caused by the high viscosity of the glycerol. The activation energy for the first step of the ethanolysis was low, 4.569 and 3.450 Kcal/mol, respectively, for the forward and reverse reactions, which indicated that the ethanolysis was less sensitive to increase in the reaction temperature.

TABLE 18.4
Activation Energies for Transesterification of Karanja Oil

	Methanolysis	Ethanolysis		
Reaction	Ea (Kcal/mol)	R^2	Ea (Kcal/mol)	R^2
TG → DG	13.579	0.9371	4.569	0.9856
DG → TG	13.251	0.9801	3.350	0.9519
DG → MG	13.015	0.9520	–	
MG → DG	13.612	0.9421	–	–
MG → GL	7.363	0.9054	–	–
GL → MG	4.592	0.9936	–	–

18.5 BIODIESEL FUEL QUALITY

The fuel characteristics of the biodiesel obtained from the karanja and jatropha oils were determined as per the ASTM method and are shown in Table 18.5. The results obtained were compared with the ASTM and EN specifications for biodiesel. The fatty acid methyl and ethyl esters of the karanja oil possessed the following fuel characteristics: acid value (mg KOH/g) 0.5, 0.5; cloud point (°C) 19, 23; pour point (°C) 15, 6; flash point (°C) 174, 148; density (g/cc at 15°C) 0.88, 0.88; viscosity (cSt) 4.77, 5.56; heating value (MJ/Kg) 40.8, 40.7, respectively. The cloud point and pour point of the karanja-based biodiesel are slightly higher, which is problematic for cold climates when pure biodiesel is to be used in the engines, but in the tropics and subtropics, this problem would not arise. When blended with diesel, the pour point is lowered to a considerable extent, 0°C for the B20 (20% karanja methyl esters) and -3°C for the B20 (20% karanja ethyl esters) biodiesel. The fuel characteristics of the methyl esters of the karanja and jatropha oils are in accordance with the ASTM 6751 specification. To satisfy the EN 14214, the storage stability needs to be improved, which is described in the following section.

TABLE 18.5
Fuel Properties of Karanja and Jatropha Methyl Esters

Parameter	Unit	KME[a]	JME[b]	ASTM D6751	EN 14214
Density at 15°C	g/cm3	0.88	0.879[c]	0.87–0.89	0.86–0.9
Viscosity at 40°C	cSt	4.77	4.84[c]	1.9–6.0	3.5–5.0
Acid value	mg KOH/g	0.5	0.24[c]	<0.8	<0.5
Flash point	°C	174	191[c]	>130	>100
Cloud point	°C	19	–	–	0/-15
Pour point	°C	15	–	–	–
Sulfur content	Wt%	0.0015	–	<0.0015	<0.0010
CCR	Wt%	0.06	0.02[c]	<0.05	–
Sulfated ash	Wt%	0.001	0.014[c]	0.02	0.02
Water	mg/kg	0.03	0.16[c]	0.05	0.05
Cu corrosion	Max. 3 h at 50°C	No. 1	–	No. 3	No. 1
Cetane number		56	51[c]	>45	>51
Ester	Wt%	98	99.6[c]	–	96.5
Free glycerol	Wt%	0.01	0.015[c]	0.02	0.02
Total glycerol	Wt%	0.19	0.088[c]	0.24	0.25
Iodine number	g/100g	86.5	–	–	<120
Oxidation stability (110°C)	h	2.24	0.56	–	6

[a] KME, karanja methyl esters.
[b] JME, jatropha methyl esters.
[c] Data from Foidl et al. (1996).

18.6 STORAGE STABILITY OF THE BIODIESEL

The induction periods of the methyl and ethyl esters of karanja and jatropha oils were estimated by the Rancimat test at 110°C using the method described by Mittelbach and Schober (2003). The methyl esters of karanja and jatropha oils have induction periods of 2.24 and 0.56 h, respectively. The smaller induction period in the case of jatropha methyl esters is due to the presence of a higher percentage of linoleic acid, 41.8% compared to 16% in the case of karanja oil (Knothe 2002). The induction period of karanja- and jatropha-based biodiesel can be improved by adding commercial natural antioxidants such as pyrogallol (PY), propylgallate (PG), *tert*-butylhydroxyquinone (TBHQ), 3-*tert*-butyl-4-hydroxyanisole (BHA), and 2,6-di-*tert*-butyl-4-methyl-phenol (BHT). The effect of antioxidants on the oxidation stability of karanja methyl esters is shown in Figure 18.4. Pyrogallol as an antioxidant at a concentration of 50 ppm improves the oxidation stability of karanja methyl esters up to 12 h. Commercial antioxidants are needed to increase the induction period of karanja- and jatropha-based biodiesel in order to satisfy the European biodiesel specifications.

18.7 CONCLUSIONS

Biodiesel is an attractive substitute for conventional petroleum-derived diesel fuel. In most of developed countries, edible-grade oils are used as feedstock for biodiesel due to the simplicity of the conventional alkali-catalyzed transesterification. The free fatty acid content of nonedible-grade oils are cheap feedstock for economic production of biodiesel. Karanja and jatropha are usually grown in degraded and waste lands and produce nonedible oils. The optimum reaction conditions for the synthesis of karanja methyl esters are 1% KOH catalyst, methanol/oil molar ratio 6:1, reaction temperature 65°C, and rate of stirring 600 rpm, which yielded 97 to 98% of methyl esters. In the case of ethanolysis, 1.4% KOH is required with 12:1 molar ratio of etha-

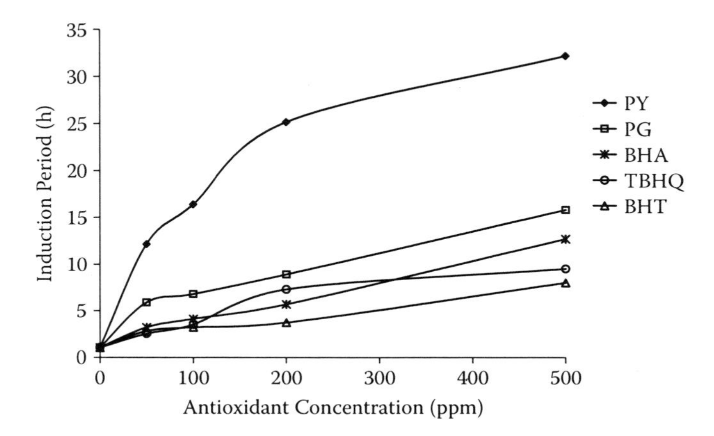

FIGURE 18.4 Effect of antioxidant concentration on the oxidation stability of karanja methyl esters.

nol to oil resulting in 95% of ethyl esters. For high-FFA oils, the dual-step process is preferred for biodiesel production. The fuel characteristics of biodiesel synthesized from karanja and jatropha oils are in accordance with biodiesel specifications, with the exception of oxidation stability. These fuels have low induction period and commercial antioxidants are recommended to improve the oxidation stability in order to satisfy the EN 14214 biodiesel specification.

REFERENCES

Bringi, N. V. 1987. *Non-traditional oilseeds and oils in India*. New Delhi: Oxford and IBH Publishing Co.

Canakci, M. and J. Von Gerpen. 2001. Biodiesel production from oils and fats with high free fatty acids. *Trans ASAE* 44: 1419–1436.

Canakci, M. and J. Von Gerpen. 1999. Biodiesel production via acid catalysis. *Trans ASAE* 42: 1203–1210.

Dorado, M. P., E. Ballesteros, F. J. Lopez, and M. Mittelbach. 2004. Optimization of alkali catalyzed transesterification of *Brassica carinata* oil for biodiesel production. *Energy Fuel* 18: 77–83.

Encinar, J. M., J. F. Gonzalez, E. Sabio, and M. J. Romiro. 1999. Preparation and properties of biodiesel from *Cynara cardunculus* L. oil. *Ind. Eng. Chem. Res.* 38: 2927–2931.

Foidl, N., G. Foidl, M. Sanchez, M. Mittelbach, and S. Hackel. 1996. *Jatropha curcas* L. as a source for the production of biofuel in Nicaragua. *Bioresource Technol.* 58: 77–82.

Freedman, B., E. H. Pryde, and T. L. Mounts. 1984. Variables affecting the yield of fatty esters from transesterified vegetable oils. *J. Am. Oil Chem. Soc.* 61: 1638–1643.

Gore, V. K. and P. Satyamoorthy. 2000. Determination of pongamol and karanjin in karanja oil by reverse phase HPLC. *Analytical Letters* 33: 337–346.

Gubitz, G. M., M. Mittelbach, and M. Trabi. 1999. Exploitation of the tropical oil seed plant Jatropha curcas L. *Bioresource Technol.* 67: 73–82.

Iwu, M. M. 1993. *Handbook of African medicinal plants*. Boca Raton, FL: CRC Press.

Karmee, S. K. and A. Chadha. 2005. Preparation of biodiesel from crude oil of *Pongamia pinnata*. *Bioresource Technol.* 96: 1425–1429.

Knothe, G. 2001. Historical perspectives on vegetable oil-based diesel fuel. *INFORM* 12: 1103–1107.

Knothe, G. 2002. Structure indices in FA chemistry. How relevant is the iodine value? *J. Am. Oil Chem. Soc.* 79: 847–854.

Knothe, G. 2005. *The biodiesel handbook*. Champaign, Illinois: AOCS Press.

Ma, F., L. D. Clements, and M. A. Hanna. 1999. The effect of mixing on transesterification of beef tallow. *Bioresource Technol.* 69: 289–293.

Meher, L. C., D. Vidya Sagar, and S. N. Naik. 2006. Optimization of alkali-catalyzed transesterification of *Pongamia pinnata* oil for production of biodiesel. *Bioresource Technol.* 97: 1392–1397.

Meher, L. C., M. G. Kulkarni, A. K. Dalai, and S. N. Naik. 2006. Transesterification of karanja (*Pongamia pinnata*) oil by solid basic catalysis. *Eur. J. Lipid Sci. Technol.* 108: 389–397.

Mittelbach, M. and S. Schober. 2003. The influence of antioxidants on the oxidation stability of biodiesel. *J. Am. Oil Chem. Soc.* 80: 817–823.

Munch and Kiefer. 1986. Die Purgiernuss. University of Hohenheim, Germany. Feb., p. 128(86.2-1).

Nagaraj, G. and N. Mukta. 2004. Seed composition and fatty acid profile of some tree borne oilseeds. *J. Oilseed Res.* 21: 117–120.

Peterson, C. L. 1986. Vegetable oil as diesel fuel: Status and research priorities. *Trans. ASAE* 29: 1412–1422.

19 Biodiesel Production from Mahua Oil and Its Evaluation in an Engine

Sukumar Puhan, Nagarajan Vedaraman, and Boppana Venkata Ramabrahmam

CONTENTS

ABSTRACT

Vegetable oils and animal fats can be transesterified to biodiesel with alcohol for use as an alternative to diesel fuel. This chapter deals mahua oil, its transfer into different esters, their performance and emission characteristics in a four-stroke, direct injection diesel engine. The results showed that the thermal efficiency was high in the case of the methyl ester compared to other esters and to diesel fuel. The tail-pipe emissions and noise levels were lower in the case of the methyl ester, compared to those of diesel and other esters. The methyl ester dominated other esters on the basis of engine performance and emissions and can be used as an alternative fuel for existing diesel engines.

19.1 INTRODUCTION

Nearly a hundred years ago, Rudolf Diesel tested vegetable oil as fuel for diesel engines. In the 1930s and 1940s vegetable oils were used as diesel fuels from time to time, but usually only in emergency situations. Recently, because of the increase in crude oil prices, dwindling resources of fossil fuel, and environmental concerns, there has been a renewed focus on vegetable oils and animal fats that can be used as biodiesel fuels in existing diesel engines. Biodiesel is, in principle, carbon dioxide (CO_2) neutral, that is, when plants grow, they absorb CO_2, and after they are harvested, converted into biofuel, and burnt, CO_2 is produced. Ideally, a closed CO_2 circuit arises. The use of biodiesel has the potential to reduce the level of pollutants and potential or probable carcinogens.

19.1.1 Worldwide Research on Vegetable Oil as Fuel

Researchers have investigated the effect of using vegetable oils alone (Barsic and Humke 1981; Frgiel and Varde 1981; Suda 1984; Murayama et al. 1984) or their blends with diesel (Ziejewski and Kaufman 1983) in a diesel engine for extended periods of time, and encountered a number of problems. Gerhard (1983) reported that the high viscosity and low volatility of pure vegetable oil reduced the fuel atomization and increased the fuel spray penetration. Higher spray penetration and polymerization of the unsaturated fatty acids at higher temperatures are partly responsible for the difficulties experienced with engine deposits and thickening of the lubricating oil. Several approaches have been undertaken to improve the physical properties of the vegetable oil, for example, (1) the addition of chemicals (additives) to improve the air-fuel mixture by decreasing the surface tension, (2) preheating to diminish the viscosity for improving the internal formation of the mixture and combustion, and (3) mixing with other fuels, to give a better internal formation of the air-fuel mixture as a consequence of a lower viscosity of the blends or to initiate better burning by easier burning components. These techniques are not suitable for long-term testing (Last and Kruger 1985), hence, the derivatives of vegetable oils in the form of alkyl esters and blends with diesel were more attractive as biodiesel. A number of studies have been carried out on the preparation and engine testing of biodiesel from various oils (canola [Spataru and Romig 1995], rapeseed [Staat and Gateau 1995], soybean [Schumacher et al. 1996], palm [Kalam and Masjuki 2002], sunflower [Da Silva, Prata, and Teixeira 2003], karanja [Raheman and Phadatare 2004], and neem oil [Nabi, Akhter, and Shahada 2006]).

19.1.2 Mahua *(Madhuca indica)* Oil for Biodiesel Production

Mahua (*M. indica*) seed oil can be used for biodiesel manufacture. Its potential is about 4,40,000 tonnes and only 10,000 tonnes are currently tapped and used, mainly by the soap industry (Roma Rao, Nanda, and Kalpana Sastry 2003). *M. indica* is a large deciduous tree with a short trunk, spreading branches, and large rounded crown. The flower is used as a vegetable and as a source of alcohol. The cake from the oil seeds is used as a fertilizer. Cattle eat the leaves, flowers, and fruits. The flowering season extends from February to April. The mature fruit falls to the ground in May to July in the north and August to September in south India. The yield of the plant depends on the climatic conditions and varies from 5 to 200 kg/plant per season, depending on the size and age of the plant. The mahua tree starts producing seeds after 10 years and continues up to 60 years. The kernel constitutes about 70% of the seed and contains 50% oil. The fats and oils are primarily water insoluble, hydrophobic substances made up of one mole of the glycerol and three moles of the fatty acids and are commonly referred to as the triglycerides. The fatty acids vary in the carbon chain length and in the number of unsaturated bonds (double bonds). The mahua oil contains approximately 47% saturated fatty acids and 53% unsaturated fatty acids. Palmitic, stearic, and oleic acids are the major constituents.

19.1.3 Transesterification

The mahua oil was used to prepare mahua oil methyl ester (MOME), mahua oil ethyl ester (MOEE), and mahua oil butyl ester (MOBE). Then their physical properties were determined and performance tested on a direct injection diesel engine to determine the engine performance and exhaust emissions in comparison with No. 2 diesel fuel.

Good quality (≤1% free fatty acid and ≤0.5% moisture content) mahua oil (5 l) was taken in a glass reactor fitted with a stirrer, external heater, and condenser for the transesterification processes. The oil was heated to 50°C in the glass reactor and NaOH dissolved in alcohol was added. The contents were heated to the required temperature (between 60 and 110°C). The reflux condenser condensed the evaporated alcohol back into the reactor. The stirring helped to achieve uniformity of the reactants and helped the reaction go faster. Methanol, ethanol, and butanol (20, 30, and 40 vol.% of oil, respectively) were used for the study. The reaction temperature was fixed in the range between 60 and 110°C at the boiling temperature of the corresponding alcohol and the reaction duration was fixed at 2 h under the reflux condition. After this, the reaction was stopped and the product was allowed to settle in two layers. The upper layer consisted of the ester and alcohol and was separated from the bottom layer (glycerin). The upper layer was distilled to remove and recover the excess alcohol and the esters were washed with hot water to remove traces of the glycerin and alkali. Finally, the product was dried for 1 h in a hot air oven at 105°C and analyzed for the fuel properties as per the standard test methods and subsequently taken for the engine test.

19.2 PROPERTIES

Table 19.1 gives a summary of the fuel properties, such as the cetane number, higher heating values, viscosity, specific gravity, flash point, pour point, sulfur content, and moisture content of different mahua oil esters and the No. 2 diesel fuel. The cetane number for butyl ester was higher compared to other esters and the diesel. The heating value increased with increase in the chain length of the fatty acid ester and decreased with increase in the number of double bonds. The increase in the heat content resulted from the increase in the number of carbons and hydrogen, as well as increase in the ratio of these elements relative to oxygen. A decrease in the heat

TABLE 19.1
Properties of Mahua Esters in Comparison with No. 2 Diesel and ASTM Standards

Properties	Diesel	MOME	MOEE	MOBE	ASTMD6751 FAME
Cetane no.	46	51	52	54	≥47
Higher heating value MJ/Kg	45	39.276	40.528	41.607	–
Kinematic viscosity (cSt)	2.4	4.2	5.4	4.7	1.9–6.0
Specific gravity	0.82	0.865	0.875	0.854	0.87–0.90
Flash point (°C)	70	157	174	164	≥130
Pour point (°C)	-10–-15	-3–-5	0–-1	-3–-1	
Sulphur content	–	0.02%	0.04%	0.03%	0.05
Moisture content	–	0.01%	0.01%	0.01%	≤0.05%

content was the result of fewer hydrogen atoms (i.e., higher unsaturation) in the molecule. The viscosity of a liquid fuel is an important parameter because the fluid has to flow through pipelines, injector nozzles, orifices, and for the atomization of the fuel in the cylinder. Proper operation of an engine depends on the accepted viscosity range of the liquid fuel. The viscosity of the mahua oil was quite high (38 cSt) and reduced to approximately one-eighth to one-tenth of the value after the transesterification. The viscosity of all three alkyl esters was within the acceptable range prescribed by the ASTM standards. The fuel consumption was significantly affected by the specific gravity of the fuel. If the specific gravity is more, the fuel is more concentrated and more fuel is likely to deliver on the mass basis, which leads to a higher fuel consumption. The MOAE has specific gravity within the range specified by ASTM standards. The flash point measures the tendency of the sample to form a flammable mixture with air under controlled conditions. This is the property that must be considered in assessing the overall flammability hazard of a material. The flash point of the MOAE was significantly higher than that of diesel fuel and thus would be quite safe for use in transportation compared to diesel. The cloud point for the MOAE was closer to that of diesel fuel.

19.3 ENGINE TESTS

The performance and emissions of the MOAE were studied in the diesel engine in comparison with the No. 2 diesel fuel. The engine used for the study was a single-cylinder, four-stroke, constant-speed, vertical, water-cooled, direct injection (DI), 3.68 kW diesel engine. The engine was coupled to a swinging field separating excited type DC generator and loaded by electrical resistance. The exhaust gas temperature was measured by an iron-constantan thermocouple. The oxides of nitrogen (NOx), carbon monoxide (CO), carbon dioxide (CO_2), and hydrocarbon (HC) were measured by the MRU emission monitoring systems DELTA 1600-L and MRU OPTRANS 1600. The fuel consumption was measured by a U-tube manometer. The engine was started in neat diesel fuel and warmed up. The warm-up periods ended when the cooling water temperature was stabilized. Then the fuel consumption, exhaust gas temperature, and different exhaust emissions were measured. The procedure was repeated for MOME, MOEE, and MOBE.

19.3.1 Brake Specific Fuel Consumption

The brake specific fuel consumption (BSFC) is the mass of fuel required to develop unit brake power. It can be seen from Figure 19.1 that the BSFC was higher for all the ester-based fuels compared to diesel. This was due to the higher specific gravity and lower heating value of the MOAE compared to the No. 2 diesel. The methyl ester showed better BSFC compared to the others. The BSFC values were 0.299, 0.319, 0.342, and 0.324 kg/kW-h correspondingly for the diesel, MOME, MOEE, and MOBE at full load.

19.3.2 Specific Energy Consumption

Figure 19.2 shows a comparison of the specific energy consumption (SEC) between the different esters and the No. 2 diesel. The reason for taking SEC into account is that comparison of thermal efficiency for different fuel becomes easier. As the thermal efficiency depends on two variables, specific fuel consumption and heating value, the comparison will be difficult unless we know the individual contribution of the variables. The product of specific fuel consumption and heating value is

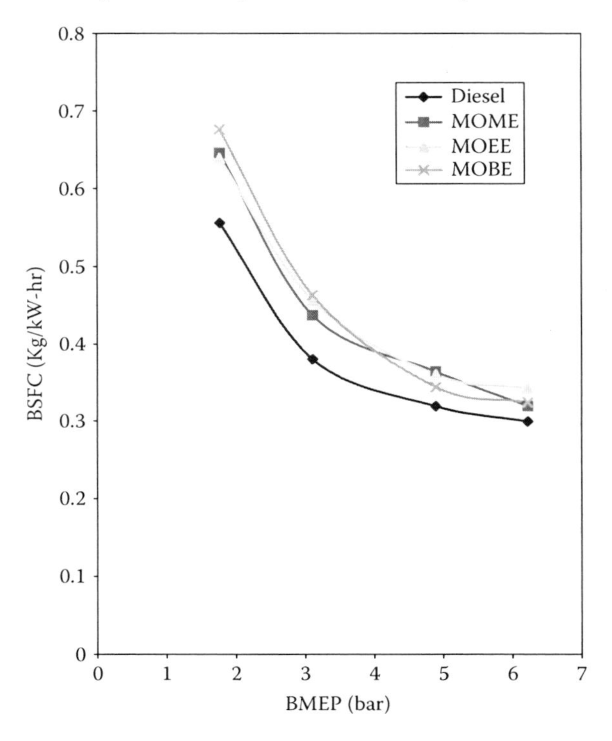

FIGURE 19.1 Variation of brake specific fuel consumption with brake mean effective pressure.

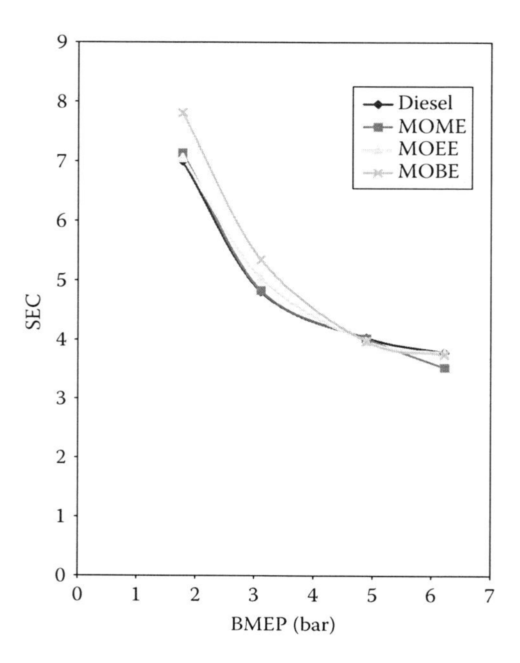

FIGURE 19.2 Variation of specific energy consumption with brake mean effective pressure.

called specific energy consumption, which does not relate to any unit. The values for the diesel, MOME, MOEE, and MOBE were 3.784, 3.525, 3.779, and 3.744, respectively, at full load. SEC was less for the MOME compared to other two esters.

19.3.3 Brake Thermal Efficiency

Figure 19.3 shows a comparison of the brake thermal efficiency (BTE) between the different esters and No. 2 diesel. The BTE is purely dependent on the engine design, type of fuel used, and the area of use. The vegetable oil-based fuel contains oxygen ranges of 10 to 12% and combustion is better in the case of the MOAE compared to the diesel. The bonded oxygen helps the fuel to burn efficiently inside the combustion chamber, thereby releasing more heat. Again, the heat release does not only depend on the oxygen content but also the heating value of the fuel. Since the vegetable oil-based fuels have 10 to 12% less heating value compared to diesel fuel, the oxygen content and heating value of the fuel are together responsible for the thermal efficiency. The data showed that the thermal efficiency for the methyl ester was high compared to those of the diesel and other esters at the full load.

19.3.4 Exhaust Gas Temperature

Figure 19.4 shows a comparison of the exhaust gas temperature (EGT) between the MOAE and diesel. In the diesel engine, there are four stages in the combustion pro-

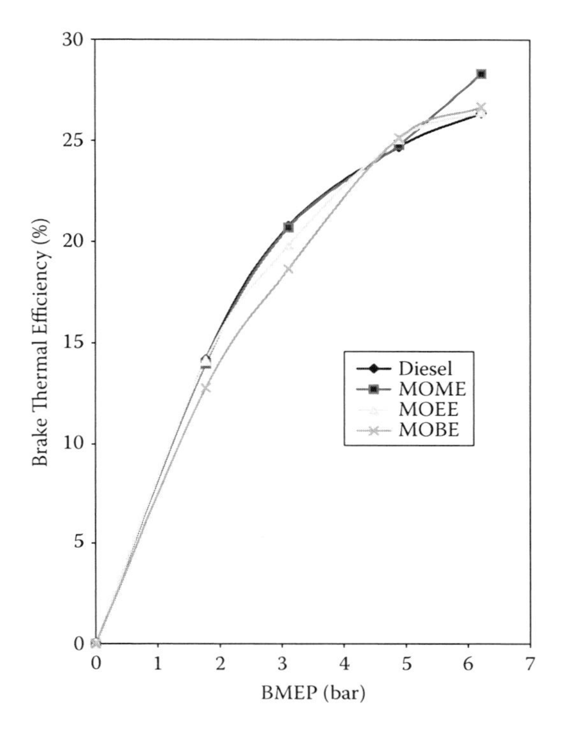

FIGURE 19.3 Variation of brake thermal efficiency with brake mean effective pressure.

cess: ignition delay, premix combustion or uncontrolled combustion, controlled combustion, and afterburning. If the afterburning phase is more or the engine misfires or the injection time is not proper, then there is every possibility for higher EGT. On the other hand, if the combustion process is perfect, then also the EGT is likely to be high. As the thermal efficiency was higher in the case of the methyl ester of the mahua oil, the combustion process was supposed to be more complete and this could be one reason for a higher EGT.

19.3.5 Noise Level

Figure 19.5 shows the variation of the noise level with load for different fuels. The noise is the indication of the sound that is created during the running of an engine. The result showed that at 100% load of the engine, the noise level for all the esters was low compared to diesel and the lowest noise level observed was 123 dB in the case of methyl ester compared to 169 dB for the No. 2 diesel at the same load condition.

19.3.6 Oxides of Nitrogen

Figure 19.6 shows a comparison of the NO_X emission between the different esters and the diesel. The oxides of nitrogen are formed inside a diesel engine due to high flame

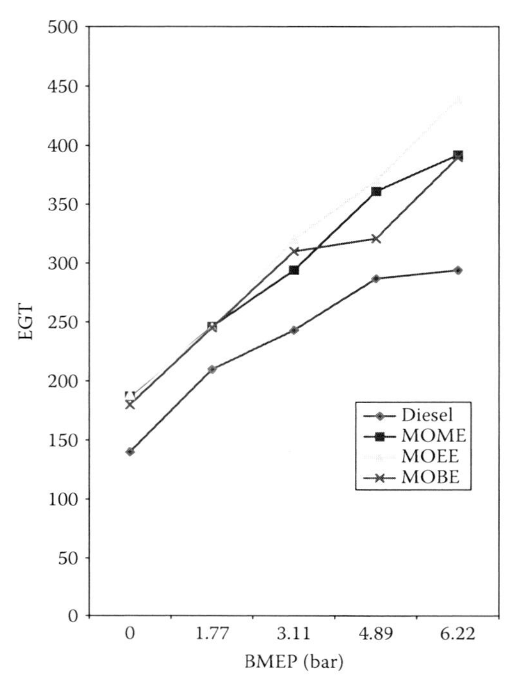

FIGURE 19.4 Variation of exhaust gas temperature with brake mean effective pressure.

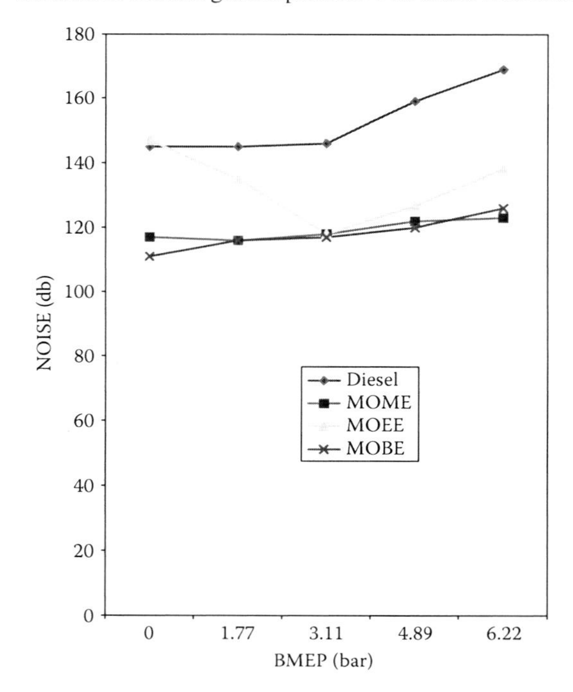

FIGURE 19.5 Variation of noise with brake mean effective pressure.

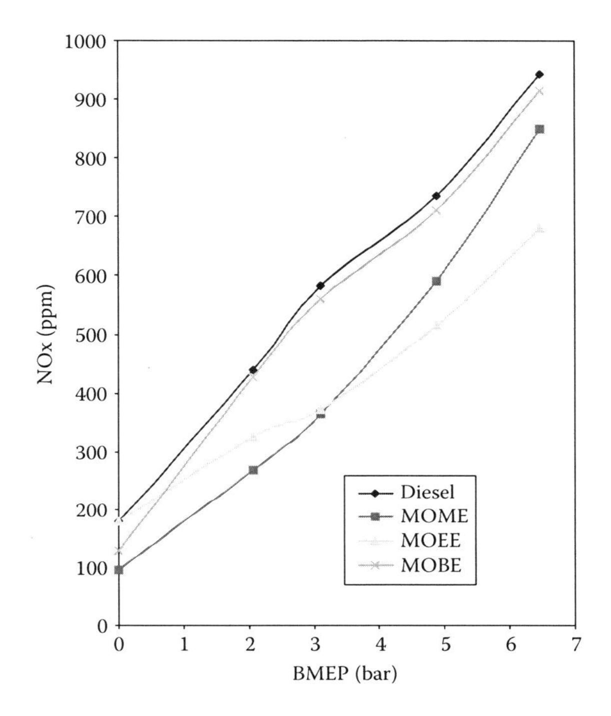

FIGURE 19.6 Variation of NOx with brake mean effective pressure.

temperature, peak pressure inside the cylinder, nitrogen content of the parent fuel, and the residence time of the fuel inside the cylinder. All these factors affect NO_X emission greatly. As the cetane number of the ester-based fuel is high compared to diesel, the residence time may be less in the case of ester-based fuel. In addition, the oxygen content of the fuel enhances the ignition quality, thereby reducing delay for esters. Hence, the MOAE is likely to produce lower heat release at the premix combustion phase, and this would lower the peak combustion temperature and reduce the NO_X emissions. In addition, other parameters such as iodine value, chemical bonding, and structure may contribute to a lower combustion temperature.

19.3.7 Carbon Monoxide Emission

Figure 19.7 shows a comparison of CO emission between the different esters and the diesel. CO emission depends on the combustion efficiency and carbon content of the fuel. This shows how efficiently the fuel is burnt inside the engine cylinder. The fuel, during combustion, undergoes a series of oxidation and reduction reactions. The carbon content of the fuel is oxidized with the oxygen available in the air to CO and subsequently to CO_2. No fuel will give 100% combustion efficiency, so the carbon that is not converted to CO_2 will come out as CO in the exhaust. The test results showed that for all the esters, the CO emission was lower than that of diesel and the methyl ester gave the lowest CO emission level compared to the other two esters (0.07% for the methyl ester; 0.34 % for the diesel at full load).

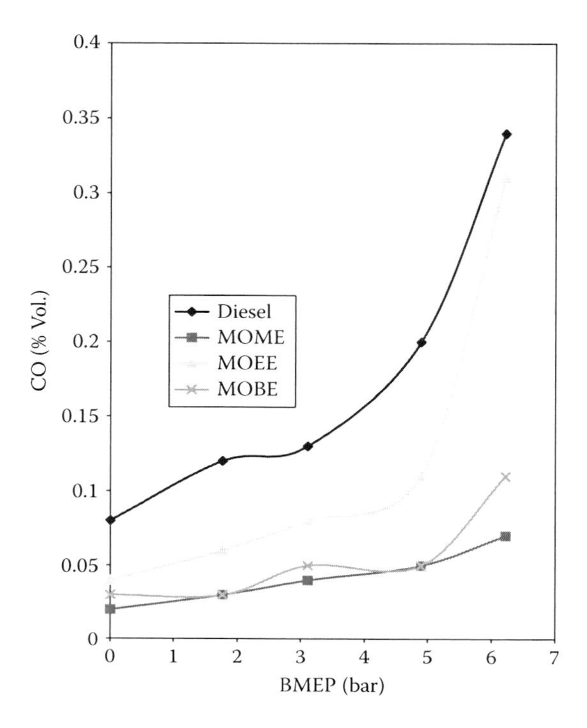

FIGURE 19.7 Variation of CO with brake mean effective pressure.

19.3.8 Carbon Dioxide Emission

Figure 19.8 shows a comparison of the CO_2 emissions between the different esters and the No. 2 diesel. Carbon dioxide emission is likely to be more for fuel with better combustion quality. The better the combustion, the more carbon as carbon dioxide is present in the exhaust. Actually, all the carbon present in the fuel cannot be converted to carbon dioxide. As the esters contained oxygen in the chemical structure, the combustion was better than with No. 2 diesel. Hence, the carbon dioxide emission in the exhaust was more than that observed for the diesel.

19.3.9 Hydrocarbon Emission

Figure 19.9 shows a comparison of the HC emission between the different esters and the diesel. The hydrocarbon present in the fuel is burnt inside the engine cylinder in the presence of air. The amount of HC that is not taking part in the combustion reaction is likely to come out as unburnt hydrocarbon. As explained earlier, due to several reasons, combustion is not 100% perfect. Hence, the HC emission is likely to occur in the exhaust system. In the case of ester-based fuels, the oxygen present in the structure helps in better combustion and hence HC emission is less than that of diesel. The results showed that the HC value for the No. 2 diesel was 89 ppm, whereas it was 35, 45, and 50 ppm for methyl, ethyl, and butyl esters, respectively, at full load.

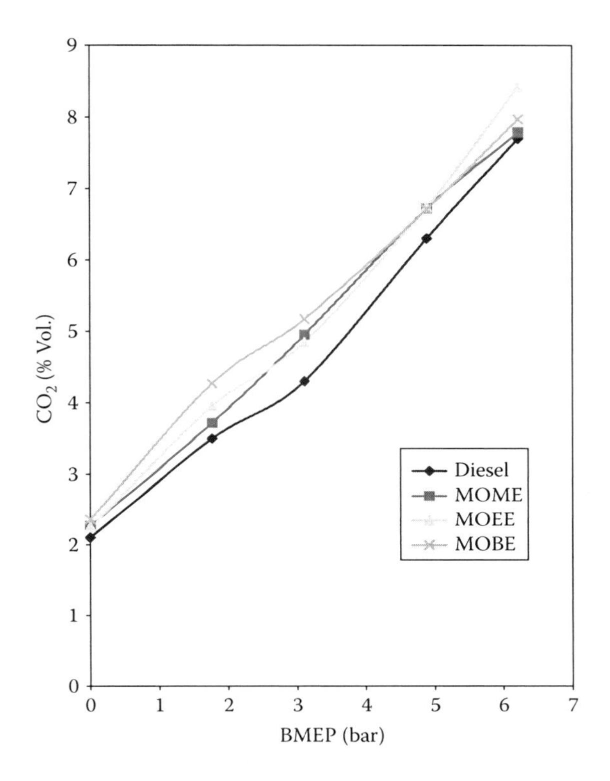

FIGURE 19.8 Variation of CO_2 with brake mean effective pressure.

19.4 CONCLUSIONS

Plant–based renewable fuels were obtained by the transesterification of mahua oil with different alcohols and tested in a single-cylinder DI diesel engine. The MOME, MOEE, and MOBE were evaluated in terms of engine performance and emissions. The CO emissions from the MOME, MOEE, and MOBE were lower than those from diesel fuel at full load. The CO_2 emission was slightly more compared to the diesel because of better combustion in the case of the MOME, MOEE, and MOBE. The NO_x emissions for all the esters were lower than that of the diesel. Hence, in terms of the performance and emission characteristics, the MOAE may be regarded as a potential substitute for diesel fuel. Among the esters, the mahua methyl ester was the best choice because of its low alcohol cost, low reaction temperature, relatively lower emission, and better engine performance compared to the other two esters.

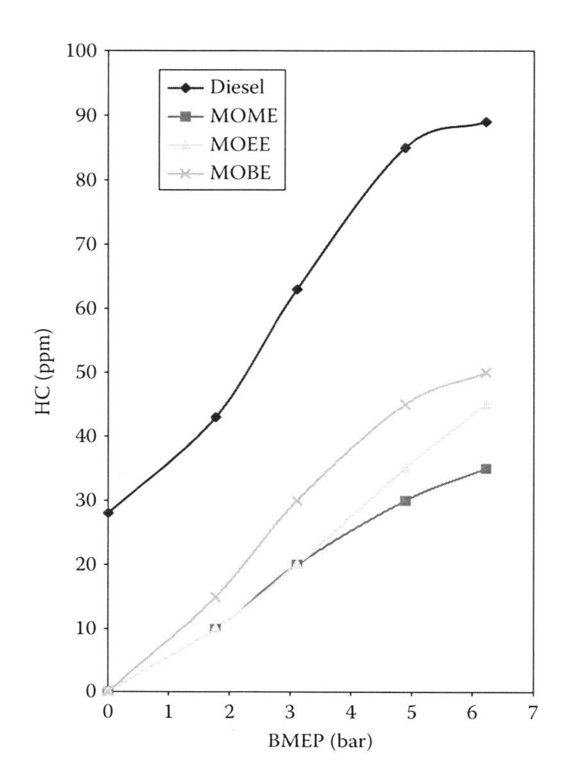

FIGURE 19.9 Variation of hydrocarbon with brake mean effective pressure.

REFERENCES

Barsic, N. J. and A. L. Humke. 1981. Performance and emission characteristics of a naturally aspirated diesel engine with vegetable oil fuels. SAE Paper No- 810262. Warrendale, PA: SAE International.

Da Silva, F. N., A. S. Prata, and J. R. Teixeira. 2003. Technical feasibility assessment of oleic sunflower methyl ester utilization in diesel bus engines. *Energy Conversion and Management* 44: 2857–2878.

Forgiel, R. and K. S. Varde. 1981. Experimental investigation of vegetable oil utilization in direct injection diesel engines. SAE Paper No. 811214. Warrendale, PA: SAE International.

Gerhard, V. 1983. Performance of vegetable oils and their monoesters as fuels for diesel engines. SAE Paper 831358. Warrendale, PA: SAE International.

Kalam, M. K. and H. H. Masjuki. 2002. Biodiesel from palm oil: An analysis of its properties and potential. *Biomass and Bioenergy* 23: 471–479.

Last, R. J. and M. D. H. Kruger. 1985. Emission and performance characteristics of a four stroke, direct injected diesel engine fuel with blends of bio diesel and low sulfur diesel fuels. SAE Paper No. 850054. Warrendale, PA: SAE International.

Murayama, T., Y.-T. Oh, N. Miyamoto, T. Chikahisa, N. Takagi, and K. Itow. 1984. Low carbon flower buildup, low smoke, and efficient diesel operation with vegetable oils by conversion to mono-esters and blending with diesel oil or alcohols. SAE Paper No. 841161. Warrendale, PA: SAE International.

Nabi, M. N., M. S. Akhter, and M. M. Z. Shahadat. 2006. Improvement of engine emissions with conventional diesel fuel and diesel-biodiesel blends. *Bioresource Technology* 97: 372–378.

Raheman, H. and A. G. Phadatare. 2004. Diesel engine emission and performance from blends of karanja methyl ester and diesel. *Biomass and Bioenergy* 27: 393–397.

Roma Rao, R. D., S. K. Nanda, and R. Kalpana Sastry. 2003. *Strategies for Augmenting Potential of Vegetable Oils as Biodiesel. Tree-Borne Oil Seeds as a Source of Energy for Decentralized Planning.* Renewable Energy Science Series XII. Ministry of Non-Conventional Energy Source, Government of India.

Schumacher, L. G., S. C. Borgelt, D. Fosseen, W. Goetz, and W. G. Hires. 1996. Heavy duty engine exhaust emission tests using methyl ester soybean oil/diesel fuel blends. *Bioresource Technology* 57: 31–36.

Spataru, A. and C. Romig. 1995. Emissions and engine performance from blends of soya and canola methyl esters with ARB#2 diesel in a DCC 6V92TA MUI engine. SAE Paper No. 952388. Warrendale, PA: SAE International.

Staat, F. and P. Gateau. 1995. The effects of rapeseed oil methyl ester on diesel engine performance exhaust emissions and long-term behavior: A summary of three years of experimentation. SAE Paper No. 950053. Warrendale, PA: SAE International.

Suda, K. J. 1984. Vegetable oil or diesel fuel: A flexible option. SAE Paper No: 840004. Warrendale, PA: SAE International.

Ziejewski, M. and K. R. Kaufman. 1983. Laboratory endurance test of a sunflower oil blend in a diesel engine. *J. Am. Oil Chem. Soc.* 60 (8): 1567–1573.

20 Biodiesel Production from Rubber Seed Oil

Arumugam Sakunthalai Ramadhas, Simon Jayaraj, and Chandrashekaran Muraleedharan

CONTENTS

ABSTRACT

Rubber seed oil is a high free fatty acid content, nonedible vegetable oil. The acid value of unrefined rubber seed oil is about 34. Neither alkaline-catalyzed transesterification nor acid-catalysed esterification alone is suitable. A two-step esterification process, that is, acid esterification followed by alkaline transesterification was developed to convert the unrefined rubber seed oil to its methyl esters and glycerol. The process parameters, such as quantity of catalyst and methanol used, reaction temperature, and reaction duration, are analysed. The properties of methyl esters of rubber seed oil are comparable to that of diesel. The performance and emission characteristics of biodiesel-diesel blends provide evidence that methyl esters of rubber seed oil are a suitable alternative fuel to diesel.

20.1 INTRODUCTION

Recent concerns over the environment, increasing fuel prices, and scarcity of supply have promoted interest in the development of alternative sources to petroleum fuels. The vegetable oils are a promising alternative fuel to diesel as their fuel properties approximate those of diesel. The sources of the vegetable oils, seeds, grow renewably in oil-yielding crops. Diesel fuel consists of saturated, nonbranched hydrocarbon molecules, with carbon ranging between 12 and 18, whereas the vegetable oil molecules are of triglycerides, generally nonbranched chains of different lengths and different degrees of saturation. Important properties such as energy density, cetane number, heat of vaporization, and stoichiometric air-fuel ratio of vegetable oil are comparable to those of diesel (Montague 1996). Currently biodiesel is produced mainly from field crop oils, such as rapeseed, sunflower, and soybean oil (Zhang 1996; Zeiejerdki and Pratt 1986). The prices of the edible oils are several-fold higher than nonedible oils. Nonedible oils also have potential for the production of biodiesel, including jatropha oil, karanji oil, rubber seed oil, etc. Biodiesel production from the nonedible oils would reduce the overall biodiesel cost. This chapter describes the biodiesel production method from unrefined rubber seed oil, its physiochemical properties, cost analysis, and evaluation of engine performance and emission characteristics with biodiesel-diesel blends.

20.2 POTENTIAL OF RUBBER SEED OIL AS AN ALTERNATIVE FUEL

The rubber tree (*Hevea brasiliensis*) is indigenous to the Amazon in Brazil. It grows quickly and is a fairly sturdy perennial tree of 25 to 30 m in height. The young plant shows its characteristic growth pattern of alternating periods of rapid elongation and consolidated development. The leaves are trifoliate with long stalks. The rubber tree may live for a hundred years or even more. However, its economic life period on plantations, is generally about 32 years, that is, seven years of immature phase and 25 years of productive phase. It flowers during the months of February and March. The fruits mature in the months between July and September, and have ellipsoidal capsules with three carpels, each containing a seed. These open up during the sunshine and the seeds fall on the ground and are normally hand picked. The rubber seeds resemble castor seeds but are slightly larger in size and each weighs 2 to 4 g. The seeds, which fall on the ground, deteriorate very rapidly due to moisture and infection. These lead to rapid increase in the free fatty acid (FFA) content of the oil. Therefore, it is essential to collect the seeds as quickly as possible and dry them, so as to reduce the moisture to a value less than 5% in order to arrest increase in the FFA. The rubber seed oil is normally obtained by expelling of the seeds. Depending on the pre-extraction history of the kernels, the color of the oil ranges from water white to pale yellow for low FFA content (about 5%) to dark color for high FFA content (about 10 to 40%). The fatty acid composition of rubber seed oil is given in Table 20.1 (Aigbodion and Pillai 2000; Aigbodion et al. 2003). The molecular formula of rubber seed oil is $C_{18}H_{32}O_2$ and its molecular weight is 278. The important physiochemical properties of rubber seed oil and diesel are shown in Table 20.2 (Ramadhas, Jayaraj, and Muraleedharan 2005a). The specific gravity of rubber seed

TABLE 20.1
Fatty Acid Composition of Rubber Seed Oil

Fatty Acid	Formula	Structure	Composition (%)
Palmitic c	$C16H32O2O_2$	16:0:0	10.2.2
Stearic	$C18H36O2O_2$	18:0:0	8.7.7
Oleic	$C18H34O2O_2$	18:1:1	24.6.6
Linoleic	$C18H32O2O_2$	18:2:2	39.6.6
Linolenic	$C18H30O2O_2$	18:3:3	13.2.2

(Reprinted from Aigbodion, A. I., Pillai, C. K. S. [2000]. Preparation, analysis and applications of rubber seed oil and its derivatives as surface coating material, *Progress in Organic Coatings*, 38, 187–192, Elsevier Publications, with permission.)

TABLE 20.2
Properties of Rubber Seed Oil in Comparison with Diesel

Property	Test Method	Rubber Seed Oil	Diesel
Specific gravity	ASTM D4052	0.91	0.835
Viscosity (mm2/s) at 40(C°)	ASTM D445	66	4.5
Flash point (°C)	ASTM D93	198	48
Fire point (°C)	ASTM D93	210	55
Calorific value (MJ/kg)	ASTM D240	37.5	42.5
Saponification value	ASTM D94	206	-
Acid value	ASTM D664	34.0	0.062

(Reprinted from Ramadhas, A. S., Jayaraj, S., Muraleedharan, C. [2005], Chacterization and effect of using rubber seed oil as fuel in the compression ignition engines, *International Journal of Renewable Energy* 30 (5), 795–803, Elsevier Publications, with permission.)

oil is higher than that of diesel; hence it has almost the same calorific value as diesel on a volumetric basis. The flash point of rubber seed oil is much higher than that of diesel and hence, from a storage point view, it is much safer than diesel. One of the undesirable properties of the oil is its viscosity, which is several times higher than that of the diesel. The calorific value of rubber seed oil is about 12% lower than that of the diesel. However, the lower calorific value of oil is compensated for by the enhanced lubrication.

20.2.1 Transesterification

For alkaline transesterification, triglycerides should have lower acid value and all reactants should be substantially anhydrous. The difficulty with processing nonedible oils and fats is that these often contain large amounts of FFA that cannot be converted to biodiesel using the alkaline catalysis method. The addition of excess sodium hydroxide catalyst with oil can compensate the higher acidity but the resulting soap would increase its viscosity or formation of the gels that interfere with the forward reaction as well as with separation of the glycerol. The yield of the transesterification process would decrease considerably with increase in FFA. The acid value of unrefined rubber seed oil is 34 mg KOH/g, that is 17% FFA content. It is known that alkaline-catalyzed transesterification does not occur if FFA content in the oil is more than 3% (Canakci and Van Gerpen 2001, 1999). Nevertheless, the refining process reduces the acid value of the oils but it increases the overall biodiesel production cost. The acid esterification process can be used to produce biodiesel from oils having FFA content higher than 3%. But this reaction is much slower than that of alkaline transesterification. A two-step esterification process is developed to produce biodiesel from unrefined rubber seed oil. The first step, acid esterification, converts FFA to esters and reduces the acid value of the oil to about 4. The second step is the alkaline-catalyzed transesterification process.

20.2.1.1 Acid Esterification

A measured quantity of the rubber seed oil is stirred and heated in the reactor to about 60°C. The calculated quantity of the methanol is mixed with the preheated rubber seed oil and the mixture is stirred vigorously for a few minutes and allowed to run at medium speed. Then a precise quantity of the concentrated sulfuric acid is added in the mixture. The heating and stirring are continued for 20 min and then the products are poured into the separating funnel. The excess alcohol with the sulfuric acid and impurities, if any, move to the upper layer and the lower layer is separated for the second step.

20.2.1.1.1 Effect of the Amount of Acid Catalyst

The quantity of the acid catalyst used in the process is an important parameter that affects the yield and quality of the biodiesel. The methanol is used in excess with varying amounts of concentrated sulfuric acid (0.25 to 2%). It was found that 0.5% concentrated sulfuric acid (v/v) gave the maximum yield (Figure 20.1). An excess amount of sulfuric acid does not increase the yield but darkens the color of the product and adds to the cost. However, an insufficient amount of the sulfuric acid lowers the yield.

20.2.1.1.2 Effect of the Amount of Methanol

The quantity of the methanol used is an important factor that affects the yield of the process and the production cost of the biodiesel. The molar ratio is defined as the ratio of number of moles of alcohol to number of moles of triglycerides. Theoretically, 3 mol of alcohol is required for the conversion of 1 mol of triglyceride to 3 mol of the ester and 1 mol of the glycerol. However, in practice, an excess methanol is required to drive the reaction towards completion. Experiments carried out with the optimal catalyst quantity (0.5% v/v) revealed the maximum yield with 20 ml of methanol for 100 ml of the rubber seed oil (Figure 20.2). With further increase in the amount of methanol, there was only little improvement in the yield. However, reduction in viscosity of the mixture was observed with increase in the quantity of methanol. Excess methanol in the biodiesel would reduce the flash point of the fuel.

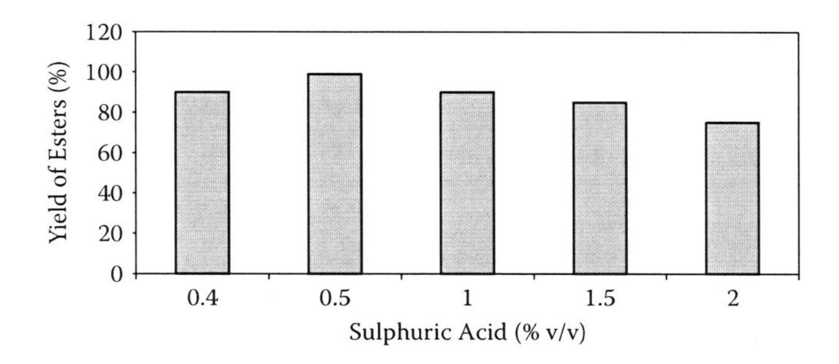

FIGURE 20.1 Effect of amount of acid catalyst on yield.

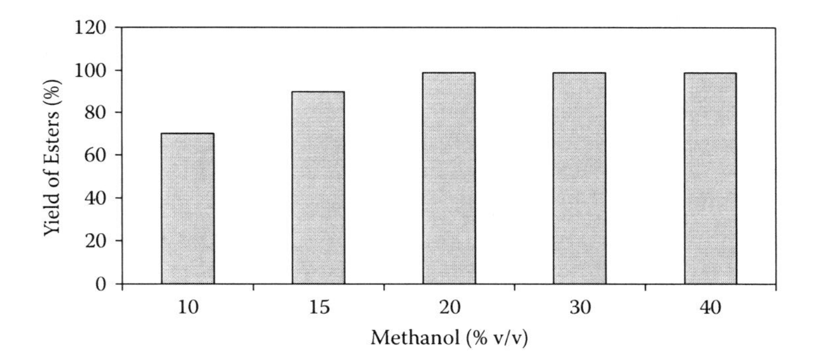

FIGURE 20.2 Effect of methanol quantity on yield of first step.

20.2.1.1.3 Effect of Reaction Temperature

The reaction temperature strongly influences the reaction rate and the yield of the process. The yield of biodiesel from the rubber seed oil was very low (about 10%) when the reaction was carried out at room temperature. The optimum temperature was in the range of 45 ± 5°C. The boiling point of the methanol is 60°C and hence higher temperature results in loss of the methanol and darkens the color of the product. Furthermore, higher reaction temperature consumes more energy and thus increases the overall production cost of the biodiesel.

20.2.1.2 Alkaline Transesterification

The product of the first step, that is, the oil-ester mixture (the lower layer in the separating funnel) was heated to the reaction temperature. The catalyst (anhydrous sodium hydroxide pellet) was dissolved in the methanol and added to the preheated mixture. Heating and stirring were continued for 30 min at the required temperature. The reaction produced two liquid phases: ester in the upper layer and crude glycerol in the lower layer. The phase separation was observed within 10 to 15 min after stirring was stopped but the complete separation required a longer time (2 to 10 h). The catalyst-glycerol mixture, settled at the bottom, was drained for further processing. The ester layer was washed with water (about 25% volume of the oil) by

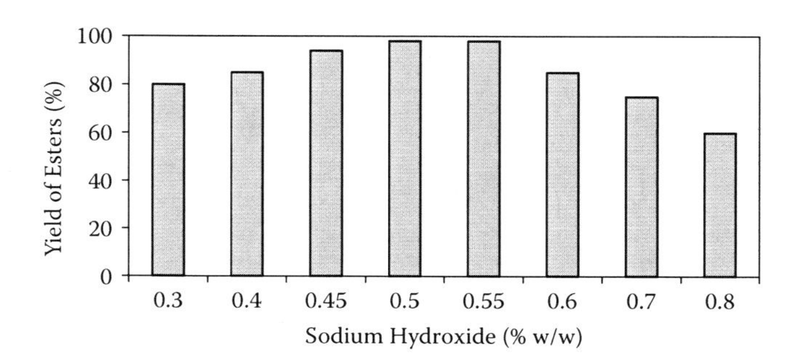

FIGURE 20.3 Effect of alkaline catalyst on yield.

gentle agitation several times until the washed water was clear, that is, the pH value was neutral.

20.2.1.2.1 Effect of the Amount of Alkaline Catalyst

In order to study the effect of the amount of alkaline catalyst on the production of biodiesel from rubber seed oil, sodium hydroxide pellets in the range of 0.3 to 1% by weight (weight of NaOH/weight of oil) were dissolved in the excess methanol. The yield of the process with respect to amount of catalyst is shown in Figure 20.3. The maximum yield was achieved with the use of 0.5% NaOH. Excess amounts of catalyst increased the viscosity of the mixture and led to the formation of soap. Also, insufficient amounts of catalyst did not initiate the reaction.

20.2.1.2.2 Effect of the Amount of Methanol

Figure 20.4 shows the yield of biodiesel with respect to the quantity of methanol used in the process. The maximum ester yield was obtained with 30% methanol by volume. With further increase in the molar ratio or methanol quantity, the yield remained almost the same. On settling of the mixture, excess methanol moved over the ester layer.

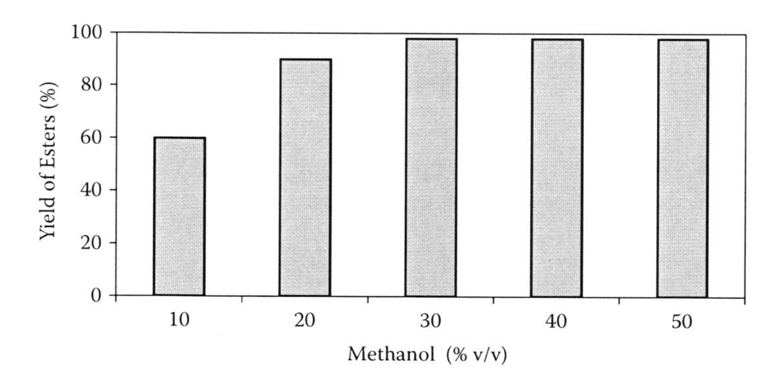

FIGURE 20.4 Effect of methanol amount on yield of second step.

TABLE 20.3
Properties of Methyl Esters of Rubber Seed Oil in Comparison with Diesel

Property	Test Procedure	Biodiesel Standard ASTM D6751-0202	Rubber Seed Oil Methyl Ester	Diesel
Specific gravity	ASTM D4052	0.87–0.90	0.874	0.835
Calorific value (MJ/kg)	ASTM D240	–	36.50	42.5
Viscosity (mm2/s) at 40°C	ASTM D445	1.9-6.0	5.81	3.8
Flash point (°C)	ASTM D93	>110	130	48
Cloud point (°C)	ASTM D2500	-3–12	4	-1
Pour point (°C)	ASTM D97	-15–10	-8	-16

20.3 PROPERTIES OF METHYL ESTERS OF RUBBER SEED OIL

The physiochemical properties of the biodiesel in comparison with the ASTM biodiesel standards, ASTM D 6751, are given in Table 20.3. The properties of the methyl esters are comparable to those of diesel and match the ASTM biodiesel standard. C, H, and O compositions of the rubber seed oil methyl esters were 76.85%, 11.82%, and 11.32%, respectively. The fuel analysis showed that the transesterification process improved the fuel properties of the oil, particularly the viscosity and flash point. The viscosity of the methyl esters of rubber seed oil was found to be closer to that of diesel, and hence, no hardware modifications are required for storage and handling of biodiesel.

20.4 ENGINE TESTS WITH BIODIESEL

The engine tests were conducted with the blends of biodiesel and diesel as fuel at the rated speed of 1500 rpm. Here, B20 represents a blend that contains 20% biodiesel and 80% diesel. The engine performance and emission characteristics obtained using biodiesel-diesel blends as fuel are described below (Ramadhas, Jayaraj, and Muraleedharan 2005b).

20.4.1 Brake Thermal Efficiency

Figure 20.5 shows the variation in the brake thermal efficiency of the engine with respect to its brake mean effective pressure (BMEP) operating with various blends of biodiesel and diesel. Increase in brake thermal efficiency of the engine with load was observed due to reduction in heat loss and increase in power. The brake thermal efficiency of 28% was achieved with B10 as compared to 25% with diesel. The lower percentage concentration of biodiesel in the blends improved the brake thermal efficiency of the engine. The additional lubricity provided by the biodiesel that reduced frictional power and the presence of the oxygen makes complete combustion. But, at the higher blends, the brake thermal efficiency of the engine decreased because of its lower calorific value.

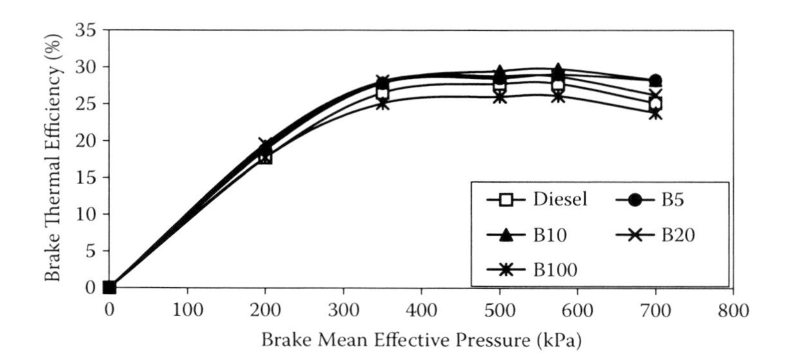

FIGURE 20.5 Comparison of brake thermal efficiency of the engine at various brake mean effective pressure (BMEP) values.

20.4.2 Specific Fuel Consumption

The variation of specific fuel consumption with respect to the BMEP for the different fuels tested is depicted in Figure 20.6. The specific fuel consumption of the engine fueled with the lower concentration of biodiesel in the blend was lower than that of diesel at all the loads. The specific fuel consumption of B50 to B100 was found to be higher as compared to diesel because of their lower calorific values. About 12% increase in fuel consumption with neat biodiesel was observed as compared to neat diesel.

20.4.3 Carbon Monoxide Emission

CO emission was found to be lower at lighter load conditions and increased with load for all the fuels tested. CO emission increased as the air-fuel ratio became lower than that of the stoichiometric air-fuel ratio (Figure 20.7). CO emission was found to be negligibly small at the stoichiometric air-fuel ratio or on the lean side of the stoichiometric. The diesel-fueled engine emitted more CO as compared to that of the biodiesel blends under all the loading conditions.

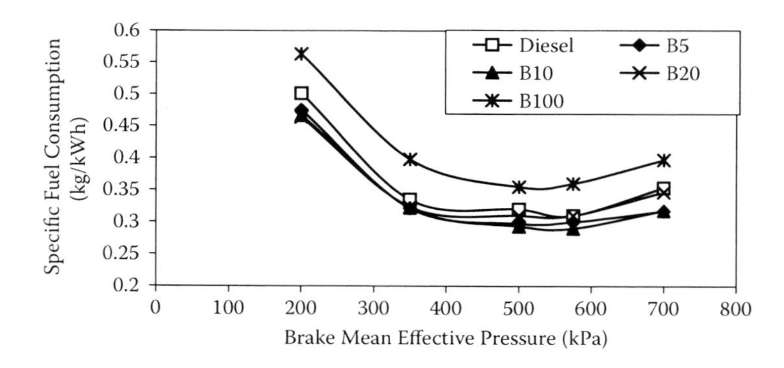

FIGURE 20.6 Comparison of specific fuel consumption of the engine at various BMEP values.

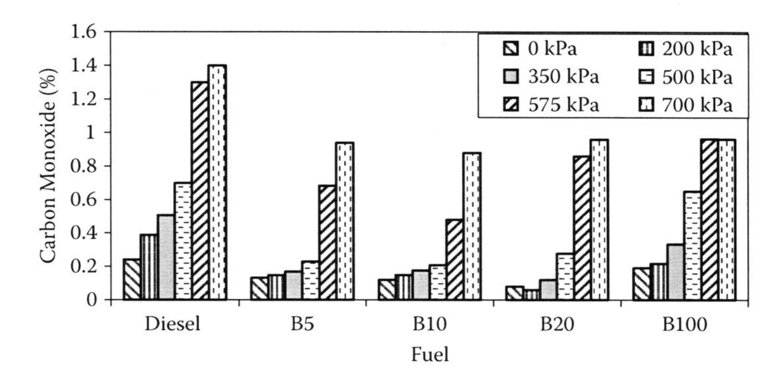

FIGURE 20.7 Comparison of CO emission of the engine at various BMEP values.

20.4.4 Carbon Dioxide Emission

CO_2 emission increased with the increase in the load, as expected. The lower percentage of biodiesel in the blends emit very low amounts of CO_2 in comparison with that of the diesel (Figure 20.8). It was observed that neat biodiesel operation emitted slightly higher amounts of the carbon dioxide as compared to that of the diesel operation. This indicated the complete combustion of the fuel and hence higher combustion chamber temperature.

20.4.5 Smoke Density

The variation of the smoke density for different fuels tested in the engine is depicted in Figure 20.9. The smoke density of the biodiesel blends was found to be lower than that of the diesel. These results support the better combustion of biodiesel blends as compared to diesel.

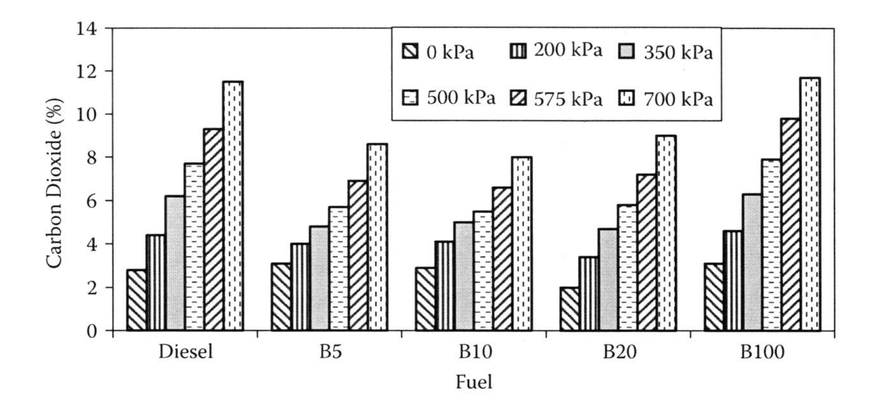

FIGURE 20.8 Comparison of CO_2 emission of the engine at various BMEP values.

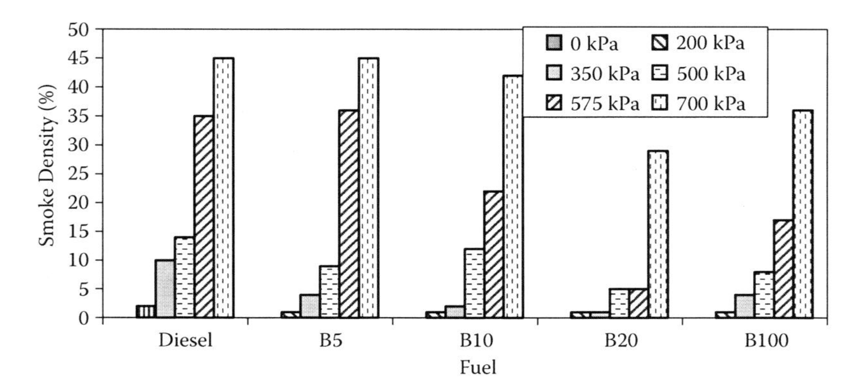

FIGURE 20.9 Comparison of smoke density of the engine at various BMEP values.

20.4.6 Exhaust Gas Temperature

The variation of the exhaust gas temperature with respect to BMEP of the engine for different fuels tested is shown in Figure 20.10. The exhaust gas temperature increased with increase in the load for all the fuels tested. It was observed that with increase in the concentration of biodiesel in the blend, the exhaust gas temperature increased marginally. The nitrogen oxides emission was directly related to the engine combustion chamber temperatures, which in turn indicated the prevailing exhaust gas temperature.

20.5 CONCLUSIONS

Low-cost, high-FFA feedstocks for the production of biodiesel were investigated. High-FFA vegetable oils such as rubber seed oil could not be transesterified with the alkaline-catalyzed transesterification process. A two-step transesterfication process was developed to convert the high-FFA vegetable oil to its methyl esters. The first step, acid-catalyzed esterification, followed by the second step, alkaline-catalyzed transesterification, converts vegetable oils into mono-esters and glycerol. This two-

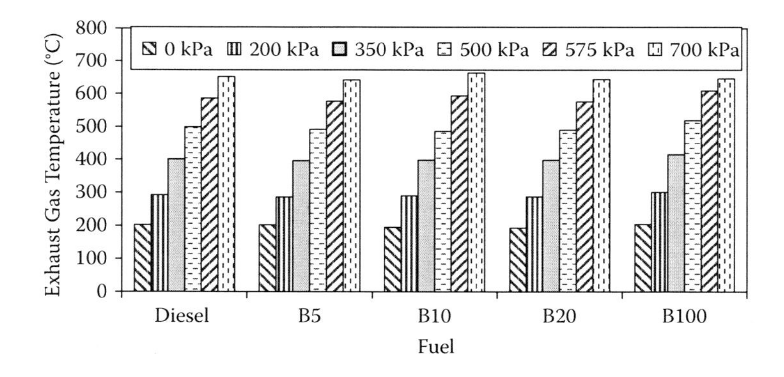

FIGURE 20.10 Comparison of exhaust gas temperature of the engine at various BMEP values.

step esterification method reduces the overall production cost of the biodiesel, as it uses low-cost, unrefined nonedible oils. The effects of alcohol to oil molar ratio, catalyst amount, reaction temperature, and reaction duration were analyzed for each step. The fuel properties of biodiesel are comparable to those of diesel. The performance and emissions characteristics support the use of biodiesel-diesel blends in engines. Use of nonedible oils for fuel purposes reduces fuel insecurity and air pollution also.

REFERENCES

Aigbodion, A. I. and C. K. S. Pillai. 2000. Preparation, analysis and applications of rubber seed oil and its derivatives as surface coating material. *Progress in Organic Coatings* 38: 187–192.

Aigbodion, A. I., F. E. Okieimen, E. O. Obazee, and I. O. Bakare. 2003. Utilization of maleinized rubber seed oil and its alkyd resin as binders in water-borne coatings. *Progress in Organic Coatings* 46: 28–31.

Canakci, M. and J. Von Gerpen. 1999. Biodiesel production via acid catalysis. *Transactions of American Society of Agricultural Engineers* 42 (5): 1203–1210.

Canakci, M. and J. Von Gerpen. 2001. Biodiesel production from oils and fats with high free fatty acids. *Transactions of American Society of Agricultural Engineers* 44: 1429–1436.

Montague, X. 1996. Introduction of rapeseed methyl ester in diesel fuel – French National program. 962065. Society of Automotive Engineers. Warrendale, PA: SAE Publication Group.

Ramadhas, A. S., S. Jayaraj, and C. Muraleedharan. 2005a. Characterization and effect of using rubber seed oil as fuel in the compression ignition engines. *International Journal of Renewable Energy* 30 (5): 795–803.

Ramadhas, A. S., S. Jayaraj, and C. Muraleedharan. 2005b. Performance and emission evaluation of a diesel engine fueled with methyl esters of rubber seed oil. *International Journal of Renewable Energy* 30: 1789–1800.

Zeiejerdki, K. and K. Pratt. 1986. Comparative analysis of the long term performance of a diesel engine on vegetable oil. 860301. Society of Automotive Engineers. Warrendale, PA: SAE Publication Group.

Zhang, Y. 1996. Combustion analysis of esters of soyabean oil in a diesel engine. 960765. Society of Automotive Engineers. Warrendale, PA: SAE Publication Group.

Index